# THE FOUNDATIONS OF ANALYSIS: A STRAIGHTFORWARD INTRODUCTION

## BOOK 2
## TOPOLOGICAL IDEAS

# THE FOUNDATIONS OF ANALYSIS: A STRAIGHTFORWARD INTRODUCTION

## BOOK 2
## TOPOLOGICAL IDEAS

K. G. BINMORE

*Professor of Mathematics*
*London School of Economics*
*and Political Science*

CAMBRIDGE UNIVERSITY PRESS

*Cambridge*
*London New York New Rochelle*
*Melbourne Sydney*

CAMBRIDGE UNIVERSITY PRESS
Cambridge, New York, Melbourne, Madrid, Cape Town, Singapore, São Paulo, Delhi

Cambridge University Press
The Edinburgh Building, Cambridge CB2 8RU, UK

Published in the United States of America by Cambridge University Press, New York

www.cambridge.org
Information on this title: www.cambridge.org/9780521233507



First published 1981
Re-issued in this digitally printed version 2008

*A catalogue record for this publication is available from the British Library*

ISBN 978-0-521-23350-7 hardback
ISBN 978-0-521-29930-5 paperback

## CONTENTS

†This material is more advanced than the remaining material and can be omitted at a first reading.

The diagram on p. x illustrates the logical structure of the books. Broken lines enclosing a chapter heading indicate more advanced material which can be omitted at a first reading. The second book depends only to a limited extent on the first. The broken arrows indicate the extent of this dependence. It will be apparent that those with some previous knowledge of elementary abstract algebra will be in a position to tackle the second book without necessarily having read the first.

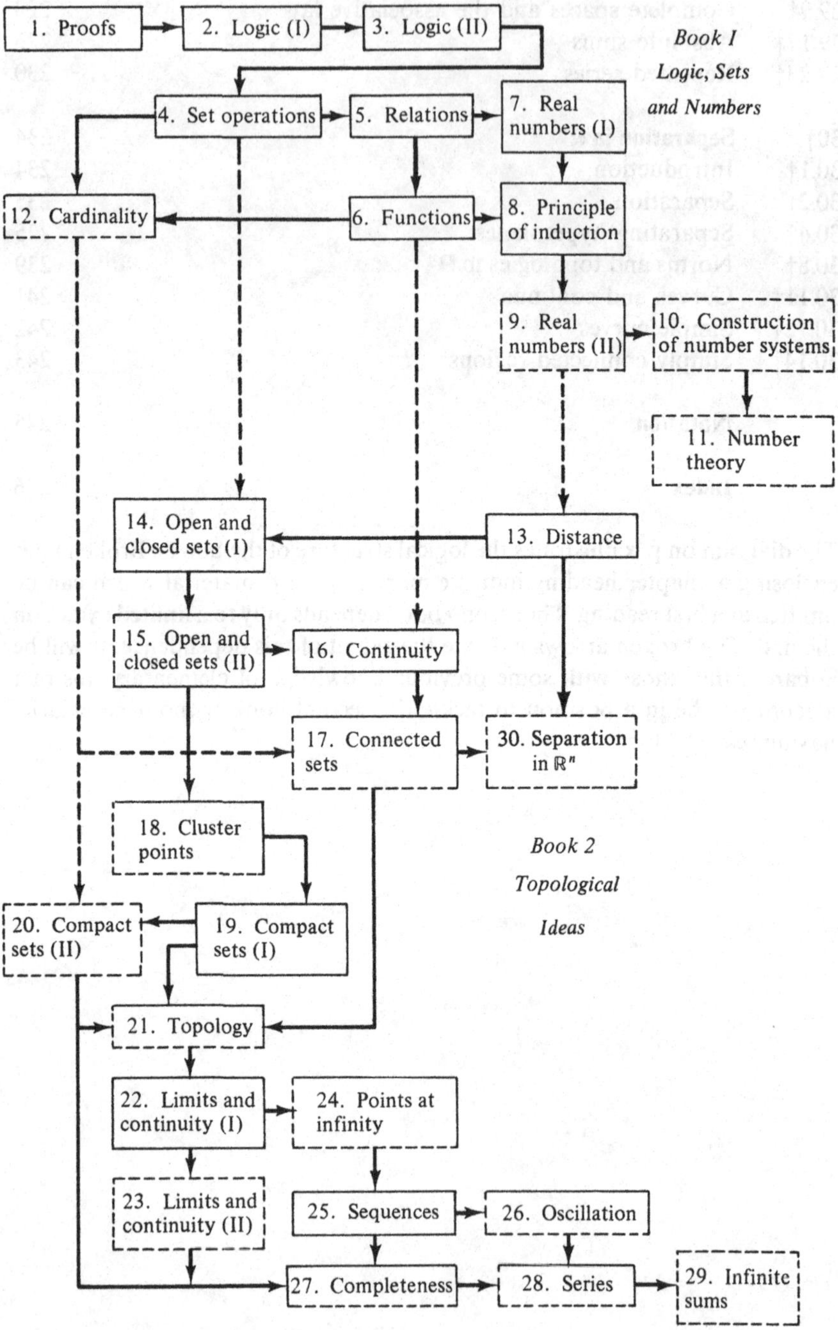
1. Proofs
2. Logic (I)
3. Logic (II)
Book I
Logic, Sets
and Numbers
4. Set operations
5. Relations
7. Real numbers (I)
12. Cardinality
6. Functions
8. Principle of induction
9. Real numbers (II)
10. Construction of number systems
11. Number theory
14. Open and closed sets (I)
13. Distance
15. Open and closed sets (II)
16. Continuity
17. Connected sets
30. Separation in ℝ^n
18. Cluster points
Book 2
Topological
Ideas
20. Compact sets (II)
19. Compact sets (I)
21. Topology
22. Limits and continuity (I)
24. Points at infinity
23. Limits and continuity (II)
25. Sequences
26. Oscillation
27. Completeness
28. Series
29. Infinite sums

# INTRODUCTION

This book is intended to bridge the gap between introductory texts in mathematical analysis and more advanced texts dealing with real and complex analysis, functional analysis and general topology. The discontinuity in the level of sophistication adopted in the introductory books as compared with the more advanced works can often represent a serious handicap to students of the subject especially if their grasp of the elementary material is not as firm as perhaps it might be. In this volume, considerable pains have been taken to introduce new ideas slowly and systematically and to relate these ideas carefully to earlier work in the knowledge that this earlier work will not always have been fully assimilated. The object is therefore not only to cover new ground in readiness for more advanced work but also to illuminate and to unify the work which will have been covered already.

Topological ideas readily admit a succinct and elegant abstract exposition. But I have found it wiser to adopt a more prosaic and leisurely approach firmly wedded to applications in the space $\mathbb{R}^n$. The idea of a relative topology, for example, is one which always seems to cause distress if introduced prematurely.

The first nine chapters of this book are concerned with open and closed sets, continuity, compactness and connectedness in metric spaces (with some fleeting references to topological spaces) but virtually all examples are drawn from $\mathbb{R}^n$. These ideas are developed independently of the notion of a limit so that this can then be subsequently introduced at a fairly high level of generality. My experience is that all students appreciate the rest from 'epsilonese' made possible by this arrangement and that many students who do not fully understand the significance of a limiting process as first explained find the presentation of the same concept in a fairly abstract setting very illuminating provided that some effort is taken to relate the abstract definition to the more concrete examples they have met before. The notion of a limit is, of course, the single most important concept in mathematical analysis. The remainder of the volume is therefore largely devoted to the application of this idea in various important special cases.

Much of the content of this book will be accessible to undergraduate students during the second half of their first year of study. This material has been indicated by the use of a larger typeface than that used for the more advanced material (which has been further distinguished by the use of the symbol †). There can be few institutions, however, with sufficient teaching time available to allow all the material theoretically accessible to first year students actually to be taught in their first year. Most students will therefore encounter the bulk of the work presented in this volume in their second or later years of study.

Those reading the book independently of a taught course would be wise to leave the more advanced sections (smaller typeface and marked with a †) for a second reading. This applies also to those who read the book during the long vacation separating their first and second years at an institute of higher education. Note, incidentally, that the exercises are intended as an integral part of the text. In general there is little point in seeking to read a mathematics book unless one simultaneously attempts a substantial number of the exercises given.

This is the second of two books with the common umbrella title *Foundations of Analysis: A Straightforward Introduction.* The first of these two books, subtitled *Logic, Sets and Numbers* covers the set theoretic and algebraic foundations of the subject. But those with some knowledge of elementary abstract algebra will find that *Topological Ideas* can be read without the need for a preliminary reading of *Logic, Sets and Numbers* (although I hope that most readers will think it worthwhile to acquire both).

A suitable preparation for both books is the author's introductory text, *Mathematical Analysis: A Straightforward Approach.* There is a small overlap in content between this introductory book and *Topological Ideas* in order that the latter work may be read without reference to the former.

Finally, I would like to express my gratitude to Mimi Bell for typing the manuscript with such indefatigable patience. My thanks also go to the students of L.S.E. on whom I have experimented with various types of exposition over the years. I have always found them to be a lively and appreciative audience and this book owes a good deal to their contributions.

*June 1980* K. G. BINMORE

# 13 DISTANCE

## 13.1 The space $\mathbb{R}^n$

Those readers who know a little linear algebra will find the first half of this chapter very elementary and may therefore prefer to skip forward to §13.18.

The objects in the set $\mathbb{R}^n$ are the $n$-tuples

$$(x_1, x_2, \ldots, x_n)$$

in which $x_1, x_2, \ldots, x_n$ are real numbers. We usually use a single symbol $\mathbf{x}$ for the $n$-tuple and write

$$\mathbf{x} = (x_1, x_2, \ldots, x_n).$$

The real numbers $x_1, x_2, \ldots, x_n$ are called the *co-ordinates* or the *components* of $\mathbf{x}$.

It is often convenient to refer to an object $\mathbf{x}$ in $\mathbb{R}^n$ as a *vector*. When doing so, ordinary real numbers are called *scalars*. If $\mathbf{x} = (x_1, x_2, \ldots, x_n)$ and $\mathbf{y} = (y_1, y_2, \ldots, y_n)$ are vectors and $\alpha$ is a scalar, we define '*vector addition*' and '*scalar multiplication*' by

$$\mathbf{x} + \mathbf{y} = (x_1 + y_1, x_2 + y_2, \ldots, x_n + y_n)$$
$$\alpha\mathbf{x} = (\alpha x_1, \alpha x_2, \ldots, \alpha x_n).$$

These definitions have a simple geometric interpretation which we shall illustrate in the case $n = 2$. An object $\mathbf{x} \in \mathbb{R}^2$ may be thought of as a point in the plane referred to rectangular Cartesian axes. Alternatively, we can think of $\mathbf{x}$ as an arrow with its blunt end at the origin and its sharp end at the point $(x_1, x_2)$.

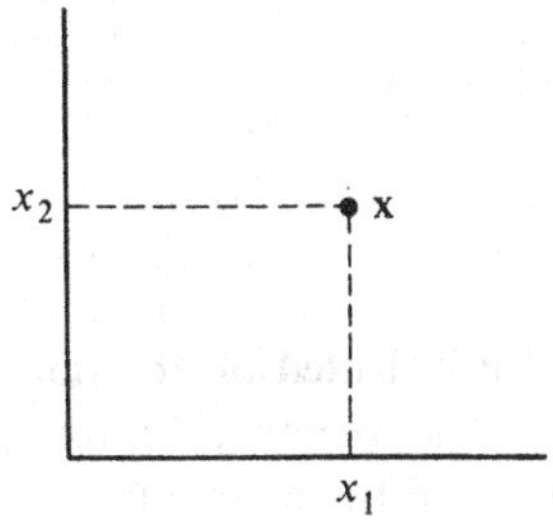

x as a point

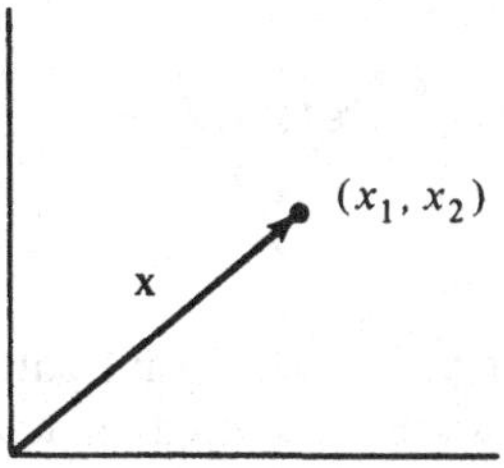

x as an arrow

Vector addition and scalar multiplication can then be illustrated as in the diagrams below. For obvious reasons, the rule for adding two vectors is called the *parallelogram law.*

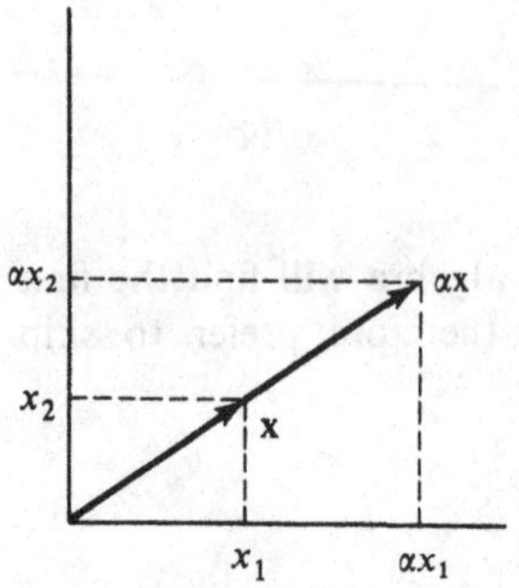


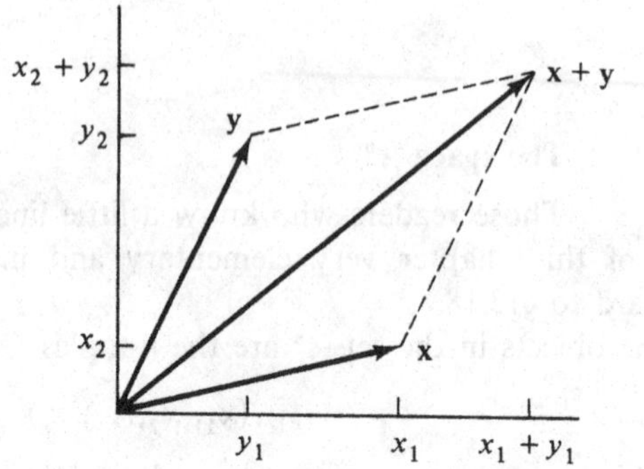


The parallelogram law is the reason that the navigators of small boats draw little parallelograms all over their charts. Suppose a boat is at $O$ and the navigator wishes to reach point $P$. Assuming that the boat can proceed at 10 knots in any direction and that the tide is moving at 5 knots in a south-easterly direction, what course should be set?

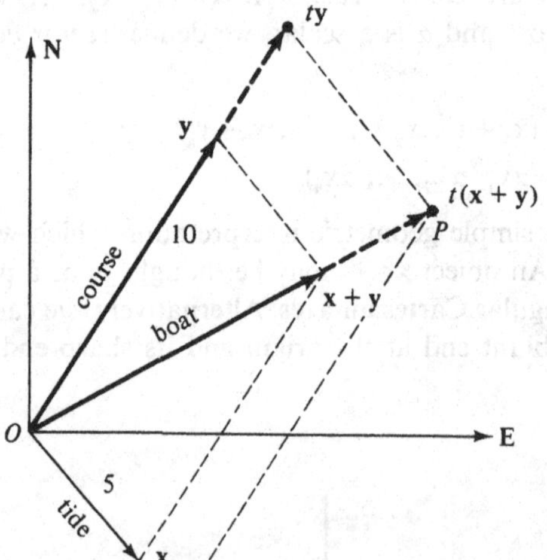


The vector **x** represents that path of the boat if it drifted on the tide for an hour (distances measured in nautical miles). The vector **y** represents the path of the boat if there were no tide and it sailed the course indicated for an hour. The vector $\mathbf{x}+\mathbf{y}$ represents the path of the boat (over the sea bed) if both influences act together. The scalar $t$ is the time it will take to reach $P$.

---

13.2 *Example* Let $\mathbf{x}=(1, 2, 3)$ and $\mathbf{y}=(2, 0, 5)$. Then

$$\mathbf{x}+\mathbf{y}=(1, 2, 3)+(2, 0, 5)=(3, 2, 8)$$
$$2\mathbf{x}=2(1, 2, 3)=(2, 4, 6).$$

---

It is very easy to check $\mathbb{R}^n$ is a commutative group under vector addition. (See §6.6.) This simply means that the usual rules for addition and subtraction are true. The zero vector is, of course,

$$\mathbf{0}=(0, 0,\ldots, 0).$$

The diagram below illustrates the vector $\mathbf{y}-\mathbf{x}=(y_1-x_1, y_2-x_2,\ldots, y_n-x_n)$ in the case $n=2$.

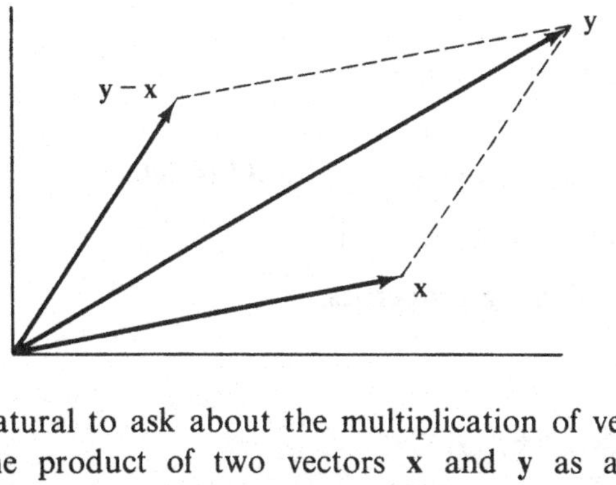


It is natural to ask about the multiplication of vectors. Is it possible to define the product of two vectors $\mathbf{x}$ and $\mathbf{y}$ as another vector $\mathbf{z}$ in a satisfactory way? There is no problem when $n=1$ since we can then identify $\mathbb{R}^1$ with $\mathbb{R}$. Nor is there a problem when $n=2$ since we can then identify $\mathbb{R}^2$ with $\mathbb{C}$ (§10.20). If $n \geqq 3$, however, there is no entirely satisfactory way of defining multiplication in $\mathbb{R}^n$. Instead we define a number of different types of 'product' none of which has all the properties which we would like a product to have.

Scalar multiplication, for example, tells us how to multiply a scalar and a vector. It does not help in multiplying two vectors. The 'inner product', which we shall meet in §13.3, tells how two vectors can be 'multiplied' to produce a scalar. In $\mathbb{R}^3$, one can introduce the 'outer product' or 'vector product' of two vectors $\mathbf{x}$ and $\mathbf{y}$. This is a vector denoted by $\mathbf{x}\wedge\mathbf{y}$ or $\mathbf{x}\times\mathbf{y}$. Unfortunately, $\mathbf{x}\wedge\mathbf{y}=-\mathbf{y}\wedge\mathbf{x}$.

Multiplication is therefore something which does not work very well with vectors. Division is almost always *meaningless*.

---

## 13.3 Length and angle in $\mathbb{R}^n$

The *Euclidean norm* of a vector **x** in $\mathbb{R}^n$ is defined by

$$\|\mathbf{x}\| = \{x_1^2 + x_2^2 + \dots + x_n^2\}^{1/2}.$$

We think of $\|\mathbf{x}\|$ as the *length* of the vector **x**. This interpretation is justified in $\mathbb{R}^2$ by Pythagoras' theorem (13.15).

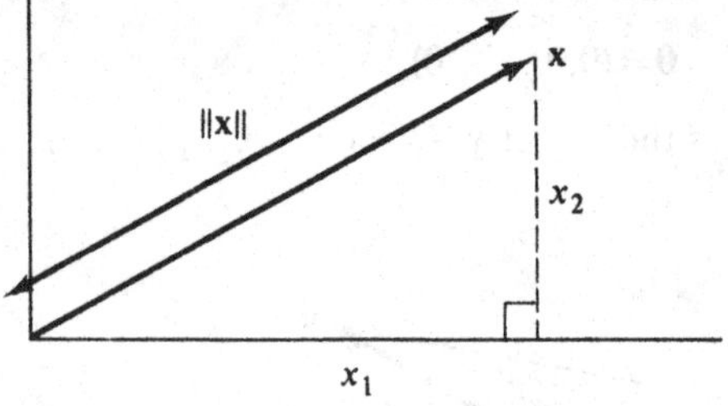


The *inner product* of two vectors **x** and **y** in $\mathbb{R}^n$ is defined by

$$\langle \mathbf{x}, \mathbf{y} \rangle = x_1 y_1 + x_2 y_2 + \dots + x_n y_n.$$

It is easy to check the following properties:

(i) $\langle \mathbf{x}, \mathbf{x} \rangle = \|\mathbf{x}\|^2$
(ii) $\langle \mathbf{x}, \mathbf{y} \rangle = \langle \mathbf{y}, \mathbf{x} \rangle$
(iii) $\langle \alpha\mathbf{x} + \beta\mathbf{y}, \mathbf{z} \rangle = \alpha\langle \mathbf{x}, \mathbf{z} \rangle + \beta\langle \mathbf{y}, \mathbf{z} \rangle$.

The geometric significance of the inner product can be discussed using the *cosine rule* (i.e. $c^2 = a^2 + b^2 - 2ab \cos \gamma$) in the diagram below.

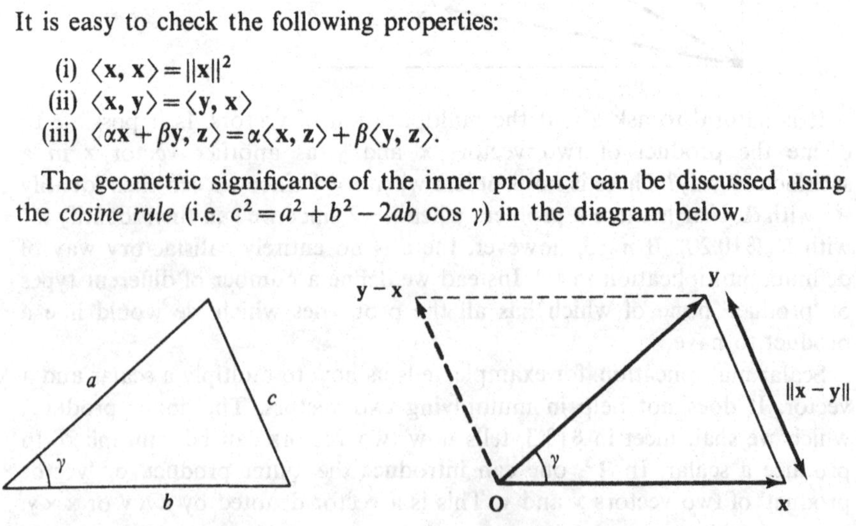


Rewriting the cosine rule in terms of the vectors introduced in the right-hand diagram, we obtain that

$$\|\mathbf{x}-\mathbf{y}\|^2=\|\mathbf{x}\|^2+\|\mathbf{y}\|^2-2\|\mathbf{x}\|\cdot\|\mathbf{y}\|\cos\gamma.$$

But,

$$\begin{aligned}\|\mathbf{x}-\mathbf{y}\|^2&=\langle\mathbf{x}-\mathbf{y},\mathbf{x}-\mathbf{y}\rangle=\langle\mathbf{x},\mathbf{x}-\mathbf{y}\rangle-\langle\mathbf{y},\mathbf{x}-\mathbf{y}\rangle\\&=\langle\mathbf{x},\mathbf{x}\rangle-2\langle\mathbf{x},\mathbf{y}\rangle+\langle\mathbf{y},\mathbf{y}\rangle=\|\mathbf{x}\|^2+\|\mathbf{y}\|^2-2\langle\mathbf{x},\mathbf{y}\rangle\end{aligned}$$

It follows that

$$\langle\mathbf{x},\mathbf{y}\rangle=\|\mathbf{x}\|\cdot\|\mathbf{y}\|\cos\gamma.$$

Of course, this argument does not *prove* anything. It simply indicates why it is helpful to think of

$$\frac{\langle\mathbf{x},\mathbf{y}\rangle}{\|\mathbf{x}\|\cdot\|\mathbf{y}\|}$$

as the cosine of the *angle* between $\mathbf{x}$ and $\mathbf{y}$.

---

13.4 *Example* Find the lengths of and the cosine of the angle between the vectors $\mathbf{x}=(1,2,3)$ and $\mathbf{y}=(2,0,5)$ in $\mathbb{R}^3$.

We have that

$$\begin{aligned}\|\mathbf{x}\|&=\{1^2+2^2+3^2\}^{1/2}=\sqrt{14},\\\|\mathbf{y}\|&=\{2^2+0^2+5^2\}^{1/2}=\sqrt{29},\\\frac{\langle\mathbf{x},\mathbf{y}\rangle}{\|\mathbf{x}\|\cdot\|\mathbf{y}\|}&=\frac{1\cdot2+2\cdot0+3\cdot5}{\sqrt{14}\sqrt{29}}=\frac{17}{\sqrt{14\times29}}.\end{aligned}$$

---

## 13.5 Some inequalities

In the previous section $\gamma$ was the angle between $\mathbf{x}$ and $\mathbf{y}$. The fact that $|\cos\gamma|\leq1$ translates into the following theorem.

---

13.6 *Theorem* (*Cauchy–Schwarz inequality*) If $\mathbf{x}\in\mathbb{R}^n$ and $\mathbf{y}\in\mathbb{R}^n$, then

$$|\langle\mathbf{x},\mathbf{y}\rangle|\leq\|\mathbf{x}\|\cdot\|\mathbf{y}\|.$$

*Proof* Let $\alpha\in\mathbb{R}$. Then

$$\begin{aligned}0\leqq\|\mathbf{x}-\alpha\mathbf{y}\|^2&=\langle\mathbf{x}-\alpha\mathbf{y},\mathbf{x}-\alpha\mathbf{y}\rangle\\&=\|\mathbf{x}\|^2-2\alpha\langle\mathbf{x},\mathbf{y}\rangle+\alpha^2\|\mathbf{y}\|^2.\end{aligned}$$

It follows that the quadratic equation $\|\mathbf{x}\|^2 - 2\alpha\langle \mathbf{x}, \mathbf{y}\rangle + \alpha^2\|\mathbf{y}\|^2$ has at most one real root (§10.10). Hence '$b^2 - 4ac \leqq 0$' – i.e.

$$4\langle \mathbf{x}, \mathbf{y}\rangle^2 - 4\|\mathbf{x}\|^2\|\mathbf{y}\|^2 \leqq 0.$$

---

It is a familiar fact in Euclidean geometry that one side of a triangle is shorter than the sum of the lengths of the other two sides.

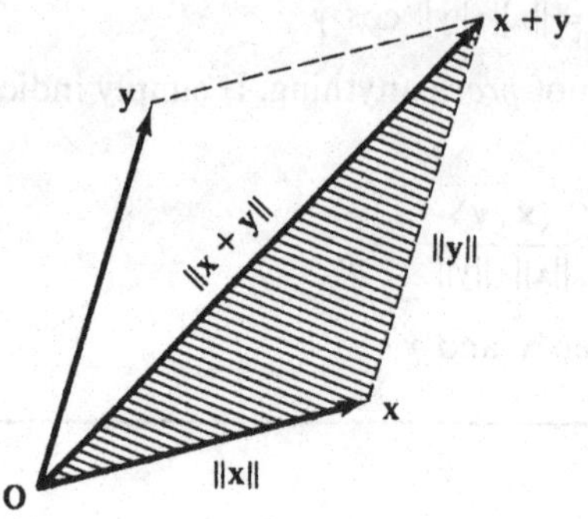


This geometric idea translates into the following theorem.

---

13.7 *Theorem* (*Triangle inequality*) If $\mathbf{x} \in \mathbb{R}^n$ and $\mathbf{y} \in \mathbb{R}^n$, then

$$\|\mathbf{x}+\mathbf{y}\| \leqq \|\mathbf{x}\| + \|\mathbf{y}\|.$$

*Proof*

$$\begin{aligned} \|\mathbf{x}+\mathbf{y}\|^2 &= \langle \mathbf{x}+\mathbf{y}, \mathbf{x}+\mathbf{y}\rangle \\ &= \|\mathbf{x}\|^2 + 2\langle \mathbf{x}, \mathbf{y}\rangle + \|\mathbf{y}\|^2 \\ &\leqq \|\mathbf{x}\|^2 + 2\|\mathbf{x}\|.\|\mathbf{y}\| + \|\mathbf{y}\|^2 \quad \text{(theorem 13.6)} \\ &= (\|\mathbf{x}\| + \|\mathbf{y}\|)^2. \end{aligned}$$

---

13.8 *Corollary* If $\mathbf{x} \in \mathbb{R}^n$ and $\mathbf{y} \in \mathbb{R}^n$, then

$$\|\mathbf{x}-\mathbf{y}\| \geqq \|\mathbf{x}\| - \|\mathbf{y}\|.$$

*Proof* It follows from the triangle inequality that

$$\|\mathbf{x}\| = \|(\mathbf{x}-\mathbf{y}) + \mathbf{y}\| \leqq \|\mathbf{x}-\mathbf{y}\| + \|\mathbf{y}\|.$$

---

**13.9 Modulus**

The *modulus* $|x|$ of a real number $x$ coincides with the Euclidean norm of $x$ thought of as a vector in $\mathbb{R}^1$. We have that

$$\|x\| = \{x^2\}^{1/2} = |x| = \begin{cases} x\ (x \geqq 0) \\ -x\ (x < 0) \end{cases}.$$

One has to remember that $y^{1/2}$ represents the *non-negative* number whose square is $y$ (§9.13).

The modulus $|z|$ of a complex number $z = x + iy$ is identified with the Euclidean norm of $(x, y)$ thought of as a vector in $\mathbb{R}^2$. We have that

$$|x + iy| = \{x^2 + y^2\}^{1/2} = \|(x, y)\|.$$

---

13.10 *Theorem* Suppose that $u$ and $v$ are real or complex numbers. Then

$$|uv| = |u| \cdot |v|.$$

*Proof* We need only consider the complex case. If $u = a + ib$ and $v = c + id$, then

$$|uv|^2 - |u|^2|v|^2 = (ac - bd)^2 + (bc + ad)^2 - (a^2 + b^2)(c^2 + d^2) = 0.$$

---

**13.11 Distance**

The *distance* $d(\mathbf{x}, \mathbf{y})$ between two vectors $\mathbf{x}$ and $\mathbf{y}$ in $\mathbb{R}^n$ is defined by

$$d(\mathbf{x}, \mathbf{y}) = \|\mathbf{x} - \mathbf{y}\|.$$

In interpreting this idea in $\mathbb{R}^n$, it is better to think of $\mathbf{x}$ and $\mathbf{y}$ as the points at the end of the arrows rather than the arrows themselves.

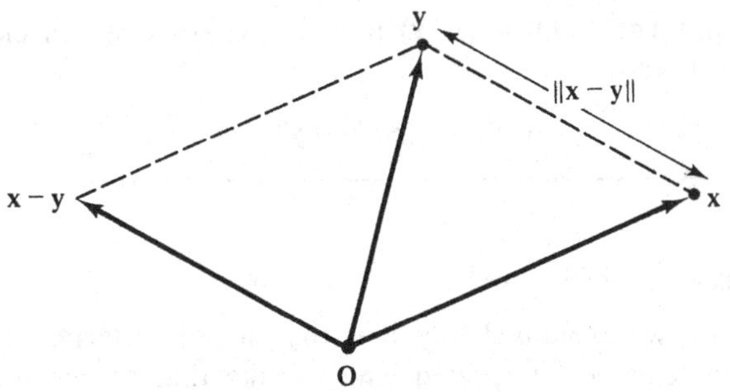


---

### 13.12 *Examples*

(i) The distance between the vectors $\mathbf{x}=(1, 2, 3)$ and $\mathbf{y}=(2, 0, 5)$ in $\mathbb{R}^3$ is

$$\|\mathbf{y}-\mathbf{x}\|=\{(2-1)^2+(0-2)^2+(5-3)^2\}^{1/2}=3.$$

(ii) The distance between 3 and 7 (regarded as vectors in $\mathbb{R}^1$) is

$$|3-7|=|-4|=4.$$

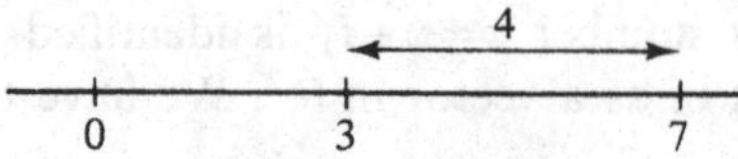


---

### 13.13 *Exercise*

(1) Let $\mathbf{x}=(0, 1, 0)$ and $\mathbf{y}=(1, 1, 0)$ be vectors in $\mathbb{R}^3$. Calculate the quantities

(i) $\mathbf{x}+\mathbf{y}$ (ii) $\mathbf{x}-\mathbf{y}$ (iii) $2\mathbf{x}$
(iv) $\|\mathbf{x}\|$ (v) $\|\mathbf{x}-\mathbf{y}\|$ (vi) $\langle\mathbf{x}, \mathbf{y}\rangle$.

What is the length of the vector **x**? What are the distance and the angle between **x** and **y**?

(2) Calculate the moduli of the following real and complex numbers:
(i) $-3$ (ii) $0$ (iii) $4$ (iv) $3+4i$ (v) $4-3i$ (vi) $i$.

(3) If $a$ and $b$ are any real numbers, prove that $|a|<b$ if and only if $-b<a<b$. If $|a|<b$ for *all* $b>0$, prove that $a=0$.

(4) Let **x** and **y** be elements of $\mathbb{R}^n$. Prove that

$$d(\mathbf{x}, \mathbf{y})\geqq|\,\|\mathbf{x}\|-\|\mathbf{y}\|\,|.$$

[*Hint*: Use question 3.]

(5) Let **x** and **y** be elements of $\mathbb{R}^n$. Prove that

$$\langle\mathbf{x}, \mathbf{y}\rangle=\tfrac{1}{4}\{\|\mathbf{x}+\mathbf{y}\|^2-\|\mathbf{x}-\mathbf{y}\|^2\}.$$

(6) Let $\mathbf{x}\in\mathbb{R}^m$ and $\mathbf{y}\in\mathbb{R}^n$. Then $(\mathbf{x}, \mathbf{y})$ may be regarded as an element of $\mathbb{R}^{m+n}$. Prove that

$$\|(\mathbf{x}, \mathbf{y})\|^2=\|\mathbf{x}\|^2+\|\mathbf{y}\|^2.$$

---

### 13.14 **Euclidean geometry and $\mathbb{R}^n$**

In §10.1, we explained that it is the models in terms of which a system of axioms can be interpreted which make that system interesting. But this is a two-way process. It is equally true that a model is interesting because of the systems of axioms which it may satisfy.

The space $\mathbb{R}^2$ is particularly interesting because it serves as a model for the axioms of Euclidean geometry in the plane. The axioms given by Euclid himself are incomplete by modern standards in that certain of his theorems cannot be deduced from his axioms alone but depend also on implicit geometrical assumptions which are systematically taken for granted but never explicitly stated. (Some of these are mentioned briefly later on.) When we refer to the axioms of Euclidean geometry, we mean the axiom system given by David Hilbert in his famous book *The foundations of geometry* published at the beginning of this century.

The geometric terms which appear in Hilbert's axioms are the words *point, line, lie on, between* and *congruent*. To show that $\mathbb{R}^2$ is a model for Euclidean plane geometry one has to give a precise definition of each of these words in terms of $\mathbb{R}^2$ and then prove each of Hilbert's axioms for Euclidean plane geometry as a theorem in $\mathbb{R}^2$. This will demonstrate, in particular, that the axioms for Euclidean geometry are *consistent*.

An attempt at describing this programme in detail is beyond the scope of this book. (Interested readers will find the book *Elementary geometry from an advanced standpoint* by E. E. Moise (Addison-Wesley, 1963) an excellent reference.) Instead, we shall merely explain how some of the very familiar ideas of elementary geometry are expressed in terms of the space $\mathbb{R}^n$.

A *point* is simply an element $\mathbf{x}\in\mathbb{R}^n$. The *line* $l$ through the distinct points $\mathbf{a}$ and $\mathbf{b}$ is the set

$$l=\{\alpha\mathbf{a}+\beta\mathbf{b}:\alpha\in\mathbb{R},\ \beta\in\mathbb{R}\text{ and }\alpha+\beta=1\}.$$

This is easier to illustrate with a picture if we rewrite it in the form

$$l=\{\mathbf{a}+\beta(\mathbf{b}-\mathbf{a}):\beta\in\mathbb{R}\}$$

and think of $l$ as the line through the point $\mathbf{a}$ in the direction of the vector $\mathbf{b}-\mathbf{a}$.

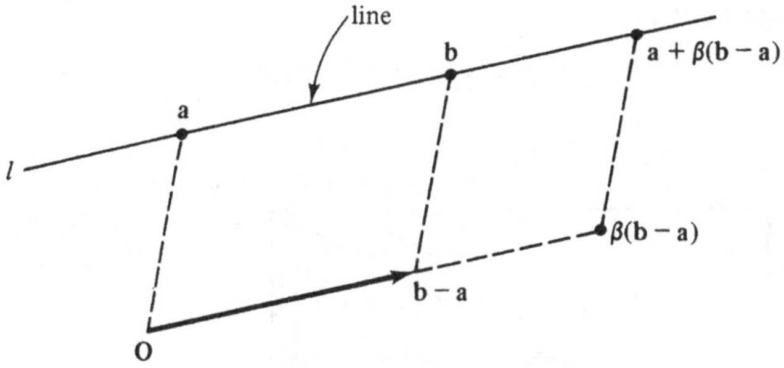


If, in the above, $\beta$ is restricted to be *non-negative* we obtain the definition of the *half-line* from $\mathbf{a}$ through $\mathbf{b}$. If $\mathbf{a}=\mathbf{0}$, such a half-line is called a *ray*. If both $\alpha$ and $\beta$ are restricted to be non-negative we obtain the definition of a line segment. Thus the *line segment* $L$ joining $\mathbf{a}$ and $\mathbf{b}$ is given by

$$L = \{\alpha\mathbf{a} + \beta\mathbf{b} : \alpha \geqq 0, \beta \geqq 0 \text{ and } \alpha + \beta = 1\}.$$

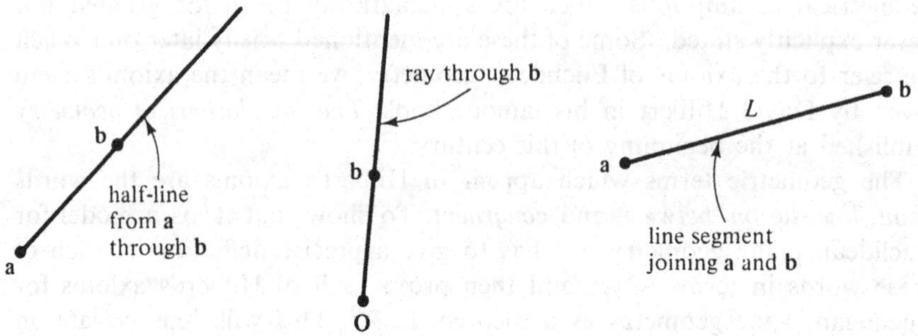


Some authors do not insist that the endpoints of a line segment belong to the line segment. Where it matters whether or not the endpoints of a line segment $L$ belong to $L$ we shall therefore sometimes stress the fact that they do by calling $L$ a *closed* line segment.

We have already discussed how *length* and *angle* are introduced into $\mathbb{R}^n$ via the norm and the inner product of the space. Two vectors $\mathbf{x}$ and $\mathbf{y}$ are said to be *orthogonal* if and only if

$$\langle \mathbf{x}, \mathbf{y} \rangle = 0.$$

Like most important ideas in mathematics, the word orthogonal has numerous synonyms of which some are *perpendicular*, *normal* and '*at right angles*'.

---

13.15 (*Pythagoras' theorem*) Let $\mathbf{x}$ and $\mathbf{y}$ be elements of $\mathbb{R}^n$. Then $\mathbf{x}$ and $\mathbf{y}$ are orthogonal if and only if

$$\|\mathbf{x} + \mathbf{y}\|^2 = \|\mathbf{x}\|^2 + \|\mathbf{y}\|^2.$$

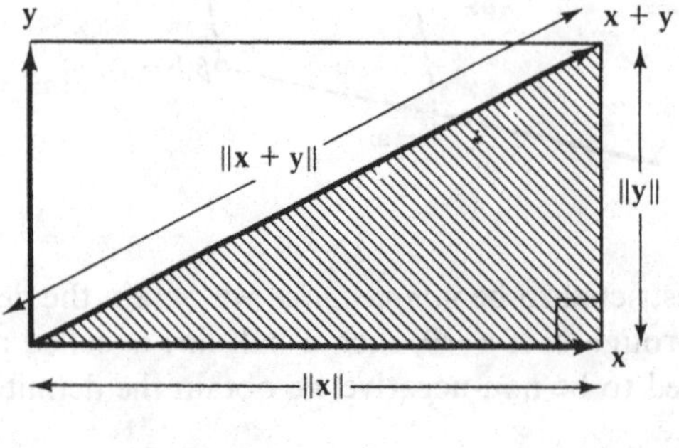

*Proof* We have that

$$\|\mathbf{x}+\mathbf{y}\|^2 = \langle \mathbf{x}+\mathbf{y}, \mathbf{x}+\mathbf{y}\rangle = \|\mathbf{x}\|^2 + 2\langle \mathbf{x}, \mathbf{y}\rangle + \|\mathbf{y}\|^2$$

and hence $\|\mathbf{x}+\mathbf{y}\|^2 = \|\mathbf{x}\|^2 + \|\mathbf{y}\|^2$ if and only if $\langle \mathbf{x}, \mathbf{y}\rangle = 0$.

---

A *circle* in $\mathbb{R}^2$ with centre $\boldsymbol{\xi}$ and radius $r>0$ is a set of the form

$$S = \{\mathbf{x} : \|\mathbf{x}-\boldsymbol{\xi}\| = r\}.$$

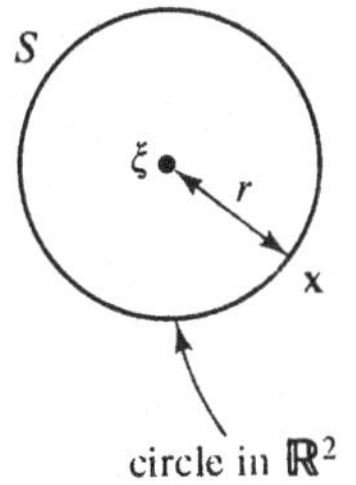


In $\mathbb{R}^3$, the set $S$ represents a *sphere* and in $\mathbb{R}^n$ we call $S$ a *hypersphere*.

The inside of a hypersphere – i.e. the set

$$B = \{\mathbf{x} : \|\mathbf{x}-\boldsymbol{\xi}\| < r\},$$

is called an *open ball*. (We should perhaps really call it a hyperball.) In $\mathbb{R}^3$, a ball looks like a ball ought to look. In $\mathbb{R}^2$ a ball is a *disc* and in $\mathbb{R}^1$ a ball is a bounded open interval $(\xi - r, \xi + r)$.

A *closed box* in $\mathbb{R}^n$ in a set $S$ of the form $I_1 \times I_2 \times \ldots \times I_n$ where each of the sets $I_k$ is a closed, bounded interval (i.e. an interval of the form $[a, b]$).

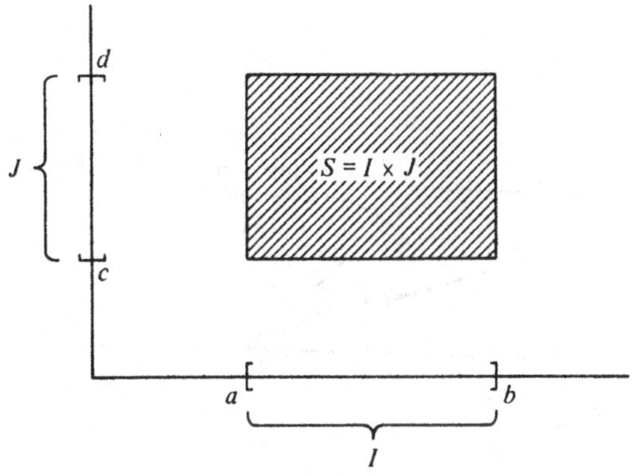

Both balls and boxes are examples of convex sets. *A* set $S$ in $\mathbb{R}^n$ is *convex* if and only if, whenever $\mathbf{x}\in S$ and $\mathbf{y}\in S$, it is true that $\alpha\mathbf{x}+\beta\mathbf{y}\in S$ provided that $\alpha\geqq 0$, $\beta\geqq 0$ and $\alpha+\beta=1$. In geometric terms, this means that, whenever two points are in the set, so is the line segment which joins them.

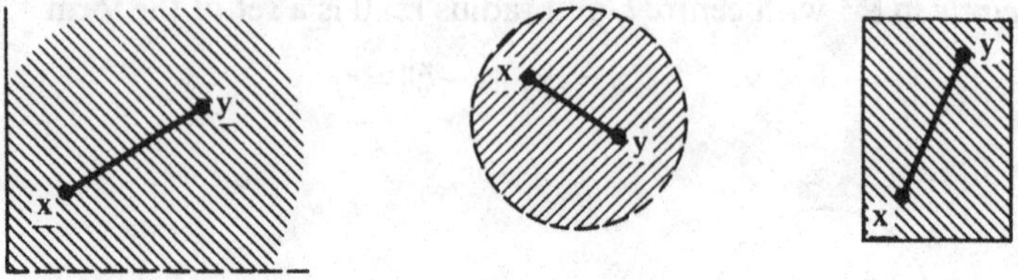


convex sets

non-convex sets

A *plane* in $\mathbb{R}^3$ through the point $\xi$ with *normal* $\mathbf{u}\neq\mathbf{0}$ is a set of the form

$$P=\{\mathbf{x}:\langle\mathbf{x}-\xi, u\rangle=0\}.$$

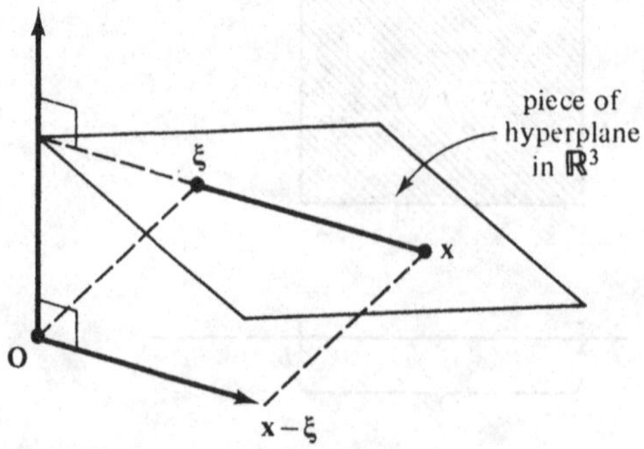

In $\mathbb{R}^n$, we call the set $P$ a hyperplane. A *hyperplane* in $\mathbb{R}^n$ with normal $\mathbf{u}\neq\mathbf{0}$ is therefore a set of the form

$$H=\{\mathbf{x}: \langle \mathbf{x}, \mathbf{u}\rangle = c\},$$

where $c$ is a constant. The hyperplane passes through any point $\xi$ which satisfies $\langle \xi, \mathbf{u}\rangle = c$. In $\mathbb{R}^3$ a hyperplane is an ordinary plane. In $\mathbb{R}^2$ a hyperplane is a line. In $\mathbb{R}^1$ a hyperplane is just a point.

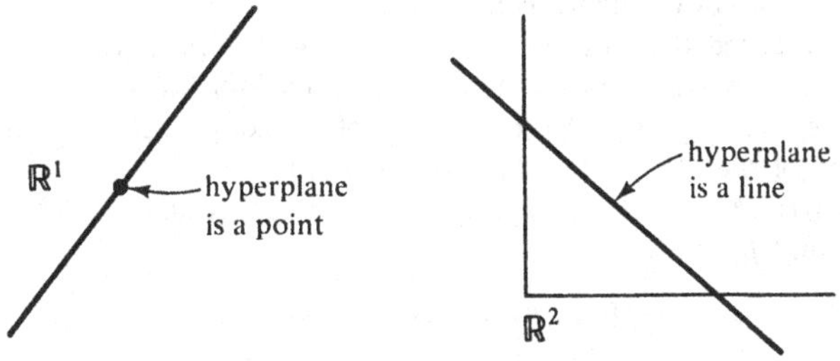


It is 'geometrically obvious' (at least in $\mathbb{R}^1$, $\mathbb{R}^2$ and $\mathbb{R}^3$) that a hyperplane splits $\mathbb{R}^n$ into two separate pieces (called half-spaces) and that any line segment joining two points from different half-spaces must cut the hyperplane. This fact is one of the assumptions that Euclid failed to give proper expression to in his axiom system. A proof of the result for $\mathbb{R}^n$ will have to wait until chapter 17 (exercise 17.28 (6)).

---

13.16 *Exercise*

(1) Let $\mathbf{x}$ and $\mathbf{y}$ be elements of $\mathbb{R}^n$. Prove that

$$\|\mathbf{x}+\mathbf{y}\|^2+\|\mathbf{x}-\mathbf{y}\|^2=2\|\mathbf{x}\|^2+2\|\mathbf{y}\|^2.$$

Interpret this result geometrically for $n=2$ in terms of the diagonals of a certain parallelogram.

(2) Let $\mathbf{x}$ and $\mathbf{y}$ be elements of $\mathbb{R}^n$. Prove that

$$\|\mathbf{x}\|=\|\mathbf{y}\| \Leftrightarrow \langle \mathbf{x}-\mathbf{y}, \mathbf{x}+\mathbf{y}\rangle = 0.$$

Interpret this result geometrically in terms of a triangle inscribed in a circle.

(3) Prove that there is a *unique* line through any two points in $\mathbb{R}^n$. Show that, if two lines meet, then their intersection consists of a single point. Prove that, in $\mathbb{R}^2$, a hyperplane is the same thing as a line.

(4) The *parallel postulate* for Euclidean plane geometry asserts that given any line $l$ and any point $P$ not on $l$, there exists a *unique* line through $P$ which does not meet $l$. *Prove* this for $\mathbb{R}^2$.

(5) Let us say that a closed box $S=I_1\times I_2\times\ldots\times I_n$ in $\mathbb{R}^n$ is proper if none

of the intervals $I_k$ consists of just a single point. Prove that every open ball $B$ contains a proper closed box $S$ and that every proper closed box $S$ contains an open ball.

(6) Prove that open balls, closed boxes, lines, half-lines and hyperplanes are all convex sets in $\mathbb{R}^n$. Prove that hyperspheres are *not* convex.

†(7) Prove that the intersection of any collection of convex sets is convex. What about unions?

†(8) A *cone* in $\mathbb{R}^n$ is the union of a collection of rays. Prove that the intersection of any collection of convex cones is a convex cone.

†(9) Taking for granted the truth of the assertion in the last paragraph of §13.15 about hyperplanes separating $\mathbb{R}^n$ into distinct 'half-spaces', explain why a line which enters a triangle in $\mathbb{R}^2$ through one of its sides must leave the triangle through one of the other sides (Pasch's theorem).

†(10) A function $f: \mathbb{R}^n \to \mathbb{R}^n$ is *linear* if and only if, for each $\mathbf{x}$ and $\mathbf{y}$ in $\mathbb{R}^n$ and all real numbers $\alpha$ and $\beta$,

$$f(\alpha\mathbf{x}+\beta\mathbf{y})=\alpha f(\mathbf{x})+\beta f(\mathbf{y}).$$

Prove that, for each $S$ in $\mathbb{R}^n$,

$$S \text{ convex} \Rightarrow f(S) \text{ convex}.$$

†(11) Prove that a function $f: \mathbb{R}^n \to \mathbb{R}^1$ is linear if and only if there exists an $\mathbf{e} \in \mathbb{R}^n$ such that $f(\mathbf{x}) = \langle \mathbf{x}, \mathbf{e} \rangle$ for each $\mathbf{x} \in \mathbb{R}^n$.

†(12) Anticipating the definition of §13.17, prove that a normed vector space $\mathcal{X}$ (with real scalars) admits an inner product satisfying (i), (ii) and (iii) of §13.3 if and only if

$$\|\mathbf{x}+\mathbf{y}\|^2+\|\mathbf{x}-\mathbf{y}\|^2=2\|\mathbf{x}\|^2+2\|\mathbf{y}\|^2$$

for each $\mathbf{x} \in \mathcal{X}$ and each $\mathbf{y} \in \mathcal{X}$.

[*Hint*: Use exercises 13.16(1) and 13.13(5).]

---

## 13.17† Normed vector spaces

The space $\mathbb{R}^n$ is an example of a normed vector space.

For a vector space, one first needs a commutative group $\mathcal{X}$ (§6.6) to serve as the *vectors*. The group operation is then called vector addition.

Next one needs a field to serve as the *scalars*. A meaning then has to be assigned to scalar multiplication in such a way that the following rules are satisfied:

(i) $\alpha(\mathbf{x}+\mathbf{y})=\alpha\mathbf{x}+\alpha\mathbf{y}$
(ii) $(\alpha+\beta)\mathbf{x}=\alpha\mathbf{x}+\beta\mathbf{x}$
(iii) $(\alpha\beta)\mathbf{x}=\alpha(\beta\mathbf{x})$
(iv) $0\mathbf{x}=\mathbf{0}$; $1\mathbf{x}=\mathbf{x}$.

For a *normed* vector space, we usually insist that the field of scalars be $\mathbb{R}$ or $\mathbb{C}$. A norm function from $\mathcal{X}$ to $\mathbb{R}$ is required which satisfies the following properties:

(i) $\|\mathbf{x}\| \geqq 0$
(ii) $\|\mathbf{x}\|=0 \Leftrightarrow \mathbf{x}=\mathbf{0}$
(iii) $\|\alpha\mathbf{x}\|=|\alpha|\,\|\mathbf{x}\|$
(iv) $\|\mathbf{x}+\mathbf{y}\| \leqq \|\mathbf{x}\|+\|\mathbf{y}\|$ (triangle inequality)

for all $\mathbf{x} \in \mathcal{X}$, $\mathbf{y} \in \mathcal{X}$ and all scalars. (Here $\|\mathbf{x}\|$ denotes the image of $\mathbf{x}$ under the norm function.)

The chief reason for giving the requirements for a normed vector space at this stage is to provide a summary of the properties of $\mathbb{R}^n$ which will be important in the next few chapters. Other normed vector spaces will be mentioned hardly at all. It should be noted, however, that the properties of infinite-dimensional normed vector spaces are often counter-intuitive. For example, a hyperplane in an infinite-dimensional normed vector space does *not*, in general, split the space into two separate half-spaces.

---

### 13.18 Metric space

A *metric space* is a set $\mathcal{X}$ and a function $d\colon \mathcal{X} \times \mathcal{X} \to \mathbb{R}$ which satisfies

(i) $d(\mathbf{x}, \mathbf{y}) \geqq 0$
(ii) $d(\mathbf{x}, \mathbf{y}) = 0 \Leftrightarrow \mathbf{x} = \mathbf{y}$
(iii) $d(\mathbf{x}, \mathbf{y}) = d(\mathbf{y}, \mathbf{x})$
(iv) $d(\mathbf{x}, \mathbf{z}) \leqq d(\mathbf{x}, \mathbf{y}) + d(\mathbf{y}, \mathbf{z})$. (triangle inequality)

The function $d$ is said to be the *metric* for the metric space and we interpret $d(\mathbf{x}, \mathbf{y})$ as the *distance* between $\mathbf{x}$ and $\mathbf{y}$.

The metric space in which we shall be most interested is the space $\mathbb{R}^n$. As we have seen in §13.11, the metric $d^2 : \mathbb{R}^n \times \mathbb{R}^n \to \mathbb{R}$ in $\mathbb{R}^n$ is defined by

$$d(\mathbf{x}, \mathbf{y}) = \|\mathbf{x} - \mathbf{y}\|.$$

Strictly speaking, one should refer to $d$ as the *Euclidean metric*. It is trivial to check that the Euclidean metric satisfies the requirements for a metric given above. In particular, the triangle inequality for a metric follows from theorem 13.7 since

$$\|\mathbf{x} - \mathbf{z}\| = \|(\mathbf{x} - \mathbf{y}) + (\mathbf{y} - \mathbf{z})\| \leqq \|\mathbf{x} - \mathbf{y}\| + \|\mathbf{y} - \mathbf{z}\|.$$

Any subset $\mathcal{Z}$ of $\mathbb{R}^n$ is also a metric space provided that we continue to use the Euclidean metric in $\mathcal{Z}$. We then say that $\mathcal{Z}$ is a *metric subspace* of $\mathbb{R}^n$. For example, the interval $(0, 1]$ is a metric subspace of $\mathbb{R}^1$ provided that the metric $d\colon (0, 1] \times (0, 1] \to \mathbb{R}$ is defined by

$$d(x, y) = |x - y|.$$

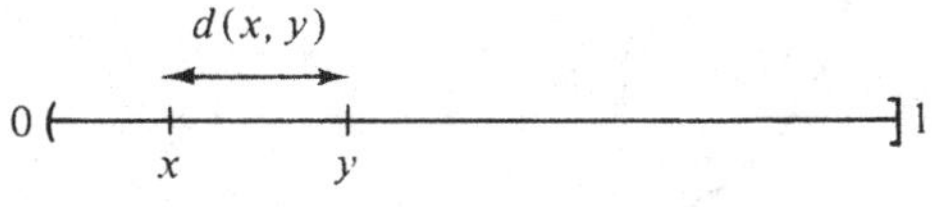


More exotic examples of metric spaces can be obtained by starting with any normed vector space $\mathcal{X}$ and defining $d\colon \mathcal{X} \times \mathcal{X} \to \mathbb{R}$ by

$$d(\mathbf{x}, \mathbf{y}) = \|\mathbf{x} - \mathbf{y}\|.$$

But there are also many interesting metric spaces which have no vector space structure at all. Since we have been discussing the fact that $\mathbb{R}^2$ (with the Euclidean metric) is a model for Euclidean geometry (§13.14), we shall give Poincaré's model for a non-Euclidean geometry as an example of such a metric space.

## 13.19† Non-Euclidean geometry

Perhaps the most well-known of Euclid's postulates is the *parallel postulate* (quoted in exercise 13.16(4)). This was always felt to be less satisfactory than Euclid's other assumptions in that it is less 'intuitively obvious' than the others. Very considerable efforts were therefore made to deduce it as a theorem from the other axioms. All these efforts were unsuccessful. Finally, it was realised that the task is impossible and Gauss, Lobachevski and Bolyai independently began to study a geometry in which the parallel postulate is *false* but all the other assumptions of Euclidean geometry are true. Gauss did not publish his work and Lobachevski published before Bolyai. Hence the non-Euclidean geometry they studied will be called *Lobachevskian geometry*. (Sometimes it is called 'hyperbolic geometry'.) This Lobachevski, incidentally, is the Nikolai Ivanovich Lobachevski of the immortal Tom Lehrer song but the scandalous suggestions made in this song are totally unfounded!

In Lobachevskian geometry there are *many* parallel lines through a given point parallel to a given line. This may seem intuitively implausible as a hypothesis about the 'real world' because we have been trained from early childhood to think of space as Euclidean. In fact, Einsteinian physics assures us that, in the vicinity of a gravitating body, space is very definitely *not* Euclidean.

The purpose of this long pre-amble is to explain the interest of the following metric space which was introduced by the great French mathematician Poincaré. This space provides a model for Lobachevskian geometry. Its existence therefore demonstrates that the axioms of Lobachevskian geometry are consistent. What is more, since the parallel postulate is *true* in $\mathbb{R}^2$ but *false* in the Poincaré model, it must be *independent* of the other axioms of Euclidean geometry. In particular, it cannot be deduced from them.

The set $\mathcal{X}$ in Poincaré's metric space is the set $\mathcal{X}=\{(x, y): x^2+y^2<1\}$ in $\mathbb{R}^2$. But the metric used in $\mathcal{X}$ is *not* the Euclidean metric.

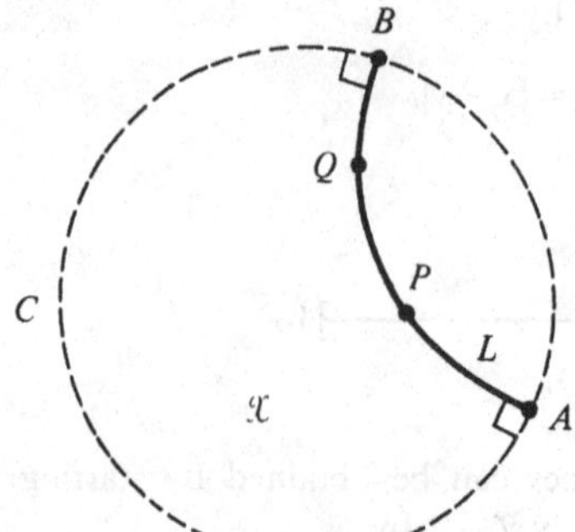

Let $P$ and $Q$ be points of $\mathscr{X}$. If $P$ and $Q$ lie on a diameter of the circle $C$ which bounds $\mathscr{X}$, let $L$ be this diameter. If $P$ and $Q$ do not lie on a diameter, let $L$ be the unique circle through $P$ and $Q$ which is orthogonal to $C$. If $A$ and $B$ are as indicated in the diagram, we then define the *Poincaré distance* between $P$ and $Q$ by

$$d(P, Q) = \left| \log\left(\frac{QB/QA}{PB/PA}\right) \right| .$$

One can think of the points in the diagram below as the footprints of a little man who tries to walk from the centre of $\mathscr{X}$ to its boundary. Each step is of equal size *relative to the Poincaré metric.* He will therefore never reach the boundary and therefore, as far as he is concerned, $\mathscr{X}$ extends indefinitely in all directions.

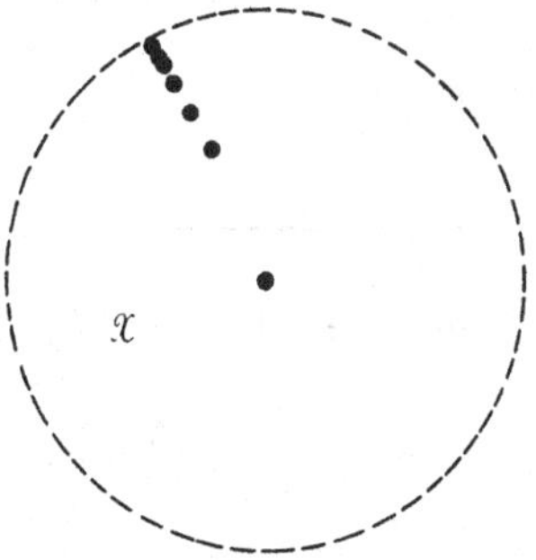


Poincaré defined a 'straight line' in $\mathscr{X}$ to be one of our curves $L$. This is perfectly reasonable since, relative to the Poincaré metric, the shortest route from $P$ to $Q$ is along $L$. He defined the angle between two 'straight lines' $L$ and $M$ to be the ordinary Euclidean angle between $L$ and $M$. With these definitions $\mathscr{X}$ is a model of Lobachevskian geometry. The diagram below shows two 'straight lines' $M$ and $N$ through the point $P$ 'parallel' to the 'straight line' $L$.

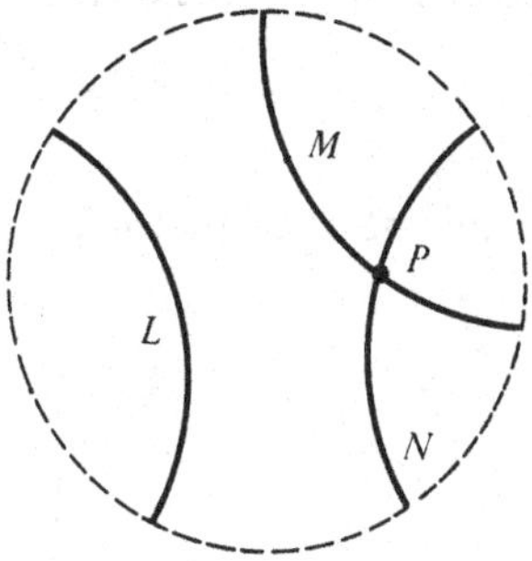


## 13.20 Distance between a point and a set

If $\xi$ is a point in a metric space $\mathscr{X}$ and $S$ is a non-empty set in $\mathscr{X}$, we define the *distance* between ξ and $S$ by

$$d(\xi, S) = \inf_{\mathbf{x} \in S} d(\xi, \mathbf{x}).$$

Obviously, $d(\xi, S)$ is always non-negative.

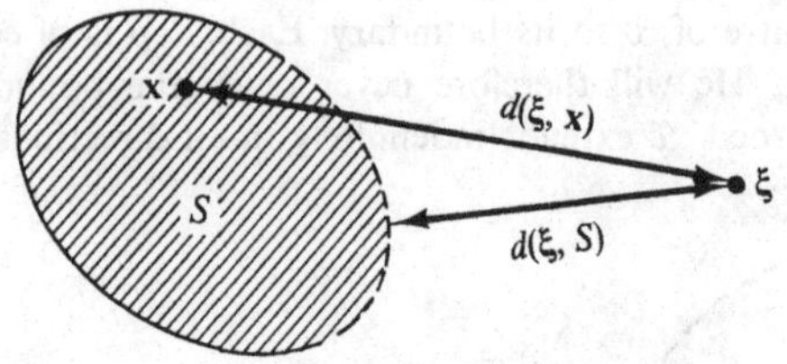


13.21 *Example* Consider the point 3 and the set $S = [0, 1]$ in $\mathbb{R}$. We have that

$$d(3, S) = d(3, 1) = 2.$$

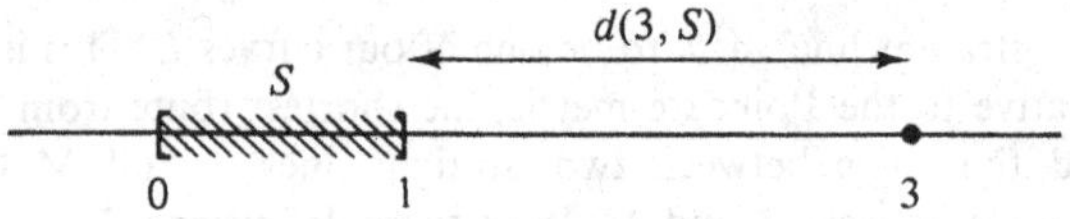


13.22 *Theorem* Let ξ be a point in a metric space $\mathcal{X}$ and let $S$ be a non-empty set in $\mathcal{X}$. Then

$$d(\xi, S) = 0$$

if and only if, for each $\varepsilon > 0$, there exists an $\mathbf{x} \in S$ such that $d(\xi, \mathbf{x}) < \varepsilon$.

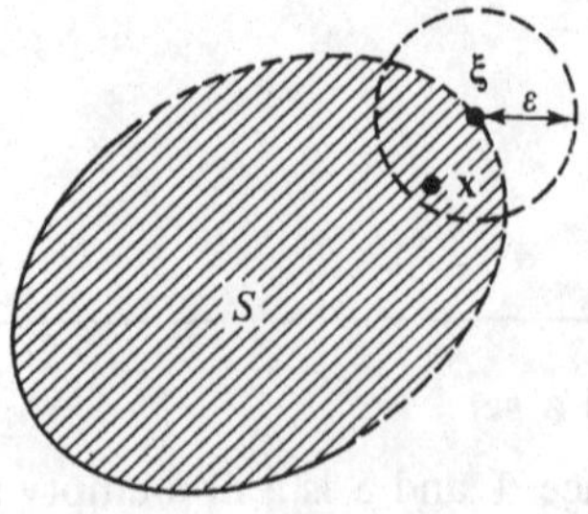

*Proof* Everything in the set

$$D = \{d(\mathbf{x}, \xi) : \mathbf{x} \in S\}$$

is non-negative. Thus 0 is a lower bound for $D$. The statement $d(\xi, S) = 0$ is therefore equivalent to the assertion that no $\varepsilon > 0$ is a lower bound for $D$ – i.e.

$$\text{not } \exists \varepsilon > 0 \ \forall \mathbf{x} \in S \ (d(\mathbf{x}, \xi) \geqq \varepsilon).$$

But this is equivalent to

$$\forall \varepsilon > 0 \ \exists \mathbf{x} \in S \ (d(\mathbf{x}, \xi) < \varepsilon)$$

as required (§3.10).

---

It is important to take note of the fact that $d(\xi, S) = 0$ does *not* imply that $\xi \in S$.

---

13.23 *Examples*

(i) Consider the point 1 and the set $S = (0, 1)$ in $\mathbb{R}$. For each $\varepsilon > 0$, we can find an $x \in S$ such that $x > 1 - \varepsilon$.

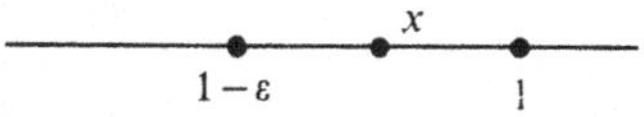


Since $1 - \varepsilon < x < 1$, $d(1, x) < \varepsilon$. It follows from theorem 13.22 that

$$d(1, S) = 0.$$

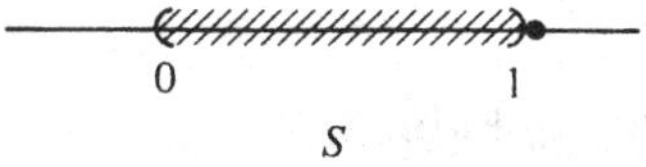


(ii) Consider the point 0 and the set $S = \{1/n : n \in \mathbb{N}\}$ in $\mathbb{R}$. Given any $\varepsilon > 0$, we can find an $n \in \mathbb{N}$ such that $n > 1/\varepsilon$ (because $\mathbb{N}$ is unbounded above). But then

$$d(0, 1/n) = 1/n < \varepsilon$$

and hence

$$d(0, S) = 0.$$

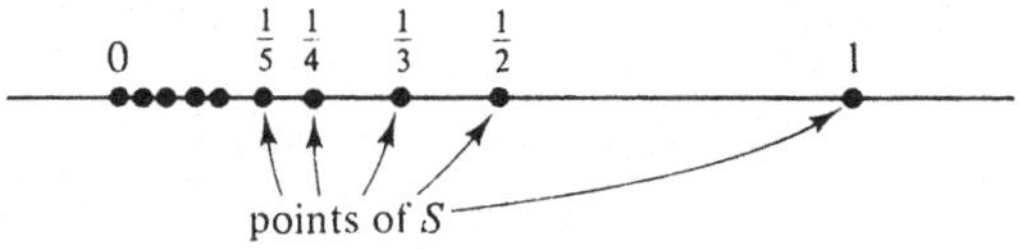


---

13.24 *Exercise*

(1) Find $d(\xi, S)$ for the following points $\xi$ and sets $S$ in $\mathbb{R}$.

(i) $\xi=1$; $S=(0, 2)$
(ii) $\xi=0$; $S=[1, 2]$
(iii) $\xi=\frac{1}{2}$; $S=\mathbb{N}$
(iv) $\xi=0$; $S=(0, 1)$
(v) $\xi=1$; $S=(2, 3)$
(vi) $\xi=2$; $S=(0, 1)\cup(3, 4)$
(vii) $\xi=\sqrt{2}$; $S=\mathbb{Q}$. [*Hint*: Use theorem 9.20.]

(2) Find $d(\boldsymbol{\xi}, S)$ for the following points $\boldsymbol{\xi}$ and sets $S$ in $\mathbb{R}^2$.

(i) $\boldsymbol{\xi}=(1, 1)$; $S=\{(x, y): 0<x<1, 0<y<1\}$
(ii) $\boldsymbol{\xi}=(2, 1)$; $S=\{(x, y): x^2+y^2\leqq 1\}$
(iii) $\boldsymbol{\xi}=(0, 0)$; $S=\{(x, y): 1<x^2+y^2\leqq 2\}$.

(3) Let $S$ be a non-empty set of real numbers which is bounded above. Prove that

$$d(\sup S, S)=0.$$

(4) Let $\boldsymbol{\xi}$ be a point in a metric space $\mathcal{X}$ and let $S$ and $T$ be non-empty sets in $\mathcal{X}$. Prove that

(i) $\boldsymbol{\xi}\in S \Rightarrow d(\boldsymbol{\xi}, S)=0$
(ii) $S\subset T \Rightarrow d(\boldsymbol{\xi}, S)\geqq d(\boldsymbol{\xi}, T)$
(iii) $d(\boldsymbol{\xi}, S\cup T) = \min\{d(\boldsymbol{\xi}, S), d(\boldsymbol{\xi}, T)\}$.

What can be said of $d(\boldsymbol{\xi}, S\cap T)$ in terms of $d(\boldsymbol{\xi}, S)$ and $d(\boldsymbol{\xi}, T)$?

(5) Let $\boldsymbol{\xi}$ and $\boldsymbol{\eta}$ be points in a metric space $\mathcal{X}$ and let $S$ be a non-empty set in $\mathcal{X}$. Prove that

$$d(\boldsymbol{\xi}, S)\leqq d(\boldsymbol{\xi}, \boldsymbol{\eta})+d(\boldsymbol{\eta}, S).$$

(6) Let $S_1$, $S_2, \ldots,$ $S_n$ be non-empty sets of real numbers. Write $S=S_1\times S_2\times\ldots\times S_n$ and suppose that $\mathbf{x}=(x_1, x_2, \ldots, x_n)\in\mathbb{R}^n$. Prove that $d(\mathbf{x}, S)=0$ if and only if

$$d(x_k, S_k)=0$$

for each $k=1, 2, \ldots, n$.

# 14 OPEN AND CLOSED SETS (I)

## 14.1 Introduction

In the next two chapters we prove a number of theorems about the properties of sets in metric spaces. Most of these theorems are true in any metric space whatsoever. Others are true only for the metric space $\mathbb{R}^n$. However, with just an occasional exception, all examples will be of sets in $\mathbb{R}^n$. A systematic attempt to give examples from a wider range of spaces at this stage would submerge the essential ideas in a sea of technicalities.

We begin by discussing open and closed sets. These generalise the idea of open and closed intervals which we met in §7.13. The intervals $(a, b)$, $(a, \infty)$, $(-\infty, b)$ are *open* and the intervals $[a, b]$, $[a, \infty)$, $(-\infty, b]$ are *closed*. (The sets $\emptyset$ and $\mathbb{R}$ count both as open and as closed intervals.) The difference between open and closed intervals lies in the treatment of their endpoints. In the general case, we shall have to talk about the boundary points of a set rather than the endpoints of an interval. Our first priority will therefore be to give a precise mathematical definition of a boundary point of a set $S$ in a metric space $\mathcal{X}$.

---

## 14.2 Boundary of a set

Let $S$ be a set in a metric space $\mathcal{X}$. A *boundary point* of the set $S$ is a point $\xi$ of $\mathcal{X}$ such that

(i) $d(\xi, S)=0$

and (ii) $d(\xi, \complement S)=0$.

Note that a boundary point of $S$ need *not* be a point of $S$.

The set of all boundary points of $S$ is denoted by $\partial S$. Observe that $S$ and $\complement S$ have the *same* boundary – i.e. $\partial S=\partial(\complement S)$.

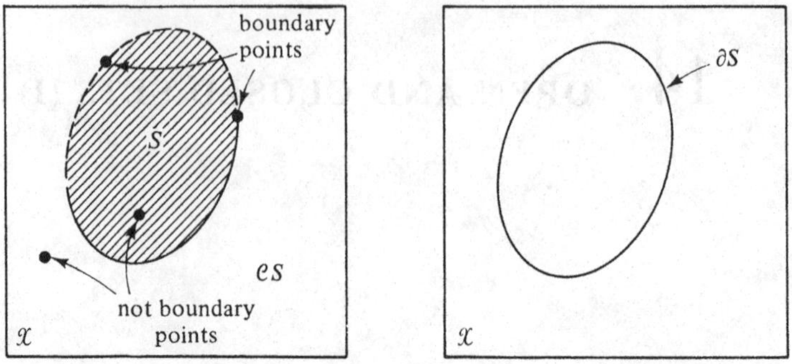


Some special consideration of the case $S=\emptyset$ is necessary. In this case $d(\xi, S)=0$ is meaningless and hence cannot be true. Thus the empty set has *no* boundary points – i.e. $\partial\emptyset=\emptyset$. Similarly, $\partial\mathcal{X}=\emptyset$.

---

### 14.3 Open balls

The open ball $B$ with centre $\xi$ and radius $r>0$ in a metric space $\mathcal{X}$ is defined by

$$B=\{\mathbf{x}: d(\xi, \mathbf{x})<r\}.$$

In $\mathbb{R}^3$, an open ball is the inside of a sphere. In $\mathbb{R}^2$, an open ball is the inside of a circle. In $\mathbb{R}^1$, an open ball is an open interval.

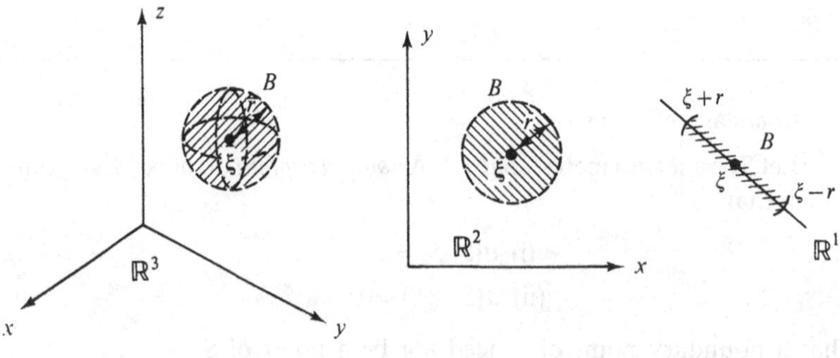


---

14.4 *Theorem* Let $S$ be a set in a metric space $\mathcal{X}$ and let $\xi$ be a point in $\mathcal{X}$. Then $\xi$ is a boundary point of $S$ if and only if each open ball $B$ with centre $\xi$ contains a point of $S$ and a point of $\mathcal{C}S$.

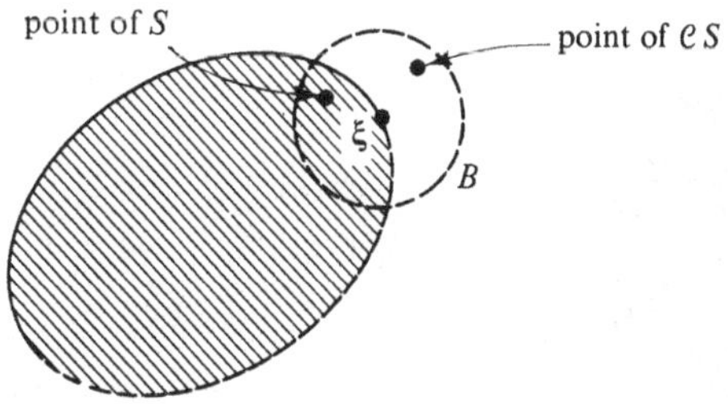


*Proof* The fact that each open ball $B$ with centre $\xi$ and radius $r>0$ contains a point of $S$ and a point of $\mathcal{C}S$ is equivalent to the assertion that, for each $r>0$, there exists $\mathbf{x}\in S$ and $\mathbf{y}\in\mathcal{C}S$ such that $d(\xi, \mathbf{x})<r$ and $d(\xi, \mathbf{y})<r$. But, from theorem 13.22, we know that this is the same as saying that $d(\xi, S)=0$ and $d(\xi, \mathcal{C}S)=0$ – i.e. $\xi$ is a boundary point of $S$ and of $\mathcal{C}S$.

---

14.5 *Examples*

(i) The boundary of the set $F$ in $\mathbb{R}^2$ defined by

$$F=\{(x, y): 0\leqq y\leqq 1\}$$

is the set

$$\partial F=\{(x, y): y=0 \text{ or } y=1\}.$$

Note that every boundary point of $F$ belongs to $F$ – i.e. $\partial F\subset F$.

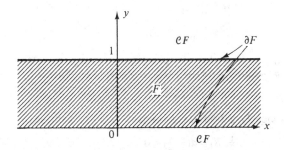

(ii) The boundary of the set $G$ in $\mathbb{R}^2$ defined by

$$G=\{(x, y): x^2+y^2<1\}$$

is the set

$$\partial G=\{(x, y): x^2+y^2=1\}.$$

Note that *no* boundary point of $G$ belongs to $G$ – i.e. $\partial G\subset\mathcal{C}G$.

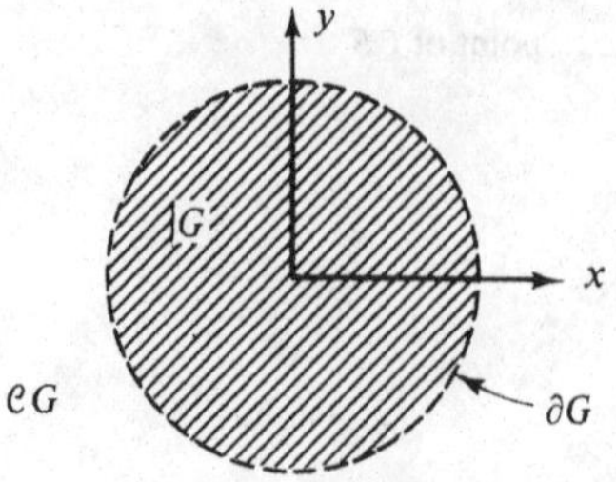


(iii) The boundary of the set $H$ in $\mathbb{R}^2$ defined by

$$H=\{(x, y)\colon x^2+y^2<1 \text{ and } y\geqq 0\}$$

is the set

$$\partial H=\{(x, y)\colon x^2+y^2=1 \text{ and } y\geqq 0\}\cup\{(x, 0)\colon -1\leqq x\leqq 1\}.$$

Note that some boundary points of $H$ belong to $H$ and some do not.

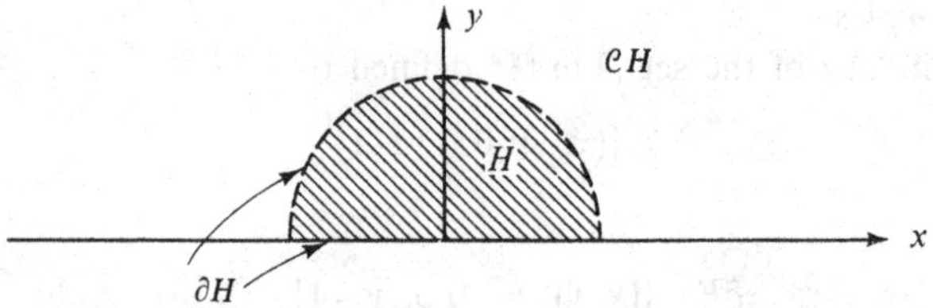


(iv) The boundary of the set $S=\{1/n\colon n\in\mathbb{N}\}$ in $\mathbb{R}^1$ is

$$\partial S=S\cup\{0\}.$$

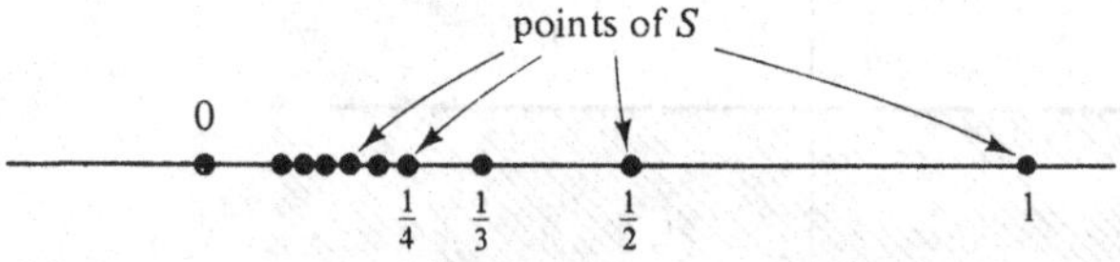


Each point of $S$ is clearly a boundary point of $S$. For example,

$$d(\tfrac{1}{2}, S)=d(\tfrac{1}{2}, \tfrac{1}{2})=0$$
$$d(\tfrac{1}{2}, \complement S)\leqq d(\tfrac{1}{2}, (\tfrac{1}{2}, 1))=0$$

and thus $\frac{1}{2}\in\partial S$. Similarly, 0 is a boundary point of $S$ because

$$d(0, S)=0 \quad \text{(example 13.23(ii))}$$
$$d(0, \complement S)=d(0, 0)=0.$$

Finally, no other point of $\mathbb{R}^1$ can be a boundary point of $S$. For example, $\frac{3}{4}$ is not a boundary point of $S$ because

$$d(\tfrac{3}{4}, S)=d(\tfrac{3}{4}, 1)=\tfrac{1}{4}\neq 0.$$

14.6 *Exercise*

(1) Sketch the following sets in $\mathbb{R}^2$ and determine their boundary points

(i) $P=\{(x, y): 0 \leqq x \leqq 1 \text{ and } 0 < y < 1\}$
(ii) $Q=\{(x, y): x < y\}$
(iii) $R=\{(x, y): y \geqq x^2\}$
(iv) $S=\{(1, 1), (2, 2), (3, 3)\}$.

(2) Sketch the following sets in $\mathbb{R}^1$ and determine their boundary points.
(i) $(0, 1)$ (ii) $[0, 1]$ (iii) $(0, 1]$ (iv) $[0, 1)$
(v) $(0, \infty)$ (vi) $[0, \infty)$ (vii) $\mathbb{N}$ (viii) $\mathbb{Q}$.

†(3) Let $S$ be a *finite* set in $\mathbb{R}^n$. Prove that $\partial S = S$.
(Note that this need not be true in a general metric space. For example, if $\mathcal{X} = S$.)

†(4) Show that the open ball $B$ in $\mathbb{R}^n$ with centre $\boldsymbol{\xi}$ and radius $r>0$ has boundary

$$\partial B = \{\mathbf{x}: d(\mathbf{x}, \boldsymbol{\xi}) = r\}.$$

(Note that this need not be true in a general metric space. For example, if $\mathcal{X}$ is the metric subspace of $\mathbb{R}^1$ consisting simply of the two points $\xi$ and $\xi + r$.)

†(5) Find the boundary points of the set $S$ in $\mathbb{R}^2$ defined by

$$S = \{(1/m, 1/n): m \in \mathbb{N} \text{ and } n \in \mathbb{N}\}.$$

†(6) Let $A$ and $B$ be sets in a metric space $\mathcal{X}$. Prove that $\partial(A \cup B) \subset \partial A \cup \partial B$. [*Hint*: Use theorem 14.4.] Deduce that $\partial(A \cap B) \subset \partial A \cap \partial B$.

---

## 14.7 Open and closed sets

Let $S$ be a set in a metric space $\mathcal{X}$. The set $S$ is *open* if and only if it contains none of its boundary points – i.e.

$$\partial S \subset \mathcal{C}S.$$

The set $S$ is *closed* if and only if it contains *all* of its boundary points – i.e.

$$\partial S \subset S.$$

It may be helpful to think of $S$ as a fenced plot of land in a city. The fence around the plot can then be thought of as the boundary of $S$. As all city dwellers know, the neighbours are not to be believed when, on moving into a new house, they explain that all of the fence is built on your land (and hence you are responsible for its upkeep). In fact, some of the fence is built on your land and some on your neighbour's land. One can think of the part of the fence on your land as the part of the boundary of $S$ which belongs to $S$. The part of the fence built on the neighbour's land can be regarded as the part of the boundary of $S$ which belongs to $\mathcal{C}S$.

An *open* set can then be thought of as a plot for which *all* the fence belongs to the neighbours. A *closed* set is a plot for which *all* of the fence belongs to the owner.

When drawing pictures, we use a dotted line to indicate the boundary of an open set and a firm line to indicate the boundary of a closed set.

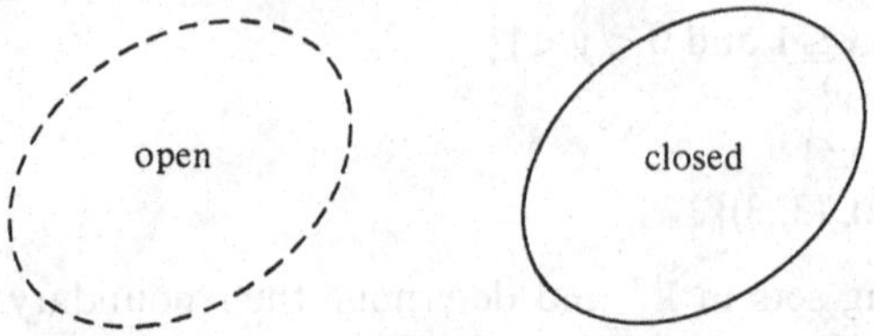


As with most city plots of land, sets are usually *neither* open *nor* closed.

---

14.8 *Example* Consider the sets of examples 14.5.
(i) The set $F$ is closed.
(ii) The set $G$ is open.
(iii) The set $H$ is neither open nor closed.
(iv) The set $S$ is neither open nor closed.

---

14.9 *Theorem* A set $S$ in a metric space $\mathscr{X}$ is open if and only if $\mathscr{C}S$ is closed.

*Proof* The theorem follows immediately from the fact that $S$ and $\mathscr{C}S$ have the same boundary.

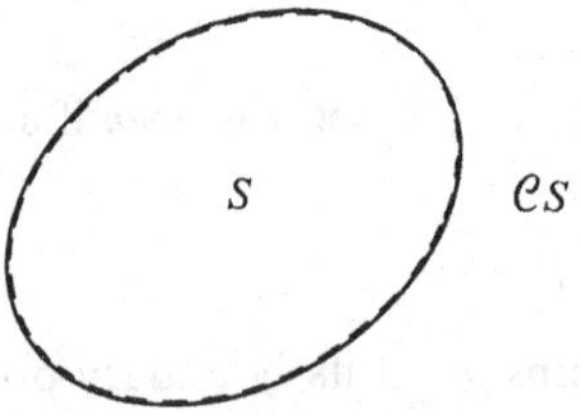


---

14.10 *Theorem* A set $F$ in a metric space $\mathscr{X}$ is closed if and only if, for each $\mathbf{x} \in \mathscr{X}$,

$$d(\mathbf{x}, F) = 0 \Rightarrow \mathbf{x} \in F.$$

*Proof* (i) Suppose that $d(\mathbf{x}, F) = 0 \Rightarrow \mathbf{x} \in F$. If $\xi$ is a boundary point of $F$, then $d(\xi, F) = 0$ and $d(\xi, \mathscr{C}F) = 0$. Since $d(\xi, F) = 0$, it follows that $\xi \in F$. Thus $F$ contains its boundary points and hence is closed.
(ii) Suppose that $F$ is closed. If $\mathbf{x} \notin F$, then $\mathbf{x} \in \mathscr{C}F$ and hence $d(\mathbf{x}, \mathscr{C}F) = 0$. If it is also true that $d(\mathbf{x}, F) = 0$, then $\mathbf{x}$ is a boundary point of $F$ and so, since

$F$ is closed, $\mathbf{x} \in F$. This is a contradiction. Hence $\mathbf{x} \notin F \Rightarrow d(\mathbf{x}, F) \neq 0$ – i.e. $d(\mathbf{x}, F) = 0 \Rightarrow \mathbf{x} \in F$.

---

14.11 *Corollary* A non-empty closed set $F$ of real members which is bounded above has a maximum.

*Proof* We have to show that $\sup F \in F$. But $d(\sup F, F) = 0$ by exercise 13.24(3). Hence the result by theorem 14.10.

---

14.12 *Theorem* A set $G$ in a metric space $\mathcal{X}$ is open if and only if each point $\xi \in G$ is the centre of an open ball $B$ entirely contained in $G$.

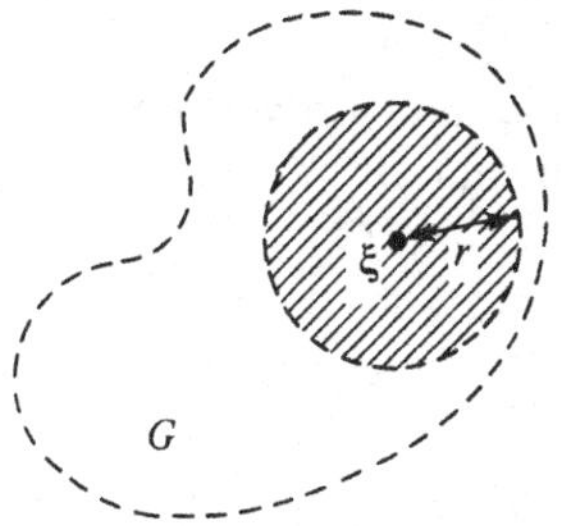


*Proof* (i) Suppose that $\xi \in G$ is the centre of an open ball $B$ entirely contained in $G$. Then $\xi$ cannot be a boundary point of $G$ because $B$ contains no point of $\mathcal{C}G$ (theorem 14.4). Thus $G$ contains none of its boundary points and so is open.

(ii) Suppose that $G$ is open. Let $\xi \in G$. *All* open balls $B$ with centre $\xi$ then contain a point of $G$. Since $\xi$ is *not* a boundary point of $G$, it follows from theorem 14.4 that at least one open ball $B$ with centre $\xi$ contains no point of $\mathcal{C}G$ – i.e. $B \subset G$.

---

14.13 *Theorem* In a metric space $\mathcal{X}$:

(i) The sets $\emptyset$ and $\mathcal{X}$ are open.

(ii) The union $S$ of any collection $\mathcal{W}$ of open sets is open.

(iii) The intersection $T$ of any *finite* collection $\{G_1, G_2, \ldots, G_k\}$ of open sets is open.

*Proof* (i) The set $\emptyset$ is open because $\partial\emptyset = \emptyset \subset \mathcal{C}\emptyset = \mathcal{X}$. The set $\mathcal{X}$ is open because $\partial\mathcal{X} = \emptyset \subset \mathcal{C}\mathcal{X} = \emptyset$.

(ii) Let $\xi \in S$. We shall use theorem 14.12 and prove that $S$ is open by finding an open ball $B$ with centre $\xi$ such that $B \subset S$.

Since $\xi \in S$, there exists a $G \in \mathcal{W}$ such that $\xi \in G$. Since $G$ is open we can find an open ball $B$ with centre $\xi$ such that $B \subset G$. But $G \subset S$ and so the result follows.

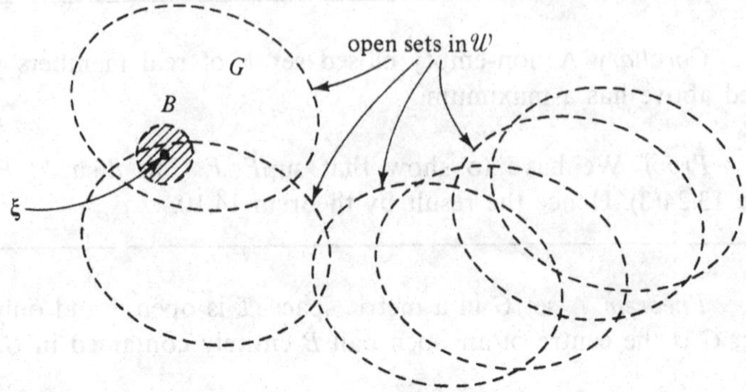


(iii) Let $\xi \in T$. We shall use theorem 14.12 and prove that $T$ is open by finding an open ball $B$ with centre $\xi$ such that $B \subset T$.

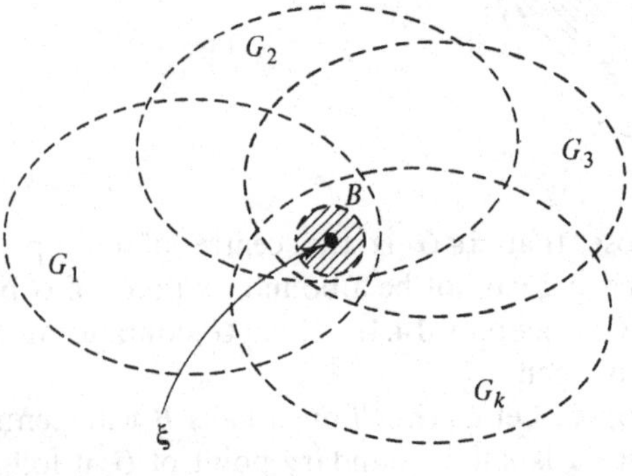


Since $\xi \in T$, we have that $\xi \in G_1, \xi \in G_2, \ldots, \xi \in G_k$. Because the sets $G_1$, $G_2$, $\ldots$, $G_k$ are all open, we can find open balls $B_1, B_2, \ldots, B_k$ each with centre $\xi$ such that

$$B_j \subset G_j \quad (j = 1, 2, \ldots, k).$$

One of the open balls $B_1, B_2, \ldots, B_k$ has minimum radius. Let this open ball be $B$. Then $B \subset B_j \subset G_j$ $(j = 1, 2, \ldots, k)$ and hence

$$B \subset T = \bigcap_{j=1}^{k} G_j.$$

Thus $B$ is an open ball with centre $\xi$ such that $B \subset T$.

---

14.14 *Theorem* In a metric space $\mathcal{X}$:

(i) The sets $\emptyset$ and $\mathcal{X}$ are closed (as well as open).

(ii) The intersection $S$ of any collection $\mathcal{W}$ of closed sets is closed.
(iii) The union $T$ of any finite collection $\{F_1, F_2, \ldots, F_k\}$ of closed sets is closed.

*Proof* The theorem follows from theorems 14.9 and 14.13.
(i) The sets $\emptyset$ and $\mathcal{X}$ are closed because $\mathcal{C}\emptyset = \mathcal{X}$ and $\mathcal{C}\mathcal{X} = \emptyset$ are open.
(ii) The set $S$ is closed because

$$\mathcal{C}S = \mathcal{C}\left(\bigcap_{F\in\mathcal{W}} F\right) = \bigcup_{F\in\mathcal{W}} \mathcal{C}F$$

is open by theorem 14.13.
(iii) The set $T$ is closed because

$$\mathcal{C}T = \mathcal{C}\left(\bigcup_{j=1}^{k} F_j\right) = \bigcap_{j=1}^{k} \mathcal{C}F_j$$

is open by theorem 14.13.

---

## 14.15 Open and closed sets in $\mathbb{R}^n$

Of the various types of sets in $\mathbb{R}^n$ introduced in §13.14, only *open balls* are open sets in the sense of §14.7. Of the other types of sets considered in §13.14:

*single points*
*lines* (and *half-lines*)
*closed line segments*
*hyperspheres*
*closed boxes*
and *hyperplanes*

are all closed sets in the sense of §14.7. We prove only that hyperplanes are closed. The proofs for the other types of set in $\mathbb{R}^n$ are left as exercises.

---

14.16 *Theorem* Any hyperplane in $\mathbb{R}^n$ is closed.

*Proof* Consider the hyperplane $H$ through the point $\boldsymbol{\xi}$ with normal $\mathbf{u} \neq \mathbf{0}$. This is given by $H = \{\mathbf{x} : \langle \mathbf{x} - \boldsymbol{\xi}, \mathbf{u}\rangle = 0\}$.
Suppose that $d(\mathbf{y}, H) = 0$. We propose to use theorem 14.10 and therefore seek to prove that $\mathbf{y} \in H$. By theorem 13.22, given any $\varepsilon > 0$, there exists an $\mathbf{x} \in H$ such that $d(\mathbf{y}, \mathbf{x}) = \|\mathbf{y} - \mathbf{x}\| < \varepsilon$. But since $\mathbf{x} \in H$,

$$\langle \mathbf{y} - \boldsymbol{\xi}, \mathbf{u}\rangle = \langle \mathbf{y} - \mathbf{x}, \mathbf{u}\rangle + \langle \mathbf{x} - \boldsymbol{\xi}, \mathbf{u}\rangle = \langle \mathbf{y} - \mathbf{x}, \mathbf{u}\rangle.$$

Hence, using the Cauchy–Schwarz inequality (13.6),

$$|\langle \mathbf{y} - \boldsymbol{\xi}, \mathbf{u}\rangle| \leqq \|\mathbf{y} - \mathbf{x}\| \cdot \|\mathbf{u}\| < \varepsilon\|\mathbf{u}\|.$$

Since this is true for *every* $\varepsilon > 0$, it follows that $\langle \mathbf{y} - \xi, \mathbf{u} \rangle = 0$ and so $\mathbf{y} \in H$. Thus $H$ is closed.

---

14.17 *Exercise*

(1) Determine which of the sets of exercise 14.6(1) and (2) are open and which are closed.

(2) Prove that the only sets in $\mathbb{R}^n$ which are *both* open *and* closed are $\emptyset$ and $\mathbb{R}^n$.

(3) Let $S$ be the set of exercise 14.6(5). Find a point $\xi \notin S$ such that $d(\xi, S) = 0$. Deduce that $S$ is not closed.

(4) Prove that open balls in a metric space $\mathcal{X}$ as defined in §14.3 are open sets as defined in §14.7.

(5) Prove that a set consisting of a single point in a metric space $\mathcal{X}$ is closed. Deduce that any *finite* set in a metric space $\mathcal{X}$ is closed.

(6) Let $S_1, S_2, \ldots, S_n$ be closed sets in $\mathbb{R}^1$. Prove that $S = S_1 \times S_2 \times \ldots \times S_n$ is a closed set in $\mathbb{R}^n$.
[*Hint*: Use exercise 13.24(6).]

(7) *Prove* that the types of set in $\mathbb{R}^n$ said in §14.15 to be closed actually are closed.

(8) A *flat* (or *affine set*) in $\mathbb{R}^n$ may be defined to be the intersection of a collection of hyperplanes. Prove that a flat is always closed.

(9) The sets $S$ and $T$ in $\mathbb{R}^n$ are defined by

$$S = \{\mathbf{x} : \langle \mathbf{x}, \mathbf{u} \rangle \leqq c\}; \; T = \{\mathbf{x} : \langle \mathbf{x}, \mathbf{u} \rangle < c\}$$

(where $\mathbf{u} \neq \mathbf{0}$ and $c$ are constant). We call $S$ a *closed half-space* of $\mathbb{R}^n$ and $T$ an *open half-space*. Prove that $S$ is closed and $T$ is open in the sense of §14.7.

†(10) Give examples to show that a convex set in $\mathbb{R}^n$ may be
(i) open
(ii) closed
(iii) neither open nor closed
(iv) both open and closed.

†(11) Prove that any open set $G$ in a metric space $\mathcal{X}$ is the union of the collection of all open balls it contains.

†(12) Find a sequence $\langle I_k \rangle$ of open intervals and a sequence $\langle J_k \rangle$ of closed intervals in $\mathbb{R}$ such that, if

$$I = \bigcap_{k=1}^{\infty} I_k; \quad J = \bigcup_{k=1}^{\infty} J_k,$$

then $I$ is *not* open and $J$ is *not* closed.

---

# 15 OPEN AND CLOSED SETS (II)

### 15.1 Interior and closure

Let $E$ be a set in a metric space $\mathcal{X}$. The *interior* $\mathring{E}$ of the set $E$ is defined by

$$\mathring{E} = E \setminus \partial E.$$

The *closure* $\bar{E}$ is defined by

$$\bar{E} = E \cup \partial E.$$

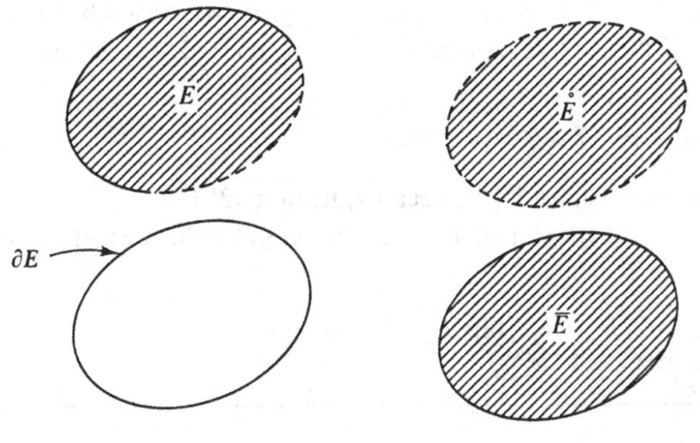


The interior $\mathring{E}$ is obtained from $E$ by removing all the boundary points of $E$ which happen to be elements of $E$. The closure $\bar{E}$ is obtained from $E$ putting in all the boundary points of $E$ which happen not to be elements of $E$.

---

15.2 *Example* Consider the set $S$ in $\mathbb{R}^2$ defined by

$$S = \{(x, y): 0 \leqq y < 1\}.$$

This set is neither open nor closed. Its interior $\mathring{S}$ is the set

$$\mathring{S} = \{(x, y): 0 < y < 1\}.$$

Its closure $\bar{S}$ is the set

$$\bar{S}=\{(x, y): 0 \leqq y \leqq 1\}.$$

---

15.3 *Exercise*

(1) Prove that, for any set $E$ in a metric space $\mathcal{X}$,
 (i) $E$ is open $\Leftrightarrow E=\mathring{E}$
 (ii) $E$ is closed $\Leftrightarrow E=\bar{E}$.

(2) Determine the interior and closure of each of the sets of exercises 14.6(1) and (2).

(3) Prove that, for any set $E$ in a metric space $\mathcal{X}$,
 (i) $\mathcal{C}(\mathring{E})=\overline{\mathcal{C}E}$ (ii) $\mathcal{C}(\bar{E})=(\mathcal{C}E)^{\circ}$.
 [*Hint*: $\mathcal{C}(\mathring{E})=\mathcal{C}(E\cap\mathcal{C}(\partial E))$ and $\mathcal{C}(\bar{E})=\mathcal{C}(E\cup\partial E)$.]

(4) Let $H$ be a hyperplane in $\mathbb{R}^n$. Prove that
 (i) $\partial H=H$ (ii) $\bar{H}=H$ (iii) $\mathring{H}=\emptyset$.
 [*Hint*: We have seen that $H$ is closed (theorem 14.16). Hence $\partial H\subset H$. If $\xi\in H$ and $H$ has normal $\mathbf{u}$, show that $d(\xi, \xi+t\mathbf{u})=|t|\cdot\|\mathbf{u}\|$. Deduce that $H\subset\partial H$.]

(5) Let $S=\{\mathbf{x}: \langle\mathbf{x}, \mathbf{u}\rangle<c\}$ and $T=\{\mathbf{x}: \langle\mathbf{x}, \mathbf{u}\rangle\leqq c\}$ (where $\mathbf{u}\neq\mathbf{0}$) be half-spaces in $\mathbb{R}^n$. Let $H$ be the hyperplane $H=\{\mathbf{x}: \langle\mathbf{x}, \mathbf{u}\rangle=c\}$. Prove that

 (i) $\partial S=H$ (ii) $\bar{S}=T$ (iii) $\mathring{S}=S$
 (iv) $\partial T=H$ (v) $\bar{T}=T$ (vi) $\mathring{T}=S$.

 [*Hint*: For (iv), (v) and (vi), recall question (3).]

†(6) Let $B_1=\{\mathbf{x}: d(\xi, \mathbf{x})<r\}$ and $B_2=\{\mathbf{x}: d(\xi, \mathbf{x})\leqq r\}$ (where $r>0$) be balls in $\mathbb{R}^n$. Prove that

 (i) $\mathring{B}_1=B_1$ (ii) $\overline{B_1}=B_2$ (iii) $\mathring{B}_2=B_1$ (iv) $\bar{B}_2=B_2$.
 [*Hint*: Use exercise 14.6(4).]

---

**15.4 Closure properties**

15.5 *Theorem* For any set $E$ in a metric space $\mathcal{X}$,

$$\mathbf{x}\in\bar{E}\Leftrightarrow d(\mathbf{x}, E)=0$$

*Proof* (i) If $\mathbf{x}\in\bar{E}=E\cup\partial E$, then $\mathbf{x}\in E$ or $\mathbf{x}\in\partial E$. In the former case $d(\mathbf{x}, E)=d(\mathbf{x}, \mathbf{x})=0$. In the latter case, we have by definition that $d(\mathbf{x}, E)=0$ and $d(\mathbf{x}, \mathcal{C}E)=0$. Thus $\mathbf{x}\in\bar{E}\Rightarrow d(\mathbf{x}, E)=0$.

(ii) If $d(\mathbf{x}, E)=0$, then either $\mathbf{x}\in E\subset\bar{E}$ or else $\mathbf{x}\in\mathcal{C}E$. In the latter case $d(\mathbf{x}, \mathcal{C}E)=d(\mathbf{x}, \mathbf{x})=0$ and so $\mathbf{x}\in\partial E\subset\bar{E}$. Thus $d(\mathbf{x}, E)=0\Rightarrow\mathbf{x}\in\bar{E}$.

---

15.6 *Corollary* Let $S$ and $T$ be sets in a metric space $\mathcal{X}$. Then

$$S \subset T \Rightarrow \bar{S} \subset \bar{T}.$$

*Proof* Let $\mathbf{x} \in \bar{S}$. By theorem 15.5, $d(\mathbf{x}, S)=0$. But $S \subset T$ implies $d(\mathbf{x}, S) \geqq d(\mathbf{x}, T)$. Hence $d(\mathbf{x}, T)=0$. Thus $\mathbf{x} \in \bar{T}$.

---

15.7† *Theorem* Let $E$ be a set in a metric space $\mathcal{X}$. Then $\bar{E}$ is the smallest closed set containing $E$.

*Proof* To show that $\bar{E}$ is closed, we use theorem 14.10. We therefore assume that $d(\xi, \bar{E})=0$ and seek to deduce that $\xi \in \bar{E}$.

By theorem 13.22, given any $\varepsilon>0$, we can find an $\mathbf{x} \in \bar{E}$ such that $d(\xi, \mathbf{x})<\varepsilon$. But $\mathbf{x} \in \bar{E} \Leftrightarrow d(\mathbf{x}, E)=0$ by theorem 15.5. Hence

$$d(\xi, E) \leqq d(\xi, \mathbf{x})+d(\mathbf{x}, E)=d(\xi, \mathbf{x}) \quad \text{(exercise 13.24(5))}.$$

Thus

$$d(\xi, E)<\varepsilon.$$

Since this is true for *any* $\varepsilon>0$, we conclude that $d(\xi, E)=0$ and hence $\xi \in \bar{E}$ by theorem 15.5.

Now suppose that $F$ is any closed set containing $E$. Since $E \subset F$, we have from corollary 15.6 that $\bar{E} \subset \bar{F}$. But $F$ is closed and therefore $F=\bar{F}$ (exercise 15.3(1ii)). It follows that $\bar{E} \subset F$ for each closed set $F$ containing $E$. Thus $\bar{E}$ is the *smallest* closed set containing $E$.

---

## 15.8† Interior properties

From exercise 15.3(3i) we know that

$$\mathcal{C}(\overline{\mathcal{C}E})=\mathring{E}.$$

Thus any result about closures leads to a corresponding result about interiors. For example,

$$S \subset T \Rightarrow \mathcal{C}S \supset \mathcal{C}T \Rightarrow \overline{\mathcal{C}S} \supset \overline{\mathcal{C}T} \Rightarrow \mathcal{C}(\overline{\mathcal{C}S}) \subset \mathcal{C}(\overline{\mathcal{C}T})$$

and so $S \subset T \Rightarrow \mathring{S} \subset \mathring{T}$.

---

15.9† *Theorem* Let $E$ be a set in a metric space $\mathcal{X}$. Then $\mathring{E}$ is the largest open set contained in $E$.

*Proof* Note first that $\mathring{E}=\mathcal{C}(\overline{\mathcal{C}E})$ and hence is open because it is the complement of a closed set. If $G \subset E$, then $\mathring{G} \subset \mathring{E}$. If $G$ is also open it follows that

$G \subset \mathring{E}$ (because then $G = \mathring{G}$). This argument shows that $\mathring{E}$ is the *largest* open set contained in $E$.

---

15.10† *Exercise*

(1) A *flat* (or affine set) $F$ is the intersection of a collection of hyperplanes in $\mathbb{R}^n$. Prove that

(i) $\bar{F} = F$ (ii) $\mathring{F} = \emptyset$.

Show that the same results hold if $F$ is a half-line or a closed line segment.

(2) Let $E$ be a set in a metric space $\mathcal{X}$. Prove that $x \in \mathring{E}$ if and only if there exists an open ball $B$ with centre $\xi$ such that $B \subset E$.

(3) Let $E$ be a set in a metric space $\mathcal{X}$. Prove that $\partial E$ is closed. [*Hint*: $\partial E = \bar{E} \cap \mathcal{C}(\mathring{E})$.]

(4) Let $E$ be a set in a metric space $\mathcal{X}$. Prove that

$$\text{(i) } \bar{E} = \bigcap_{\substack{E \subset F \\ F \text{ closed}}} F \quad \text{(ii) } \mathring{E} = \bigcup_{\substack{G \subset E \\ G \text{ open}}} G.$$

(5) Let $A$ and $B$ be sets in a metric space $\mathcal{X}$. Prove that

$$\text{(i) } \overline{A \cup B} = \bar{A} \cup \bar{B};\ (A \cap B)^{\circ} = \mathring{A} \cap \mathring{B}$$
$$\text{(ii) } A \subset B \Rightarrow \bar{A} \subset \bar{B};\ A \subset B \Rightarrow \mathring{A} \subset \mathring{B}$$
$$\text{(iii) } \overline{A \cap B} \subset \bar{A} \cap \bar{B};\ (A \cup B)^{\circ} \subset \mathring{A} \cup \mathring{B}.$$

Give examples to show that $\subset$ cannot be replaced in general by $=$ in (iii).

(6) Let $E$ be a set in a metric space $\mathcal{X}$. Prove that

$$\partial(\partial E) = \partial E.$$

Show also that:

(i) $\partial(\bar{E}) \subset \partial E$ (ii) $\partial(\mathring{E}) \subset \partial E$.

Give examples to show that $\subset$ cannot be replaced in general by $=$ in either case.

(7) Let $C$ be a convex set in $\mathbb{R}^n$ with a non-empty interior. Prove that

(i) $\mathring{\bar{C}} = \mathring{C}$ (ii) $\bar{\mathring{C}} = \bar{C}$.

Give examples of (non-convex) sets $C$ for which these equations are false. Give an example of a closed convex set $C$ for which $\mathring{C} = \emptyset$.

(8) Let $C$ be a convex set in $\mathbb{R}^n$. Prove that $\mathring{C}$ and $\bar{C}$ are also convex.

(9) Let $E$ be a set in a metric space $\mathcal{X}$. Let $\mathcal{W}$ denote the collection of all sets which can be obtained from $E$ by applying the operations $\mathcal{C}$, $^-$ and $^\circ$ to $E$ any finite number of times in any order. Show that the maximum number of distinct elements that $\mathcal{W}$ can contain is 14.

---

## 15.11 Contiguous sets

We shall say that two sets $A$ and $B$ in a metric space $\mathcal{X}$ are *contiguous* (i.e. 'next to each other') if a point of the closure of one of the sets belongs to the other. Thus $A$ and $B$ are contiguous if and only if

$$\bar{A} \cap B \neq \emptyset \quad \text{or} \quad A \cap \bar{B} \neq \emptyset.$$

We say that the sets $A$ and $B$ are *separated* if and only if they are *not* contiguous.

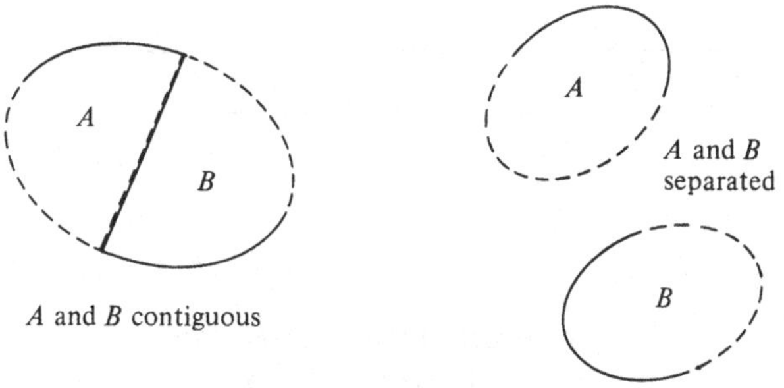


To say that $A$ and $B$ are contiguous, means that either the sets overlap or else a boundary point of one set belongs to the other. Using the metaphor of §14.7 which interprets sets as plots of land in a city, the owners of $A$ and $B$ are neighbours.

---

15.12 *Theorem* Two sets $A$ and $B$ are contiguous if and only if there exists a point in one of the sets which is at zero distance from the other.

*Proof* This follows from theorem 15.5. We have that

$$\mathbf{a} \in A \cap \bar{B} \Leftrightarrow \mathbf{a} \in A \text{ and } d(\mathbf{a}, B) = 0.$$

Similarly,

$$\mathbf{b} \in \bar{A} \cap B \Leftrightarrow \mathbf{b} \in B \text{ and } d(\mathbf{b}, A) = 0.$$

---

15.13 *Examples*

(i) The intervals $A = (0, 1)$ and $B = [1, 2]$ in $\mathbb{R}^1$ are *contiguous*. We have that $d(1, A) = 0$. Alternatively one can note that

$$\bar{A} \cap B = [0, 1] \cap [1, 2] = \{1\} \neq \emptyset.$$

The diagram illustrates the owner of set $A$ walking from his house $a \in A$ to visit his neighbour whose house is at $b \in B$. Observe that this visit does not involve crossing any point outside $A \cup B$.

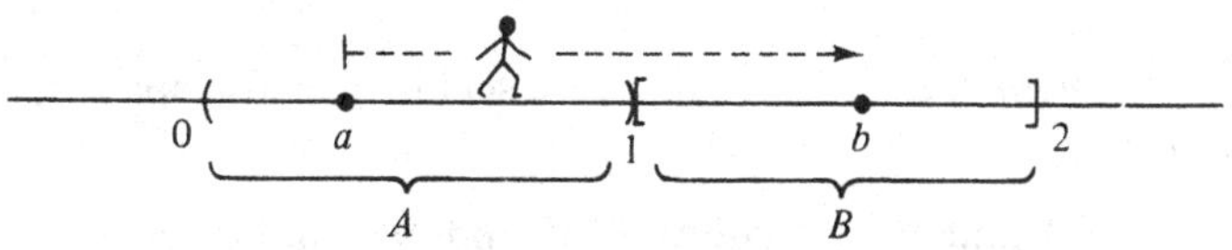

(ii) The intervals $A=(0, 1)$ and $B=(1, 2)$ in $\mathbb{R}^1$ are *separated.* The diagram illustrates the owner of $A$ jumping over the point 1 in order to get to $B$.

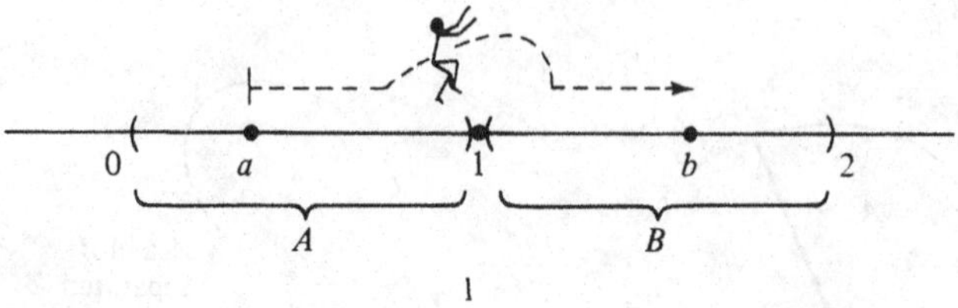


In formal terms

$$\bar{A}\cap B=[0, 1]\cap(1, 2)=\emptyset$$
$$A\cap\bar{B}=(0, 1)\cap[1, 2]=\emptyset.$$

---

15.14 *Theorem* Two *open* sets $G$ and $H$ in a metric space $\mathcal{X}$ are contiguous if and only if they overlap.

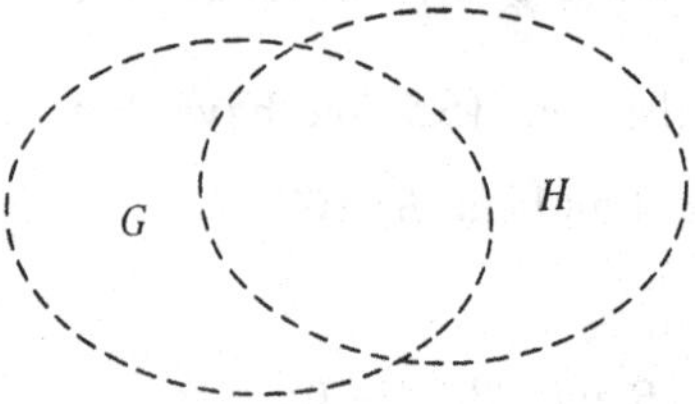


*Proof* Two sets which overlap are obviously contiguous. We therefore need only prove that, if $G$ and $H$ are contiguous open sets, then $G\cap H\neq\emptyset$.

Suppose that $G\cap H=\emptyset$. Then $G\subset \mathcal{C}H$. Hence $\bar{G}\subset\overline{\mathcal{C}H}=\mathcal{C}H$ (corollary 15.6) because $\mathcal{C}H$ is closed. Thus $\bar{G}\cap H=\emptyset$. Similarly, $G\cap\bar{H}=\emptyset$. Thus $G$ and $H$ are not contiguous.

---

15.15 *Theorem* Let $A$, $B$, $G$ and $H$ be sets in a metric space $\mathcal{X}$. If $A\subset G$ and $B\subset H$, then

$$A \text{ and } B \text{ contiguous} \Rightarrow G \text{ and } H \text{ contiguous.}$$

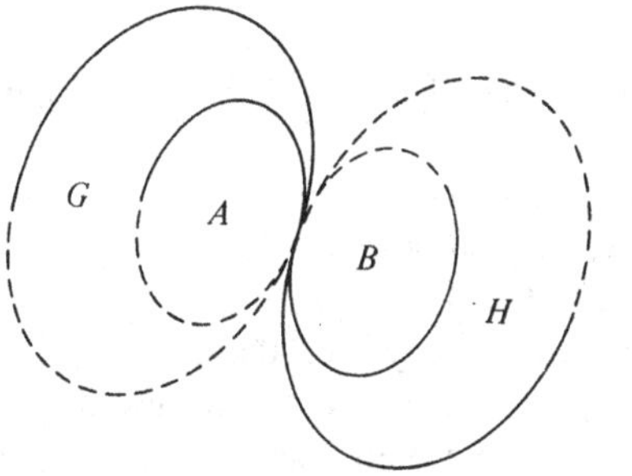


*Proof* Suppose that $\bar{A}\cap B\neq\emptyset$. Then $\bar{G}\cap H\supset\bar{A}\cap B\neq\emptyset$. Similarly, $A\cap\bar{B}\neq\emptyset\Rightarrow G\cap\bar{H}\neq\emptyset$.

---

15.16† *Theorem* Two sets $A$ and $B$ in a metric space $\mathscr{X}$ are separated if and only if there exist *disjoint* (i.e. non-overlapping) *open* sets $G$ and $H$ such that $A\subset G$ and $B\subset H$.

*Proof* (i) Suppose that $G$ and $H$ are disjoint open sets for which $A\subset G$ and $B\subset H$. Then $A$ and $B$ are separated by theorems 15.14 and 15.15.

(ii) Suppose that $A$ and $B$ are separated. Define the sets $G$ and $H$ by

$$G=\{\mathbf{x}:d(\mathbf{x},A)<d(\mathbf{x},B)\}$$
$$H=\{\mathbf{x}:d(\mathbf{x},B)<d(\mathbf{x},A)\}.$$

The sets $G$ and $H$ are obviously disjoint. We prove that $G$ and $H$ are open using theorem 14.12. Let $\xi\in G$ and write

$$r=\tfrac{1}{2}\{d(\xi,B)-d(\xi,A)\}>0.$$

Then the open ball with centre $\xi$ and radius $r>0$ lies entirely inside $G$. To prove this, we suppose that $d(\xi,\mathbf{x})<r$. We then need to show that $d(\mathbf{x},B)-d(\mathbf{x},A)>0$.

Observe that

$$d(\xi,B)\leqq d(\xi,\mathbf{x})+d(\mathbf{x},B)$$

– i.e.

$$d(\mathbf{x},B)\geqq d(\xi,B)-d(\xi,\mathbf{x}).$$

Also

$$d(\mathbf{x},A)\leqq d(\mathbf{x},\xi)+d(\xi,A).$$

Hence

$$\begin{aligned} d(\mathbf{x},B)-d(\mathbf{x},A) &\geqq d(\xi,B)-2d(\xi,\mathbf{x})-d(\xi,A)\\ &>d(\xi,B)-2r-d(\xi,A)\\ &=0.\end{aligned}$$

Thus $\mathbf{x}\in G$ and so $G$ is open. Similarly, $H$ is open.

It remains to show that $A\subset G$ and $B\subset H$. Since $A$ and $B$ are separated, $A\cap\bar{B}=\emptyset$. Hence, given any $\mathbf{a}\in A$, $d(\mathbf{a},B)>0$ (theorem 15.12). Thus, for each $\mathbf{a}\in A$,

$$d(\mathbf{a},A)=0<d(\mathbf{a},B)$$

and hence $A\subset G$. Similarly, $B\subset H$.

---

15.17 *Exercise*

(1) Decide which of the following pairs of sets in $\mathbb{R}^2$ are contiguous and which are separated.
 (i) $A=\{(x, y): x>0\}$; $B=\{(x, y): x<0\}$
 (ii) $A=\{(x, y): x>0\}$; $B=\{(x, y): y=0 \text{ and } x \leqq 0\}$
 (iii) $A=\{(x, y): x^2+y^2<1\}$; $B=\{(x, y): x^2+y^2>0\}$
 (iv) $A=\{(x, y): x>0 \text{ and } y \geqq 0\}$; $B=\{(x, y): x \leqq 0 \text{ and } y<0\}$
 (v) $A=\{(x, y): x \geqq 0 \text{ and } y \geqq 0\}$; $B=\{(x, y): x<0 \text{ and } y<0\}$.

(2) Prove that two *closed* sets $A$ and $B$ in a metric space $\mathcal{X}$ are contiguous if and only if they overlap.

(3) Suppose that $A$ and $B$ are sets in a metric space $\mathcal{X}$ such that $A \cup B = \mathcal{X}$. If $A$ and $B$ are separated, prove that each set is *both* open *and* closed. What does this imply about $A$ and $B$ in the metric space $\mathbb{R}^n$?

# 16 CONTINUITY

## 16.1 Introduction

A child takes a lump of modelling clay and by moulding it and pressing it, without ripping or tearing, shapes it into the figure of a man. We say that the original shape is transformed '*continuously*' into the final shape.

What is it that distinguishes a continuous transformation from any other? Most people would agree that the essential feature of a continuous transformation is that pieces of clay which were 'next to each other' at the beginning should still be 'next to each other' at the end. In this chapter, we shall express this idea in a precise mathematical definition and explore some of its immediate consequences.

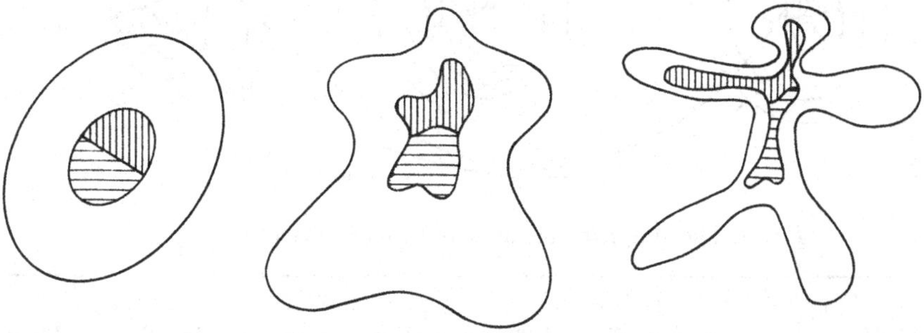

## 16.2 Continuous functions

Let $\mathcal{X}$ and $\mathcal{Y}$ be metric spaces and let $S \subset \mathcal{X}$. A function $f\colon S \to \mathcal{Y}$ is *continuous on the set* $S$ if and only if, for each pair $A$ and $B$ of subsets of $S$,

$$A \text{ and } B \text{ contiguous} \Rightarrow f(A) \text{ and } f(B) \text{ contiguous.}$$

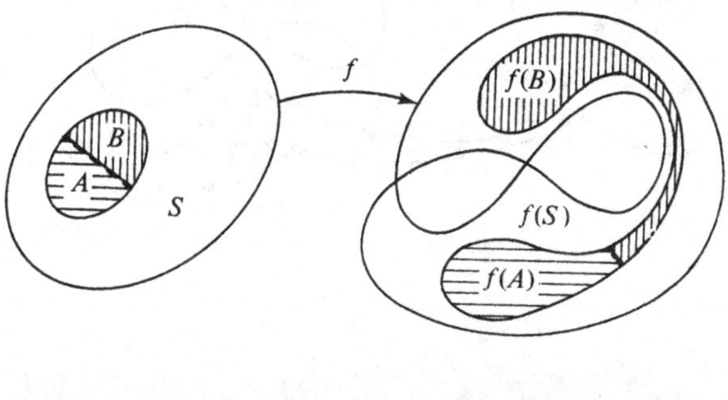

If $f\colon S\to\mathcal{Y}$, then $f(f^{-1}(T))=T$ for each $T\subset f(S)$ (exercise 6.4(5)). It follows that an equivalent formulation of the definition is that $f$ is continuous on $S$ if and only if, for each pair $C$ and $D$ of subsets of $f(S)$,

$$C \text{ and } D \text{ separated} \Rightarrow f^{-1}(C) \text{ and } f^{-1}(D) \text{ separated.}$$

---

Our first theorem records the unremarkable fact that, if one continuous transformation is followed by another, then the result will be a continuous transformation.

---

16.3 *Theorem* Let $\mathcal{X}$, $\mathcal{Y}$ and $\mathcal{Z}$ be metric spaces and let $S\subset\mathcal{X}$, $T\subset\mathcal{Y}$. Suppose that $g\colon S\to\mathcal{Y}$ and $f\colon T\to\mathcal{Z}$ are continuous on $S$ and $T$ respectively and that $g(S)\subset T$. Then the composite function $f\circ g\colon S\to\mathcal{Z}$ is continuous on $S$.

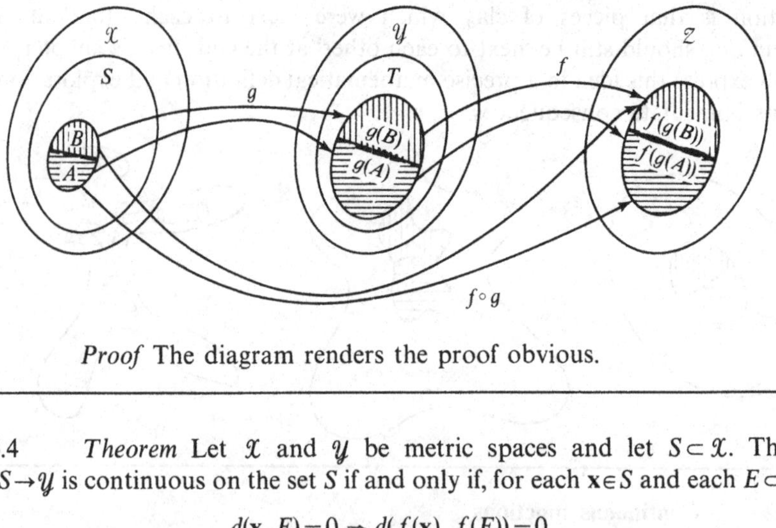


*Proof* The diagram renders the proof obvious.

---

16.4 *Theorem* Let $\mathcal{X}$ and $\mathcal{Y}$ be metric spaces and let $S\subset\mathcal{X}$. Then $f\colon S\to\mathcal{Y}$ is continuous on the set $S$ if and only if, for each $\mathbf{x}\in S$ and each $E\subset S$,

$$d(\mathbf{x}, E)=0 \Rightarrow d(f(\mathbf{x}), f(E))=0.$$

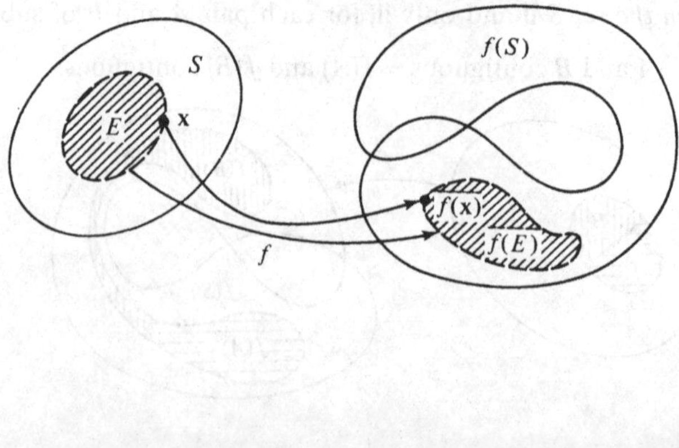

*Proof* (i) Suppose that $f$ is continuous on $S$. If $d(\mathbf{x}, E)=0$, then $\{\mathbf{x}\}$ and $E$ are contiguous by theorem 15.12. Hence $\{f(\mathbf{x})\}$ and $f(E)$ are contiguous. Thus $d(f(\mathbf{x}), f(E))=0$ by theorem 15.12 again.

(ii) Suppose that $d(\mathbf{x}, E)=0 \Rightarrow d(f(\mathbf{x}), f(E))=0$. Let $A$ and $B$ be contiguous subsets of $S$. Then a point of one set is at zero distance from the other by theorem 15.12. Suppose $\mathbf{a} \in A$ and $d(\mathbf{a}, B)=0$. Then $d(f(\mathbf{a}), f(B))=0$ and hence $f(A)$ and $f(B)$ are contiguous by theorem 15.12 yet again.

---

If $f: S \to \mathbb{R}^n$, the formula

$$f(\mathbf{x})=(f_1(\mathbf{x}), \ldots, f_n(\mathbf{x}))$$

defines $n$ real-valued functions $f_1: S \to \mathbb{R}, f_2: S \to \mathbb{R}, \ldots, f_n: S \to \mathbb{R}$. We call these functions the component functions of $f$ and write

$$f=(f_1, f_2, \ldots, f_n).$$

---

16.5 *Theorem* Let $\mathcal{X}$ be a metric space and let $S \subset \mathcal{X}$. A function $f: S \to \mathbb{R}^n$ is continuous on the set $S$ if and only if each of its component functions is continuous on $S$.

*Proof* This follows immediately from theorem 16.4 and exercise 13.24(6).

---

16.6 *Corollary* The projection function $P_k: \mathbb{R}^n \to \mathbb{R}$ defined by

$$P_k(x_1, x_2, \ldots, x_n)=x_k$$

is continuous on $\mathbb{R}^n$.

*Proof* The identity function $I: \mathbb{R}^n \to \mathbb{R}^n$ defined by $I(\mathbf{x})=\mathbf{x}$ is obviously continuous. But $I=(P_1, P_2, \ldots, P_n)$ and hence $P_k$ is continuous by theorem 16.5.

---

**16.7 The continuity of algebraic operations**

It is easy to see that constant functions and identity functions are continuous. We shall obtain some slightly less trivial examples of continuous functions from these by using the operations of addition, subtraction, multiplication and division. But first we must prove that these operations are themselves continuous.

We begin with a simple lemma.

16.8 *Lemma* Let $\mathcal{X}$ and $\mathcal{Y}$ be metric spaces and let $S \subset \mathcal{X}$. Then $f: S \to \mathcal{Y}$ is continuous on $S$ provided that for each $\mathbf{x} \in S$ there exists a $\gamma > 0$ and a $c > 0$ such that

$$d(\mathbf{x}, \mathbf{y}) < \gamma \Rightarrow d(f(\mathbf{x}), f(\mathbf{y})) \leqq c d(\mathbf{x}, \mathbf{y})$$

for each $\mathbf{y} \in S$.

*Proof* Suppose that $d(\mathbf{x}, E) = 0$. Given any $\varepsilon > 0$, we shall prove that there exists a $\mathbf{y} \in E$ such that $d(f(\mathbf{x}), f(\mathbf{y})) < \varepsilon$. Thus $d(f(\mathbf{x}), f(E)) = 0$ by theorem 13.22.

Choose $\varepsilon_0 = \min \{\varepsilon c^{-1}, \gamma\}$. Then $\varepsilon_0 > 0$ and hence there exists $\mathbf{y} \in E$ such that $d(\mathbf{x}, \mathbf{y}) < \varepsilon_0$ by theorem 13.22. But $\varepsilon_0 \leqq \gamma$. It follows that

$$d(f(\mathbf{x}), f(\mathbf{y})) \leqq c d(\mathbf{x}, \mathbf{y}) < c\varepsilon_0 \leqq \varepsilon \text{ as required.}$$

16.9 *Theorem* Let $a$ and $b$ be real numbers and let $L: \mathbb{R}^n \times \mathbb{R}^n \to \mathbb{R}$, $M: \mathbb{R} \times \mathbb{R} \to \mathbb{R}$ and $D: \mathbb{R} \times (\mathbb{R} \setminus \{0\}) \to \mathbb{R}$ be defined by

(i) $L(\mathbf{x}_1, \mathbf{x}_2) = a\mathbf{x}_1 + b\mathbf{x}_2$
(ii) $M(x_1, x_2) = x_1 x_2$
(iii) $D(x_1, x_2) = x_1 / x_2$.

Then $L$, $M$ and $D$ are continuous on their domains of definition.

*Proof* (i) We have that $\|\mathbf{x}_1 - \mathbf{y}_1\| \leqq \|(\mathbf{x}_1, \mathbf{x}_2) - (\mathbf{y}_1, \mathbf{y}_2)\|$ and

$\|\mathbf{x}_2 - \mathbf{y}_2\| \leqq \|(\mathbf{x}_1, \mathbf{x}_2) - (\mathbf{y}_1, \mathbf{y}_2)\|$ because

$$\|(\mathbf{x}_1, \mathbf{x}_2) - (\mathbf{y}_1, \mathbf{y}_2)\|^2 = \|\mathbf{x}_1 - \mathbf{y}_1\|^2 + \|\mathbf{x}_2 - \mathbf{y}_2\|^2.$$

Hence

$$\begin{aligned} \|L(\mathbf{x}_1, \mathbf{x}_2) - L(\mathbf{y}_1, \mathbf{y}_2)\| &= \|a(\mathbf{x}_1 - \mathbf{y}_1) + b(\mathbf{x}_2 - \mathbf{y}_2)\| \\ &\leqq |a| \cdot \|\mathbf{x}_1 - \mathbf{y}_1\| + |b| \cdot \|\mathbf{x}_2 - \mathbf{y}_2\| \\ &\leqq \{|a| + |b|\} \cdot \|(\mathbf{x}_1, \mathbf{x}_2) - (\mathbf{y}_1, \mathbf{y}_2)\|. \end{aligned}$$

Thus $L$ is continuous on $\mathbb{R}^{2n}$ by lemma 16.8.

(ii) We use lemma 16.8 with $\gamma = 1$. Assume that

$$\|(x_1, x_2) - (y_1, y_2)\| < 1.$$

Then

$$|x_1 - y_1| \leqq \|(x_1, x_2) - (y_1, y_2)\| < 1$$

and

$$|x_2 - y_2| \leqq \|(x_1, x_2) - (y_1, y_2)\| < 1.$$

In particular,

$$|y_2| = |y_2 - x_2 + x_2| \leqq |y_2 - x_2| + |x_2| < 1 + |x_2|.$$

Put $c = 1 + |x_1| + |x_2|$. Then

$$\begin{aligned} |M(x_1, x_2) - M(y_1, y_2)| &= |x_1x_2 - y_1y_2| = |x_1x_2 - x_1 y_2 + x_1 y_2 - y_1 y_2| \\ &\leqq |x_1| \cdot |x_2 - y_2| + |y_2| \cdot |x_1 - y_1| \\ &\leqq c\|(x_1, x_2) - (y_1, y_2)\|. \end{aligned}$$

(iii) We use lemma 16.8. If $(x_1, x_2) \in \mathbb{R} \times \mathbb{R} \setminus \{0\}$, then $x_2 \neq 0$ and we take $\gamma = \frac{1}{2}|x_2|$. Assume that

$$\|(x_1, x_2) - (y_1, y_2)\| < \tfrac{1}{2}|x_2|.$$

Then

$$|x_1 - y_1| \leqq \|(x_1, x_2) - (y_1, y_2)\|$$

and

$$|x_2 - y_2| < \|(x_1, x_2) - (y_1, y_2)\| < \tfrac{1}{2}|x_2|.$$

In particular,

$$|y_2| = |y_2 - x_2 + x_2| \geqq |x_2| - |y_2 - x_2| > \tfrac{1}{2}|x_2|.$$

Put $c = 2|x_2|^{-2}(|x_1| + |x_2|)$. Then

$$\begin{aligned} |D(x_1, x_2) - D(y_1, y_2)| &= \left| \frac{x_1}{x_2} - \frac{y_1}{y_2} \right| \\ &= \left| \frac{x_1 y_2 - x_1x_2 + x_1x_2 - y_1x_2}{x_2 y_2} \right| \\ &< 2|x_2|^{-2}\{|x_1| \cdot |y_2 - x_2| + |x_2| \cdot |y_1 - x_1|\} \\ &\leqq c\|(x_1, x_2) - (y_1, y_2)\| \end{aligned}$$

and the result follows from lemma 16.8.

---

16.10 *Corollary* Let $\mathcal{X}$ be a metric space and let $S \subset \mathcal{X}$. If $f_1 \colon S \to \mathbb{R}^n$ and $f_2 \colon S \to \mathbb{R}^n$ are continuous on the set $S$, then so is the function $F \colon S \to \mathbb{R}^n$

defined by

$$F(\mathbf{x})=af_1(\mathbf{x})+bf_2(\mathbf{x}).$$

*Proof* This follows immediately from theorems 16.3 and 16.9 since $F=L\circ(f_1,f_2)$.

---

16.11 *Corollary* Let $\mathcal{X}$ be a metric space and let $S\subset\mathcal{X}$. If $f_1: S\to\mathbb{R}$ and $f_2: S\to\mathbb{R}$ are continuous on the set $S$, then so is the function $F: S\to\mathbb{R}$ defined by

$$F(\mathbf{x})=f_1(\mathbf{x})f_2(\mathbf{x}).$$

*Proof* We have that $F=M\circ(f_1,f_2)$.

---

16.12 *Corollary* Let $\mathcal{X}$ be a metric space and let $S\subset\mathcal{X}$. If $f_1: S\to\mathbb{R}$ and $f_2: S\to\mathbb{R}\setminus\{0\}$ are continuous on the set $S$, then so is the function $F: S\to\mathbb{R}$ defined by

$$F(\mathbf{x})=f_1(\mathbf{x})/f_2(\mathbf{x}).$$

*Proof* We have that $F=D\circ(f_1,f_2)$.

---

**16.13 Rational functions**

Suppose that $P(x_1, x_2,\ldots, x_n)$ is an expression obtained from the variables $x_1, x_2,\ldots, x_n$ and a finite number of real constants using a finite number of multiplications and additions. Then we say that $P:\mathbb{R}^n\to\mathbb{R}$ is a *polynomial*. If $Q:\mathbb{R}^n\to\mathbb{R}$ is a polynomial and

$$Z=\{\mathbf{x}:Q(\mathbf{x})=0\}$$

then the function $R:\mathbb{R}^n\setminus Z\to\mathbb{R}$ defined by

$$R(\mathbf{x})=\frac{P(\mathbf{x})}{Q(\mathbf{x})}$$

is called a *rational function.*

---

16.14 *Example* The function $P:\mathbb{R}^2\to\mathbb{R}$ defined by

$$P(x, y)=x^2+3xy+x^2y^3+1$$

is a polynomial. The function $R:\mathbb{R}^2\setminus\{(0, 0)\}\to\mathbb{R}$ defined by

$$R(x, y)=\frac{x^2+3xy+x^2y^3+1}{x^2+xy+y^2}$$

is a rational function.

---

16.15 *Theorem* All rational functions are continuous on their domains of definition.

*Proof* Constant functions and the identity function are continuous. So are the projection functions (corollary 16.6). The fact that rational functions are continuous therefore follows from corollaries 16.10, 16.11 and 16.12.

---

16.16 *Exercise*

(1) For each of the following functions $f: \mathbb{R}\to\mathbb{R}$, find non-empty contiguous sets $A$ and $B$ in $\mathbb{R}$, for which $f(A)$ and $f(B)$ are *not* contiguous.

(i) $f(x)=\begin{cases} x & (x\leqq 1) \\ x-1 & (x>1) \end{cases}$

(ii) $f(x)=\begin{cases} 1 & (x\in\mathbb{Q}) \\ 0 & (x\notin\mathbb{Q}). \end{cases}$

Deduce that in neither case is $f$ continuous.

(2) Let $\mathbf{e}\in\mathbb{R}^n$. Prove that the functions $f:\mathbb{R}^n\to\mathbb{R}^1$ and $g:\mathbb{R}^n\to\mathbb{R}^1$ defined by (i) $f(\mathbf{x})=\|\mathbf{x}\|$ (ii) $g(\mathbf{x})=\langle\mathbf{x}, \mathbf{e}\rangle$ (see §13.3) are continuous.

(3) A function $f: \mathbb{R}^n\to\mathbb{R}^m$ is said to be *linear* if and only if $f(\alpha\mathbf{x}+\beta\mathbf{y})=\alpha f(\mathbf{x})+\beta f(\mathbf{y})$ for all real $\alpha$ and $\beta$ and all $\mathbf{x}$ and $\mathbf{y}$ in $\mathbb{R}^n$.
Prove that a linear function $f_1: \mathbb{R}^n\to\mathbb{R}^1$ satisfies $f_1(\mathbf{x})=\langle\mathbf{x}, \mathbf{e}\rangle$ for some $\mathbf{e}\in\mathbb{R}^n$. Hence show that any linear function $f: \mathbb{R}^n\to\mathbb{R}^m$ is continuous.

(4) Let $\mathcal{X}$ be a metric space and let $S$ be a non-empty set in $\mathcal{X}$. Prove that the function $f: \mathcal{X}\to\mathbb{R}$ defined by

$$f(\mathbf{x})=d(\mathbf{x}, S)$$

is continuous on $\mathcal{X}$.
[*Hint*: Begin by showing that $|d(\mathbf{x}, S)-d(\mathbf{y}, S)|\leqq d(\mathbf{x}, \mathbf{y})$ using exercise 13.24(5).]

(5) Let $\mathcal{X}$ and $\mathcal{Y}$ be metric spaces and let $S\subset T\subset\mathcal{X}$. If $f: T\to\mathcal{Y}$ and $g: T\to\mathcal{Y}$ satisfy $f(\mathbf{x})=g(\mathbf{x})$ for each $\mathbf{x}\in S$, prove that $g$ continuous on $T$ implies $f$ continuous on $S$.

†(6) Prove theorem 16.9 with $\mathbb{R}$ replaced everywhere by $\mathbb{C}$.

---

## 16.17† Complex-valued functions

In the above discussion we have confined our attention to *real* polynomials and *real* rational functions. However, it is worth noting that *all* of the results of this chapter remain valid if $\mathbb{R}$ is replaced throughout by $\mathbb{C}$ and the word 'real' replaced by 'complex'. No other change in the proofs is necessary.

# 17 CONNECTED SETS

## 17.1 Introduction

The characteristic feature of an interval in $\mathbb{R}^1$ is that it is 'all in one lump'. A little man standing on any point of an interval $I$ would be able to walk to any other point of $I$ without having to cross any points of the complement of $I$.

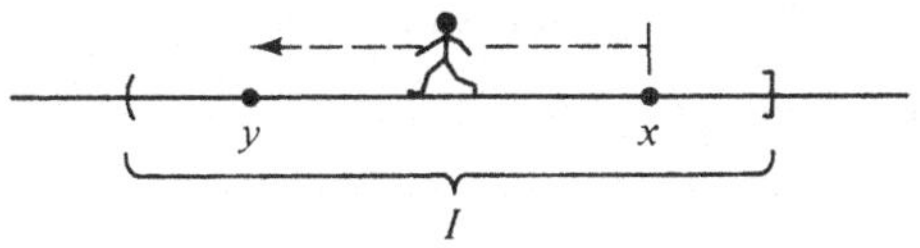


What sets in $\mathbb{R}^n$ have the same property? One can think, for example, of a set $S$ in $\mathbb{R}^2$ as a country entirely surrounded by water. When does $S$ consist of a single island and when does it consist of a whole archipelago of islands?

Before trying to answer this question, we must of course first express it in precise mathematical terms.

---

## 17.2 Connected sets

A set $S$ in a metric space $\mathcal{X}$ is *connected* if and only if $S$ cannot be split into two non-empty separated subsets. This means that if $A$ and $B$ are any two non-empty subsets of $S$ satisfying $A \cup B = S$, then $A$ and $B$ are contiguous.

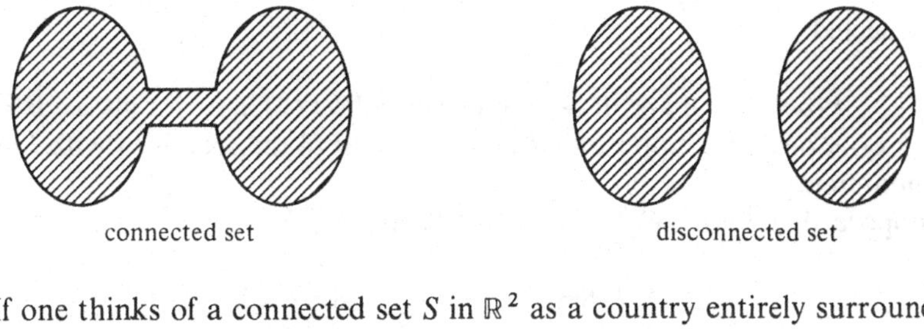


If one thinks of a connected set $S$ in $\mathbb{R}^2$ as a country entirely surrounded by water, the definition says that, however one divides $S$ into two provinces

$A$ and $B$, these two provinces will be 'joined up'. A little man will be able to go from $A$ to $B$ without getting his feet wet.

---

17.3 *Theorem* A metric space $\mathcal{X}$ is connected if and only if the only sets in $\mathcal{X}$ which are *both* open *and* closed are $\emptyset$ and $\mathcal{X}$.

*Proof* Observe that

$$\text{(i) } A\cup B=\mathcal{X} \text{ and } A\cap\bar{B}=\emptyset \Rightarrow A\cup B=\mathcal{X} \text{ and } A\cap B=\emptyset$$
$$\Rightarrow B=\mathcal{C}A$$
$$\Rightarrow A\cup B=\mathcal{X}.$$
$$\text{(ii) } A\cup B=\mathcal{X} \text{ and } A\cap\bar{B}=\emptyset \Rightarrow A\cup\bar{B}=\mathcal{X} \text{ and } A\cap\bar{B}=\emptyset$$
$$\Rightarrow \bar{B}=\mathcal{C}A$$
$$\Rightarrow A\cap\bar{B}=\emptyset.$$

Taking these two results together we obtain that

$$A\cup B=\mathcal{X} \text{ and } A\cap\bar{B}=\emptyset \Leftrightarrow B=\bar{B} \text{ and } B=\mathcal{C}A$$
$$\Leftrightarrow B \text{ is closed and } A=\mathcal{C}B.$$

It follows that $\mathcal{X}$ can be split into two disjoint, separated sets $A$ and $B$ if and only if $A=\mathcal{C}B$ and $B$ is simultaneously open and closed. The theorem is an immediate consequence.

---

17.4 *Corollary* The metric space $\mathbb{R}^n$ is connected.

*Proof* See exercise 14.17(2).

---

17.5 *Theorem* Let $\mathcal{W}$ be a collection of connected sets in a metric space $\mathcal{X}$ all of which contain a common point $\xi$. Then

$$T=\bigcup_{S\in\mathcal{W}} S$$

is connected.

*Proof* Let $A$ and $B$ be non-empty subsets of $T$ such that $A\cup B=T$. Either $\xi\in A$ or $\xi\in B$. Suppose that $\xi\in A$. Then all of the sets $S\cap A$ with $S\in\mathcal{W}$ are non-empty.

Suppose that $S\cap B=\emptyset$ for all $S\in\mathcal{W}$ Then

$$\emptyset=\bigcup_{S\in\mathcal{W}} S\cap B=B\cap\bigcup_{S\in\mathcal{W}} S=B\cap T=B$$

because $B\subset T$. This is a contradiction because $B\neq\emptyset$. Hence there exists an

$S_1 \in \mathcal{W}$ such that $S_1 \cap B \neq \emptyset$. Also $S_1 \cap A \neq \emptyset$ and

$$(S_1 \cap A) \cup (S_1 \cap B) = S_1 \cap (A \cup B) = S_1 \cap T = S_1.$$

Because $S_1$ is connected, it follows that $S_1 \cap A$ and $S_1 \cap B$ are contiguous. Thus $A$ and $B$ are contiguous by theorem 15.15.

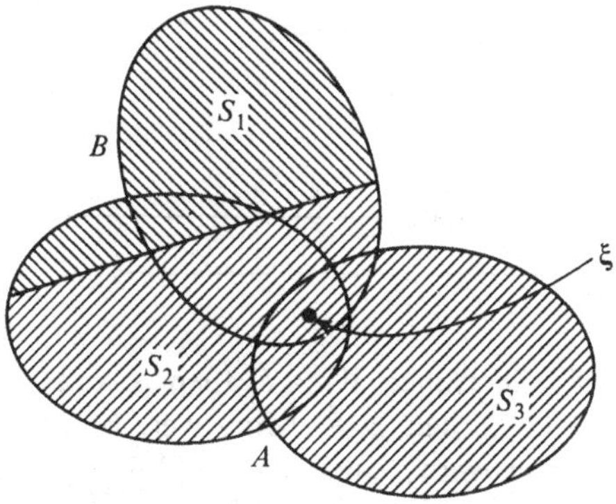


## 17.6 Connected sets in $\mathbb{R}^1$

An *interval* $I$ in $\mathbb{R}^1$ is a set with the property that, if $a \in I$ and $b \in I$, then

$$a < c < b \Rightarrow c \in I.$$

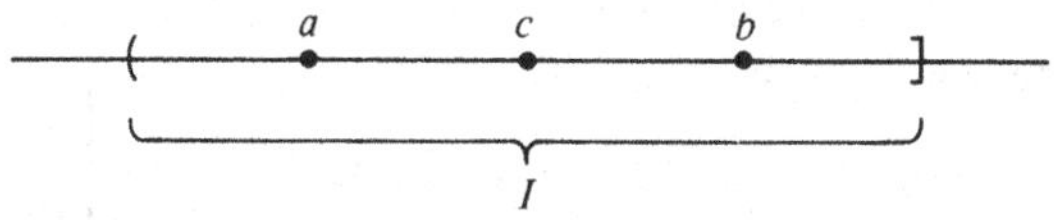


It seems 'geometrically obvious' that a connected set in $\mathbb{R}^1$ is the same thing as an interval. However, the proof is rather more intricate than one would expect.

17.7 *Theorem* A set $E$ in $\mathbb{R}^1$ is connected if and only if it is an interval.

*Proof* (i) Suppose $E$ is connected. If $a \in E$, $b \in E$ and $a < c < b$, we have to prove that $c \in E$. If $c \notin E$, then the sets $A = E \cap (-\infty, c)$ and $B = E \cap (c, \infty)$ are non-empty and $A \cup B = E$. But $A$ and $B$ are separated by theorem 15.14. This is a contradiction and it follows that $E$ is an interval.

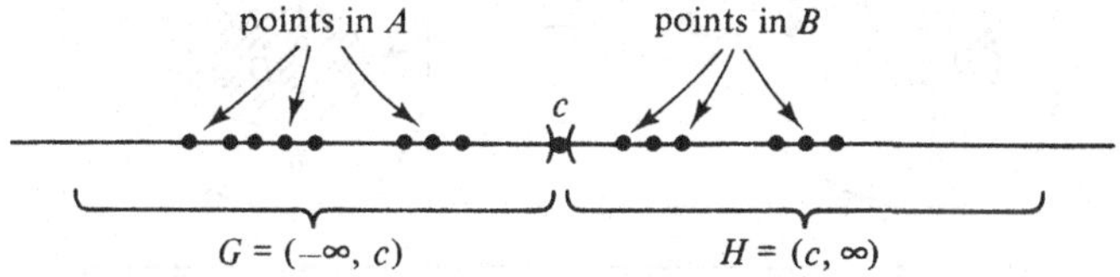

(ii) Suppose that $E$ is an interval. Split $E$ into two non-empty subsets $A$ and $B$ with $A \cup B = E$. We need to show that $A$ and $B$ are contiguous.

Let $a \in A$ and $b \in B$. We shall assume that $a \leqq b$. (If not, reverse the roles of $A$ and $B$.) Let $S = A \cap [a, b]$ and $T = B \cap [a, b]$. If $S$ and $T$ are contiguous, then so are $A$ and $B$ by theorem 15.15.

Since $E$ is an interval, $[a, b] \subset E$ and thus $S \cup T = [a, b]$. In particular, $c = \sup S$ satisfies $a \leqq c \leqq b$ and hence $c \in S$ or $c \in T$.

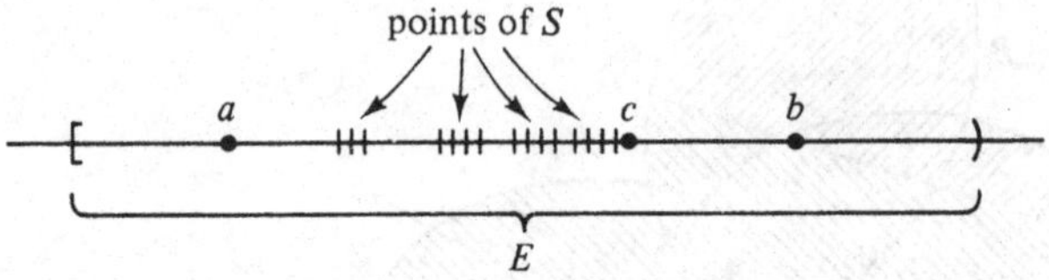


By exercise 13.24(3), $d(c, S) = 0$. Hence, if $c \in T$, then $S$ and $T$ are contiguous by theorem 15.12. If $c \notin T$, then $c \in S$. Moreover, because $c$ is an upper bound of $S$, $(c, b]$ is a non-empty subset of $T$. Thus $d(c, T) = 0$ and hence we arrive again at the conclusion that $S$ and $T$ are contiguous.

## 17.8 Continuity and connected sets

> 17.9 *Theorem* Let $\mathfrak{X}$ and $\mathcal{Y}$ be metric spaces and let $S \subset \mathfrak{X}$. If $f: S \to \mathcal{Y}$ is continuous on the set $S$, then
>
> $$S \text{ connected} \Rightarrow f(S) \text{ connected}.$$

*Proof* Let $C$ and $D$ be non-empty subsets of $f(S)$ such that $C \cup D = f(S)$. We need to show that $C$ and $D$ are contiguous.

If $C$ and $D$ are separated, then so are $f^{-1}(C)$ and $f^{-1}(D)$ (§16.2). But this contradicts the hypothesis that $S$ is connected because $f^{-1}(C) \cup f^{-1}(D) = S$ and neither of the sets $f^{-1}(C)$ or $f^{-1}(D)$ is empty.

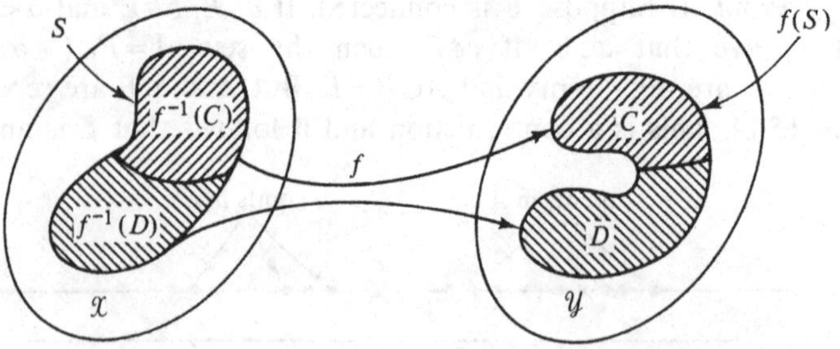

17.10 *Theorem* Suppose that $I$ is an interval in $\mathbb{R}$ and that $f: I \to \mathbb{R}$ is continuous on $I$. Then $f(I)$ is an interval.

*Proof* This is an immediate consequence of the previous two theorems.

---

17.11 *Corollary* (*Intermediate value theorem*) Suppose that $f:[a, b] \to \mathbb{R}$ is continuous on $[a, b]$ and $\lambda$ lies between $f(a)$ and $f(b)$. Then there exists a $\xi \in [a, b]$ such that

$$\lambda = f(\xi).$$

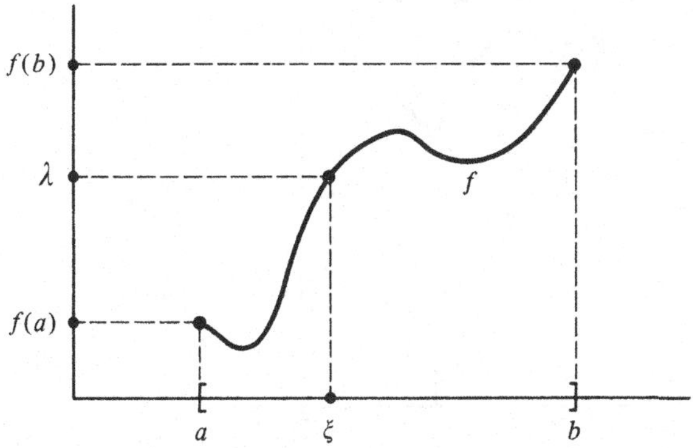


---

17.12 *Example* Show that the equation

$$17x^7 - 19x^5 - 1 = 0$$

has a solution $\xi$ which satisfies $-1 < \xi < 0$.

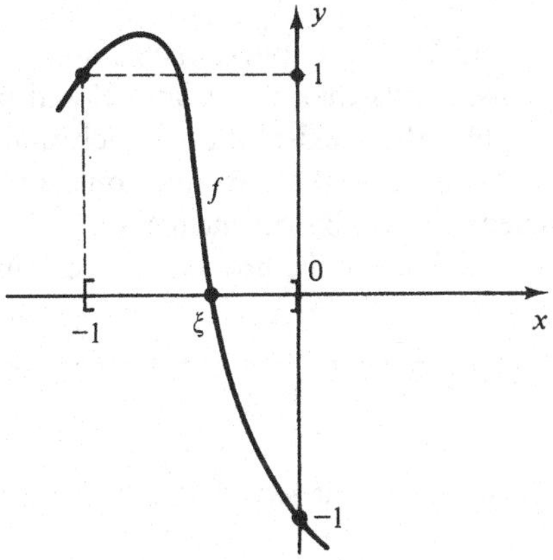

*Proof* The function $f\colon \mathbb{R}\to\mathbb{R}$ defined by $f(x)=17x^7-19x^5-1$ is a polynomial and hence is continuous on $\mathbb{R}$. Thus it is continuous on $[-1, 0]\subset\mathbb{R}$. Now $f(-1)=1$ and $f(0)=-1$. Since $-1<0<1$, it follows from the intermediate value theorem that $f(\xi)=0$ for some $\xi$ between $-1$ and 0.

---

17.13 *Example* Let $f\colon[a, b]\to[a, b]$ be continuous on $[a, b]$. Then, for some $\xi\in[a, b]$, $f(\xi)=\xi$. Thus a continuous function from $[a, b]$ to $[a, b]$ 'fixes' some point of $[a, b]$.

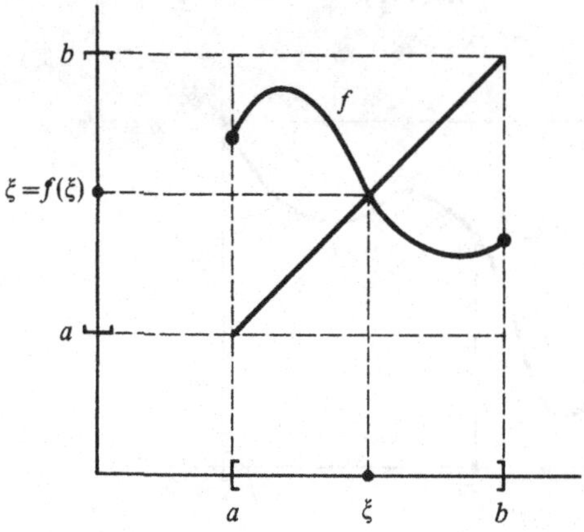


*Proof* The image of $[a, b]$ under $f$ is a subset of $[a, b]$. Thus $f(a)\geqq a$ and $f(b)\leqq b$. The function $g\colon [a, b]\to\mathbb{R}$ defined by $g(x)=f(x)-x$ is continuous on $[a, b]$ by corollary 16.10. But $g(a)\geqq 0$ and $g(b)\leqq 0$. By the intermediate value theorem, it follows that there exists a $\xi\in[a, b]$ such that $g(\xi)=0$

---

17.14 *Note* The previous example is the one-dimensional version of *Brouwer's fixed point theorem*. This asserts that, if $B$ is any closed ball in $\mathbb{R}^n$ and $f\colon B\to B$ is continuous on $B$, then there exists a $\boldsymbol{\xi}\in B$ such that $f(\boldsymbol{\xi})=\boldsymbol{\xi}$. Thus, if a spherical tank full of water is shaken around, one tiny drop of water will end up where it started. This is by no means 'obvious' and the proof in the general case is quite difficult. It is, however, a very important result with many applications.

---

**17.15 Curves**

What is a curve? A physicist might say that a curve is the path traced out by a particle as it changes its position with time. This idea leads us to the following definition.

Let $\mathcal{X}$ be any metric space. We shall say that a *path* in $\mathcal{X}$ is a function $f$: $[0, 1]\to\mathcal{X}$ which is continuous on $I=[0, 1]$. One can think of $\mathbf{x}=f(t)$ as the position in $\mathcal{X}$ of a particle at time $t$. The diagram illustrates the case $\mathcal{X}=\mathbb{R}^2$.

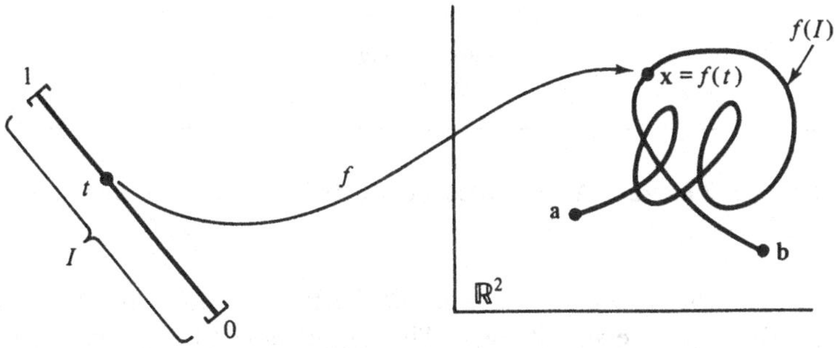


We shall call the set $f(I)$ a *curve*. From the intuitive point of view, this is not a very satisfactory definition since it classifies as curves all sorts of sets which do not look at all like curves 'ought to look'. Many mathematicians therefore prefer not to use the word 'curve' except with some qualifying adjective.

If $f(0)=\mathbf{a}$ and $f(1)=\mathbf{b}$, we shall say that the curve $f(I)$ *joins* $\mathbf{a}$ and $\mathbf{b}$.

---

17.16 *Example* The closed line segment in $\mathbb{R}^n$ given by

$$L=\{t\mathbf{a}+(1-t)\mathbf{b} : 0\leqq t\leqq 1\}$$

is a curve which joins the points $\mathbf{a}$ and $\mathbf{b}$.

We have that $L=f(I)$ where $f$: $I\to\mathbb{R}^n$ is defined by

$$f(t)=t\mathbf{a}+(1-t)\mathbf{b}.$$

---

17.17 *Theorem* A curve in a metric space $\mathcal{X}$ is connected.

*Proof* This is an immediate consequence of theorems 17.7 and 17.9.

---

**17.18 Pathwise connected sets**

A *pathwise connected set* is a set with the property that each pair of points in the set can be joined by a curve which lies entirely in the set.

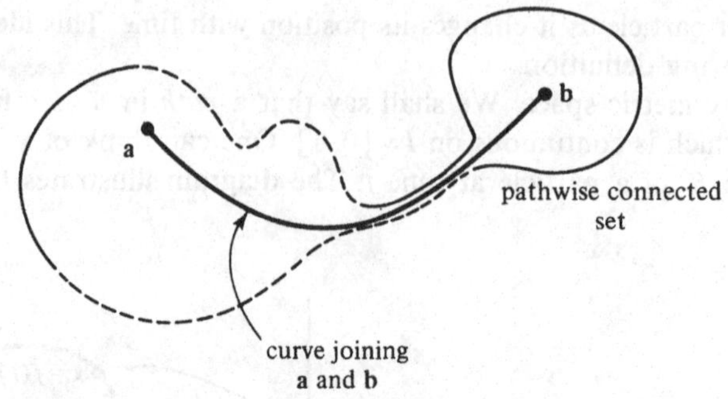


17.19 *Theorem* A pathwise connected set $E$ in a metric space $\mathcal{X}$ is connected.

*Proof* Let $\xi \in E$ and let $\mathcal{W}$ be the collection of all curves which contain $\xi$ and are subsets of $E$. The hypothesis of the theorem ensures that

$$E = \bigcup_{S \in \mathcal{W}} S.$$

Hence $E$ is connected by theorems 17.5 and 17.17.

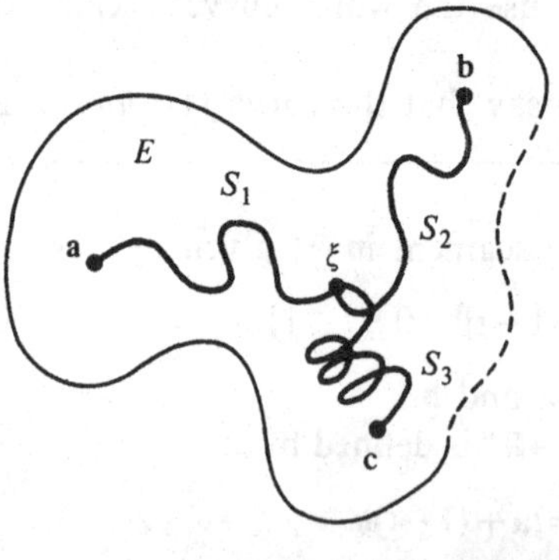


It may be helpful to think of $E$ in the previous theorem as an island and of the paths emanating from $\xi$ as railway lines linking the capital $\xi$ to the rest of the island. The theorem supplies us with a useful means of generating examples of connected sets. But, before considering these examples, there is a natural question which needs to be answered. Are *all* connected sets pathwise connected?

The diagram below illustrates an example due to Brouwer of a set $E$ in $\mathbb{R}^2$ which shows the answer to be negative. The set $E$ consists of the union of a circle and a spiral which winds around the circle infinitely often. The circle and the spiral are contiguous and so $E$ is connected. But no *curve* lying within the set $E$ joints the points **a** and **b**.

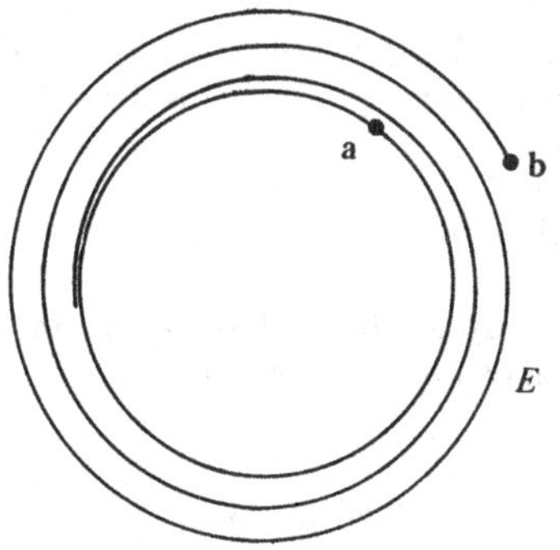


17.20 *Example* Consider the set $S$ in $\mathbb{R}^2$ defined by

$$S = \{(x, y) : x \leqq 0\}.$$

For each point $\mathbf{u} \in S$, it is true that the closed line segment joining $\mathbf{0}$ and $\mathbf{u}$ lies in $S$. Hence $S$ is connected.

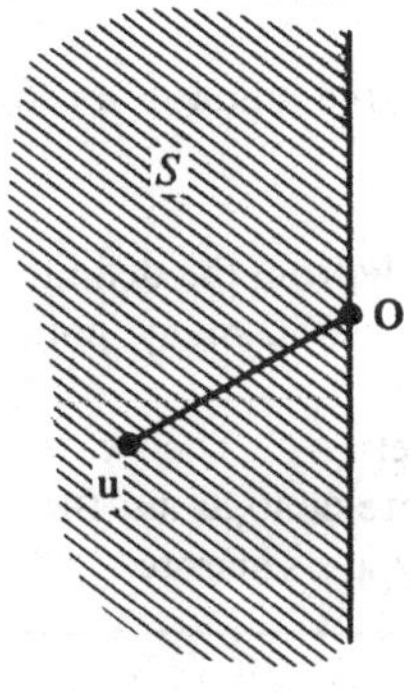


**17.21† Components**

17.22† *Theorem* Let $C$ and $D$ be contiguous sets in a metric space $\mathcal{X}$. If $C$ and $D$ are connected, then so is $C \cup D$.

*Proof* Suppose that $A$ and $B$ are non-empty subsets of $C \cup D$ such that $A \cup B = C \cup D$. We must show that $A$ and $B$ are contiguous.

If $A \cap C \neq \emptyset$ and $B \cap C \neq \emptyset$, then $A \cap C$ and $B \cap C$ are contiguous because $C$ is connected. Hence $A$ and $B$ are contiguous by theorem 15.15. If $B \cap C = \emptyset$, then either $A \cap D \neq \emptyset$ (in which case $A$ and $B$ are contiguous because $D$ is connected) or else

$A \cap D = \emptyset$ (in which case $A = C$ and $B = D$ and so $A$ and $B$ are contiguous by hypothesis). Similarly if $A \cap C = \emptyset$.

---

Let $E$ be a set in a metric space $\mathcal{X}$. If $\xi \in E$, then the *component* of $E$ containing $\xi$ is the largest connected subset $S$ of $E$ which contains $\xi$ – i.e. if $\mathcal{U}$ is the collection of all connected subsets of $E$ containing $\xi$, then

$$S = \bigcup_{T \in \mathcal{U}} T.$$

If one thinks of $E$ as a country entirely surrounded by water, then it is an archipelago of which the constituent islands are the components of $E$.

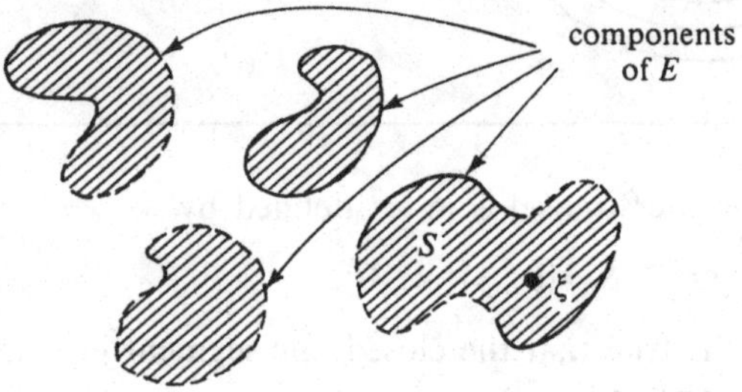


---

17.23† *Theorem* Any set $E$ in a metric space $\mathcal{X}$ is the union of its components. Any two distinct components of $E$ are separated.

*Proof* The first sentence follows immediately from the definition because each $\xi \in S$ determines a component. Of course, in general, many points will determine the *same* component.

Suppose that $C$ and $D$ are *distinct* components. Let $y \in C$. Then $C$ is the largest connected subset of $E$ containing $y$. But, if $C$ and $D$ are contiguous, then $C \cup D$ is a connected subset of $E$ larger than $C$. Hence $C$ and $D$ are separated.

---

17.24 *Examples*

(i) The components of the set $E$ in $\mathbb{R}^2$ defined by

$$E = \{(x, y) : x \leqq 0 \text{ or } x > 1\}$$

are

$$E_1 = \{(x, y) : x \leqq 0\}$$

and

$$E_2 = \{(x, y) : x > 1\}.$$

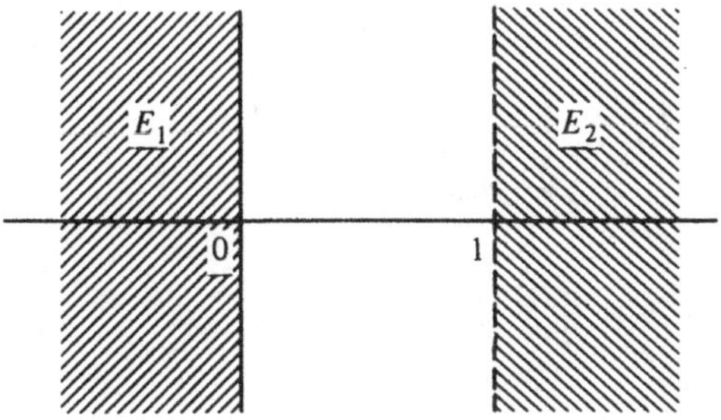


(ii) The components of the set $S=\{1/n: n\in\mathbb{N}\}$ in $\mathbb{R}^1$ are the sets $\{1\}$, $\{\frac{1}{2}\}$, $\{\frac{1}{3}\},\ldots$. Each point of $S$ therefore constitutes a separate component of $S$. We say that such a set is '*totally disconnected*'.

(iii) The components of the set $T=\{0\}\cup S$ are the sets $\{0\}$, $\{1\}$, $\{\frac{1}{2}\}$, $\{\frac{1}{3}\},\ldots$. Notice that any two distinct components are separated but that $\{0\}$ is *not* separated from the *union* of the other components.

---

## 17.25† Structure of open sets in $\mathbb{R}^n$

Theorem 14.13 asserts that the union of any collection of open sets is open. It follows that the set

$$G=(0,\ 1)\cup(1,2)\cup(4,\ \infty)$$

is an open set in $\mathbb{R}^1$.

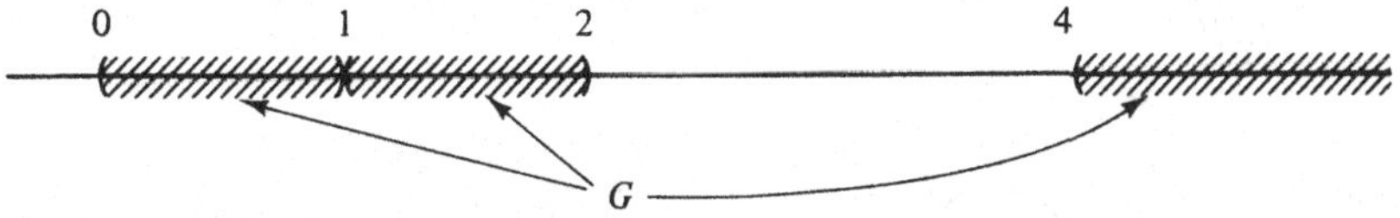


We shall prove that *all* open sets in $\mathbb{R}^1$ are of this general type. We begin with the following theorem.

---

17.26† *Theorem* Any component of an open set $G$ in $\mathbb{R}^n$ is itself an open set.

*Proof* Let $S$ be a component of $G$. Let $\xi\in S$. Since $S\subset G$, $\xi\in G$. Because $G$ is open there exists an open ball $B$ with centre $\xi$ such that $B\subset G$ (theorem 14.12).

Observe that it must also be true that $B\subset S$. Otherwise $B\cup S$ would be a connected subset of $G$ larger than the largest connected subset containing $\xi$ (i.e. $S$). Thus $S$ is open by theorem 14.12.

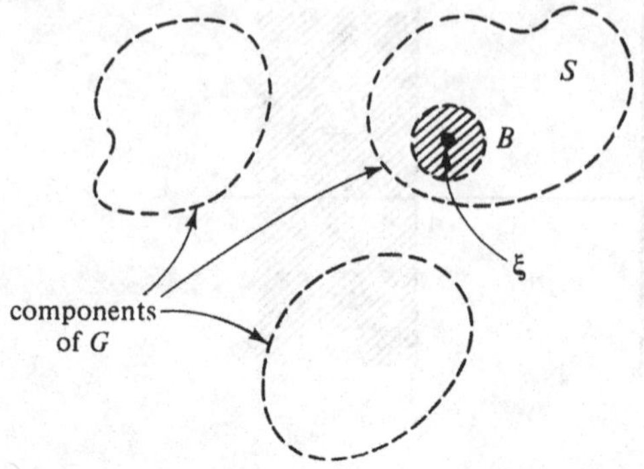


17.27† *Theorem* Any open set $G$ in $\mathbb{R}^1$ is the union of a countable collection of disjoint open intervals.

*Proof* We know from theorem 17.23 that

$$G = \bigcup_{I \in \mathcal{W}} I$$

where $\mathcal{W}$ is the collection of components of $G$. These components are connected by definition and hence are intervals by theorem 17.7. By theorem 17.26, the components are open. The components are therefore open intervals. Moreover, distinct components are separated. In particular, distinct components are disjoint. (See theorem 15.14.)

The only thing left to prove is that $\mathcal{W}$ is a *countable* collection. This is very easy. From theorem 9.20, we know that each open interval contains a rational number. We may therefore construct a function $f: \mathcal{W} \to \mathbb{Q}$ which has the property that $f(I) \in I$ for each $I \in \mathcal{W}$.

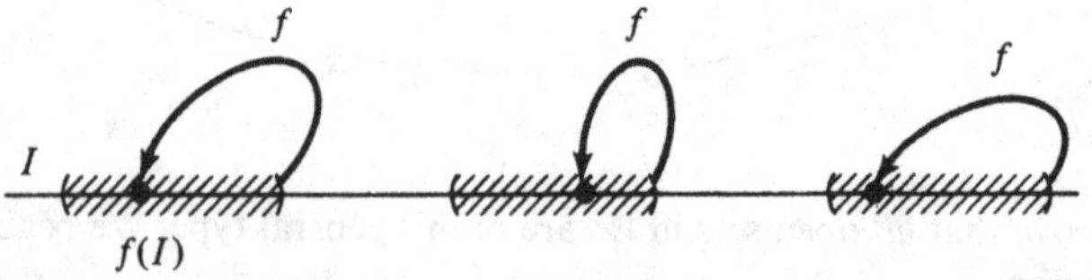


The function $f$ is *injective* because the intervals $I$ in $\mathcal{W}$ are disjoint. Since $\mathbb{Q}$ is countable, it follows from theorem 12.7 that $\mathcal{W}$ is countable.

17.28 *Exercise*

(1) Which of the following sets in $\mathbb{R}^2$ are connected?

(i) $\{(x, y): 0 < x \leqq 1 \text{ and } 0 \leqq y < 1\}$
(ii) $\{(x, y): x > 0 \text{ or } y \leqq 0\}$

(iii) $\{(x, y): x \geqq 1 \text{ or } x < 0\}$
(iv) $\{(x, y): 1 < x^2 + y^2 < 2\}$
(v) $\{(\frac{1}{n}, \frac{1}{n}): n \in \mathbb{N}\}$
(vi) $\{(x, y): x \neq y \text{ and } x \neq -y\}$
(vii) $\{(x, y): xy < 1\}$
(viii) $\{(x, y): xy > 1\}$
(ix) $\{(x, y): xy = 1\}$
(x) $\{(x, y): xy = 0\}$.

(2) Prove that any convex set in $\mathbb{R}^n$ is connected. (See §13.14.) Deduce that open balls, closed boxes, lines, half-lines and hyperplanes are all connected sets in $\mathbb{R}^n$. [*Hint*: Use exercise 13.16(6).] Show also that half-spaces (exercise 14.17(9)) are connected.

(3) Let $I$ be an interval in $\mathbb{R}^1$ and suppose that $f: I \to \mathbb{R}^1$ is an injection which is continuous on $I$. Prove that $f$ is either strictly decreasing on $I$ or else strictly increasing on $I$. [*Hint*: Intermediate value theorem.]

(4) Let $S$ be the set in $\mathbb{R}^2$ defined by

$$S = \{(x, y): 1 \leqq x^2 + y^2 \leqq 2\}.$$

Give an example of a continuous function $f: S \to S$ which has *no* fixed point.

(5) The sets $A$ and $B$ in $\mathbb{R}^2$ are defined by

$$A = \{(x, y): y^2 \leqq 1\};\ B = \{(x, y): x^2 \geqq 1\}.$$

Explain why a function $f: \mathbb{R}^2 \to \mathbb{R}^2$ cannot be continuous on the set $A$ if $f(A) = B$.

†(6) Let $H$ be a hyperplane in $\mathbb{R}^n$. Prove that $\mathbb{R}^n \backslash H$ has two components and that these are the open half-spaces determined by H. (See exercise 14.17(9).) Explain why any line segment which joins two points from different half-spaces contains a point of $H$. (See §13.14.)

# 18 CLUSTER POINTS

## 18.1 Cluster points

A *cluster point* of a set $E$ in a metric space $\mathcal{X}$ is a point $\xi \in \mathcal{X}$ such that

$$\xi \in \overline{E \setminus \{\xi\}}.$$

A cluster point of a set $E$ may or may not be an element of $E$. Equally, an element of $E$ may or may not be a cluster point of $E$. Those elements of $E$ which are *not* cluster points of $E$ are called *isolated points* of E.

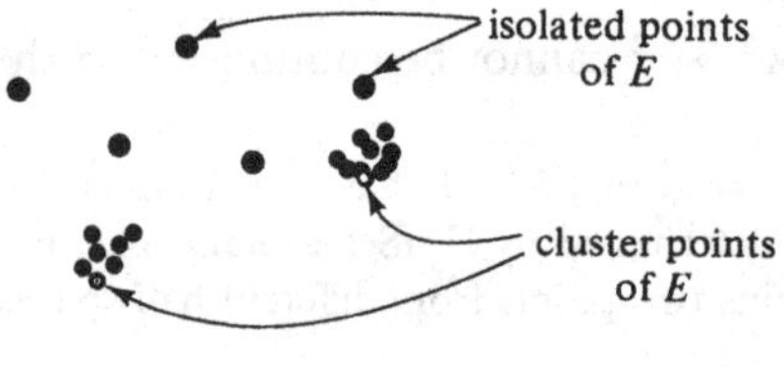


Cluster points are often referred to as 'points of accumulation' or as 'limit points' of the set. The latter usage is unfortunate and we prefer not to employ the word 'limit' except when discussing convergence. (See §26.2.)

The tables below list a number of equivalent assertions about cluster points and isolated points. The proofs of these equivalences provide useful practice in the techniques of chapters 14 and 15 and so are the subject of one of the problems in the next set of exercises.

TABLE I

| |
|---|
| (i) $\xi$ is a cluster point of $E$. |
| (ii) $\xi \in \bar{E}$ but is not an isolated point of $E$. |
| (iii) $d(\xi, E \setminus \{\xi\}) = 0$. |
| (iv) Each open ball with centre $\xi$ contains a point of $E$ other than $\xi$. |
| (v) Each open ball with centre $\xi$ contains an *infinite* subset of $E$. |

TABLE II

| |
|---|
| (i) $\xi$ is an isolated point of $E$. |
| (ii) $\xi \in E$ but is not a cluster point of $E$. |
| (iii) $\xi \in E$ and $d(\xi, E \setminus \{\xi\}) > 0$. |
| (iv) There exists an open ball containing no point of $E$ except $\xi$. |

To deduce Iv from Iiv, consider a sequence of smaller and smaller open balls with centre $\xi$. Note that Iv implies that every point of a *finite* set is an isolated point.

---

## 18.2 *Examples*

(i) The set $S = \{1/n : n \in \mathbb{N}\}$ in $\mathbb{R}^1$ has a single cluster point, namely 0. Note that $0 \notin S$. Every point of $S$ is an isolated point of $S$.

(ii) The set $\mathbb{N}$ in $\mathbb{R}^1$ has *no* cluster points. All its points are isolated.

(iii) The cluster points of the set $T$ in $\mathbb{R}^2$ defined by

$$T = \{(x, y) : x^2 + y^2 < 1\} \cup \{(2, 2)\}$$

are the points of $\{(x, y) : x^2 + y^2 \leqq 1\}$. The point (2, 2) is an isolated point of $T$.

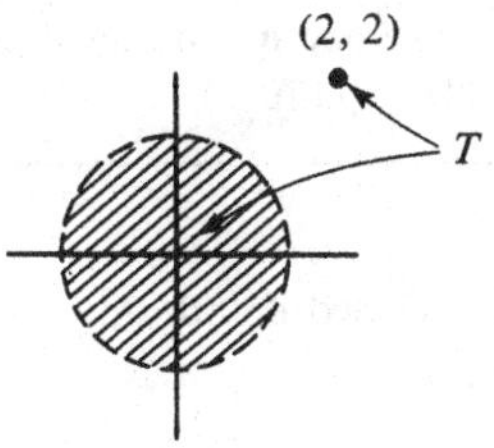


---

## 18.3 *Exercise*

(1) Find the cluster points and the isolated points of the following sets in $\mathbb{R}^1$

(i) $(0, 1)$ (ii) $[0, 1)$ (iii) $(-\infty, 0]$ (iv) $\mathbb{N}$ (v) $\mathbb{Q}$
(vi) $\mathbb{R}$ (vii) $\emptyset$ (viii) $\{1, 2, 3\}$ (ix) $[0, 1]\cup\{2\}$ (x) $[0, 1]\cup(1, 2]$

(2) Repeat the above question for the following sets in $\mathbb{R}^2$.

(i) $A=\{(x, y): |x|+|y|\leqq 1\}$
(ii) $B=[(x, y): x\geqq 0\}$
(iii) $C=\{(x, y): x^2+y^2<1\}$
(iv) $D=\mathbb{N}^2$
(v) $E=\{(1/m, 1/n): m\in\mathbb{N} \text{ and } n\in\mathbb{N}\}$
(vi) $F=\{(x, y): x^2+(y-2)^2\leqq 1\}\cup\{(x, y): (x-2)^2+y^2\leqq 1\}$
(vii) $G=\{(x, y): x^2<1\}\cup\{(x, y): y^2>1\}$.

(3) Prove the equivalences given in tables I and II of §18.1.

---

## 18.4† Properties of cluster points

18.5† *Theorem* Any isolated point $\xi$ of a set $E$ in $\mathbb{R}^n$ is a boundary point of $E$.

*Proof* By §18.1 (IIiv), there exists an open ball $B$ with centre $\xi$ all of whose points are in $\mathcal{C}E$ except for $\xi\in E$. By theorem 14.4, $\xi$ is a boundary point of $E$.

(Note that this theorem need not be true in a general metric space. For example, if $\mathcal{X}=E$.)

---

18.6† *Theorem* Let $E$ be a set in a metric space $\mathcal{X}$. A boundary point $\eta$ of $E$ which is not a cluster point of $E$ is an isolated point of $E$.

*Proof* If $\eta$ is not a cluster point of $E$, then Iiv does not hold and hence there exists an open ball $B$ with centre $\eta$ containing no point of $E$ except (possibly) $\eta$. Since $\eta$ is a boundary point, any open ball $B$ with centre $\eta$ contains a point of $E$. Hence $\eta\in E$. Thus $\eta$ is an isolated point of $E$ by §18.1 (IIii).

---

18.7† *Theorem* A set $E$ in a metric space $\mathcal{X}$ is closed if and only if it contains all its cluster points.

*Proof* (i) Suppose $E$ is closed and $\xi$ is a cluster point of $E$. Then $\xi\in\overline{E\setminus\{\xi\}}\subset\bar{E}=E$.

(ii) Suppose that $E$ contains all its cluster points. Since it contains all its isolated points (by definition), it follows from theorem 18.6 that $E$ contains all its boundary points and hence is closed.

---

18.8† *Exercise*

(1) Let $E$ be a set in a metric space $\mathscr{X}$ and let $r$ be a positive real number. If

$$d(\mathbf{x}, \mathbf{y}) \geqq r$$

for each distinct pair of points $\mathbf{x}$ and $\mathbf{y}$ in $E$, prove that $E$ has no cluster points.

(2) Show that any set $E$ in a metric space $\mathscr{X}$ with no cluster points is closed. [*Hint*: §2.10.] Show also that every point of $E$ is isolated. Give an example of a closed set with a cluster point and a set with only isolated points which is not closed. Show that a closed set $F$ in a metric space $\mathscr{X}$ whose elements are all isolated points has no cluster points.

(3) Use the previous results to show that $\mathbb{N}$ is a closed set in $\mathbb{R}^1$ and that any finite set in a metric space $\mathscr{X}$ is closed.

---

## 18.9† The Cantor set

In mathematics, an example which shows that unexpected things can happen is said to be pathological. The Cantor set is a magnificent instance of a pathological example in that it shows that lots of unexpected things can happen simultaneously.

From the point of view of this chapter, the Cantor set is interesting because it is a *perfect set*. This means that it is a closed set in $\mathbb{R}^1$ with *no* isolated points. The fact that perfect sets exist is not particularly surprising. Any closed interval in $\mathbb{R}^1$ is a perfect set. However, the Cantor set is also totally disconnected (see example 17.24(ii)). This means that each of its components consists of just a single point.

The Cantor set is constructed from the closed interval $[0, 1]$ as follows.

Delete from $[0, 1]$ its open middle third $(\frac{1}{3}, \frac{2}{3})$. This leaves the set

$$F_1 = [0, \tfrac{1}{3}] \cup [\tfrac{2}{3}, 1]$$

which is closed, being the finite union of closed sets (theorem 14.14).

Delete the open middle thirds of each of the closed intervals $[0, \frac{1}{3}]$ and $[\frac{2}{3}, 1]$ which make up $F_1$. This leaves the set

$$F_2 = [0, \tfrac{1}{9}] \cup [\tfrac{2}{9}, \tfrac{1}{3}] \cup [\tfrac{2}{3}, \tfrac{7}{9}] \cup [\tfrac{8}{9}, 1]$$

which is closed and satisfies $F_2 \subset F_1$.

Now delete the open middle thirds of each of these intervals and so on. We thus obtain a sequence $\langle F_n \rangle$ of closed sets which satisfy

$$F_1 \supset F_2 \supset F_3 \supset \ldots.$$

The Cantor set is defined by

$$F = \bigcap_{n=1}^{\infty} F_n.$$

Observe that $F$ is closed by theorem 14.14.

Obviously $F$ contains all the endpoints of the closed intervals which make up the sets $F_1, F_2, F_3, \ldots$. Thus $F$ contains all the points

$$0, 1, \tfrac{1}{3}, \tfrac{2}{3}, \tfrac{1}{9}, \tfrac{2}{9}, \tfrac{7}{9}, \tfrac{8}{9}, \tfrac{1}{27}, \ldots.$$

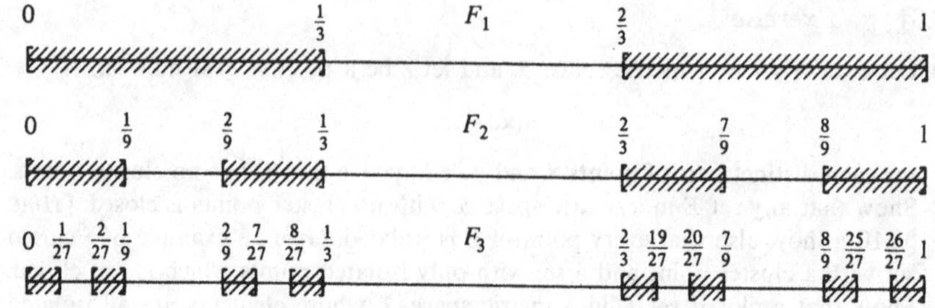


At first sight it may seem that $F$ contains only this countable set of rational numbers. But $F$ is in fact an *uncountable* set. It contains not only the points listed above but all cluster points of the set of these points.

Some of the properties of the Cantor set are listed below. The proofs are left for the next set of exercises.

(i) $F$ is closed
(ii) $F$ is uncountable
(iii) $F$ is totally disconnected
(iv) $F$ is nowhere dense – i.e. $\bar{F}$ contains no open ball
(v) $F$ is perfect
(vi) $F$ has 'zero length'.

The last property can be interpreted as meaning that, for each $\varepsilon>0$, intervals $I_1$, $I_2$, ... exist such that $F\subset I_1\cup I_2\cup\ldots$ and

$$\sum_{k=1}^{\infty} l(I_k)<\varepsilon,$$

where $l(I_k)$ denotes the length of the interval $I_k$.

---

## 18.10† *Exercise*

(1) Let $F$ be the Cantor set as constructed in §18.9. Prove that $x\in F$ if and only if

$$x=\sum_{n=1}^{\infty}\frac{a_n}{3^n}$$

where each of the coefficients $a_n$ is 0 or 2. Use exercise 12.22(2) to deduce that $F$ is uncountable. [*Hint*: Observe that the left-hand endpoints of the closed intervals which make up $F_N$ are the numbers of the form

$$\sum_{n=1}^{N}\frac{a_n}{3^n}$$

where each $a_n$ is 0 or 2. Moreover, each of these closed intervals is of length $3^{-N}$ and

$$\sum_{n=N+1}^{\infty}\frac{a_n}{3^n}\leq\sum_{n=N+1}^{\infty}\frac{2}{3^n}=3^{-N}.]$$

(2) Prove assertions (i) – (vi) of §18.9. [*Hint* $\bar{F}=F\subset F_n(n=1, 2,\ldots)$.]

(3) By theorem 14.9 the complement of the Cantor set is a collection of disjoint open intervals. Obtain explicit formulae for the endpoints of these intervals and check directly that there is only a countable number of these. (See theorem 17.27.)

# 19 COMPACT SETS (I)

## 19.1 Introduction

Of the different types of interval in $\mathbb{R}^1$, perhaps the most important in analysis are those of the form

$$[a, b]=\{x: a \leqq x \leqq b\}.$$

Such intervals are said to be compact. We shall always assume in discussing a compact interval $[a, b]$ that $a \leqq b$ – i.e. that compact intervals are non-empty.

Why are compact intervals so important? This is not an easy question to answer without anticipating the theorems which we shall prove in this and later chapters. However, some readers will already know of the vital theorem which asserts that any continuous function $f$: $[a, b] \rightarrow \mathbb{R}$ attains a maximum and a minimum value on $[a, b]$. This theorem is fundamental for a rigorous development of the differential and integral calculus.

In this chapter we shall generalise the idea of a compact interval by introducing the notion of a compact set. In seeking such a generalisation, the natural way to begin is by asking: what distinguishes a compact interval from other intervals? The obvious answer is that an interval $I$ is compact if and only if it is closed and bounded. Should we therefore define a set $S$ in a metric space $\mathfrak{X}$ to be a closed and bounded set?

The meaning of 'closed' was discussed in chapter 14 and it is easy to provide an appropriate definition of 'bounded'. We say that a set $S$ in a metric space $\mathfrak{X}$ is *bounded* if and only if there exists an open ball $B$ of which $S$ is a subset.

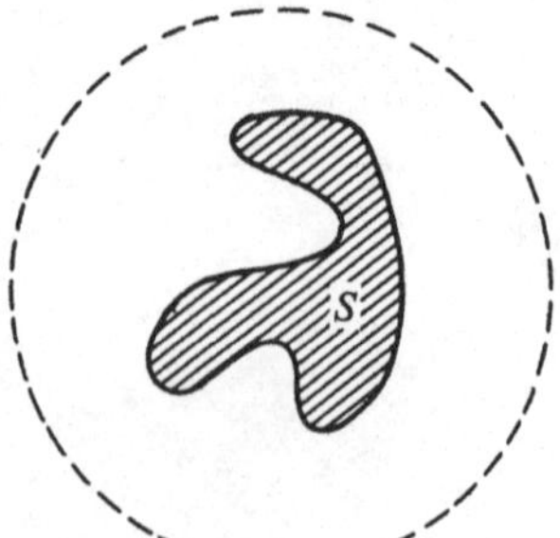


*S bounded*

As we shall see, it would be adequate to define a compact set $S$ to be a closed and bounded set *provided* that we stick very firmly to the space $\mathbb{R}^n$. But this definition would not be of any use in more general metric spaces. The second, and more important reason from our point of view, is that it is *NOT* the fact that compact intervals are closed and bounded which makes them so useful. What makes them important is another more subtle property which we discuss in the next section. This will require some preliminary remarks on oriental culture.

---

## 19.2 Chinese boxes

The next theorem is named after the works of art produced by Chinese and Indian artists consisting of innumerable filligree boxes, one inside another, all carved from a single block of ivory.

Recall from §13.14 that a *closed box* in $\mathbb{R}^n$ is a set $S$ of the form $I_1 \times I_2 \times \ldots \times I_n$, where each of the sets $I_k$ is a compact interval in $\mathbb{R}^1$. The diagram illustrates a nested sequence of closed boxes in $\mathbb{R}^2$.

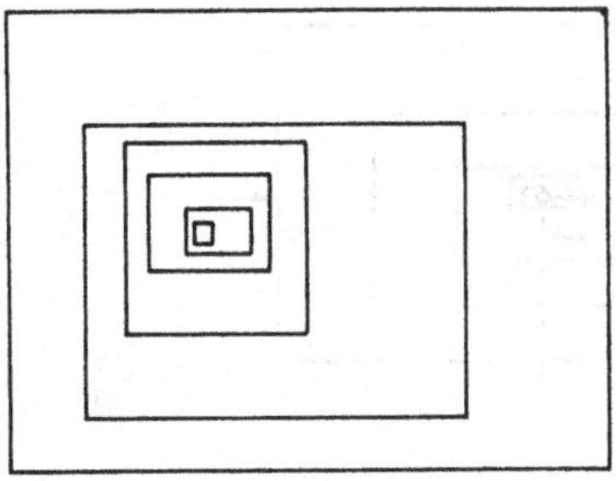

A sequence $\langle S_k \rangle$ of sets is said to be *nested* if and only if

$$S_{k+1} \subset S_k$$

for each $k \in \mathbb{N}$.

---

19.3 *Theorem* (*Chinese box theorem*) Let $\langle S_k \rangle$ be a nested sequence of closed boxes in $\mathbb{R}^n$. Then

$$\bigcap_{k=1}^{\infty} S_k \neq \emptyset.$$

*Proof* We begin with the proof for $\mathbb{R}^1$. Suppose therefore that $\langle I_k \rangle$ is a nested sequence of compact intervals $I_k = [a_k, b_k]$. Put $A = \{a_k : k \in \mathbb{N}\}$ and $B = \{b_k : k \in \mathbb{N}\}$. Since $\langle I_k \rangle$ is nested, each element of $B$ is an upper bound for $A$.

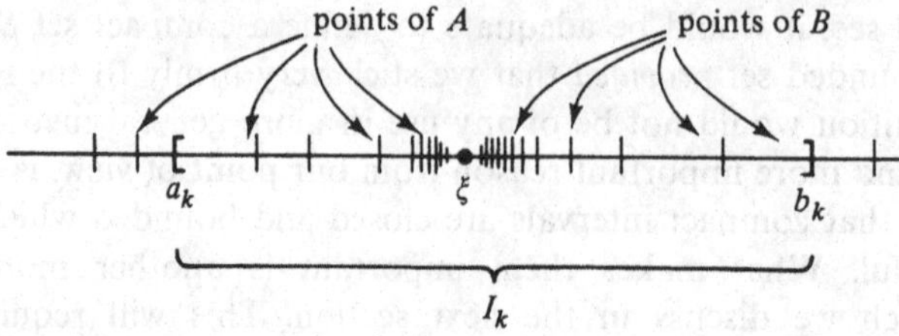


Let $\xi = \sup A$. Then $a_k \leq \xi \leq b_k$ for each $k \in \mathbb{N}$. Hence $\xi \in I_k$ for all $k \in \mathbb{N}$. This proves the theorem in the case $n=1$.

Now let $\langle S_k \rangle$ be any nested sequence of closed boxes of the form $S_k = I_k^{(1)} \times I_k^{(2)} \times \ldots \times I_k^{(n)}$ in $\mathbb{R}^n$. The first part of the proof demonstrates the existence of a real number $\xi_l \in I_k^{(l)}$ for each $k \in \mathbb{N}$. But then

$$(\xi_1, \xi_2, \ldots, \xi_n) \in I_k^{(1)} \times I_k^{(2)} \times \ldots \times I_k^{(n)}$$

for each $k \in \mathbb{N}$. Hence there exists a $\xi \in S_k$ for each $k \in \mathbb{N}$.

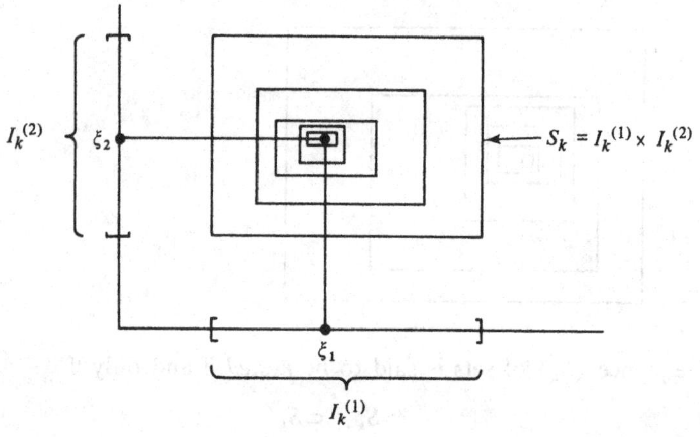


19.4 *Example* We give another proof of the fact that $\mathbb{R}$ is uncountable using the Chinese box theorem.

Suppose that [0, 1] is countable and that $f: \mathbb{N} \to [0, 1]$ is a surjection. Split [0, 1] into three congruent compact subintervals $[0, \frac{1}{3}]$, $[\frac{1}{3}, \frac{2}{3}]$, $[\frac{2}{3}, 1]$. At least one of these subintervals does not contain $f(1)$. Call this subinterval $I_1$. Split the compact interval $I_1$ into three congruent compact subintervals in the same way. At least one of these subintervals does not contain $f(2)$. Call this subinterval $I_2$.

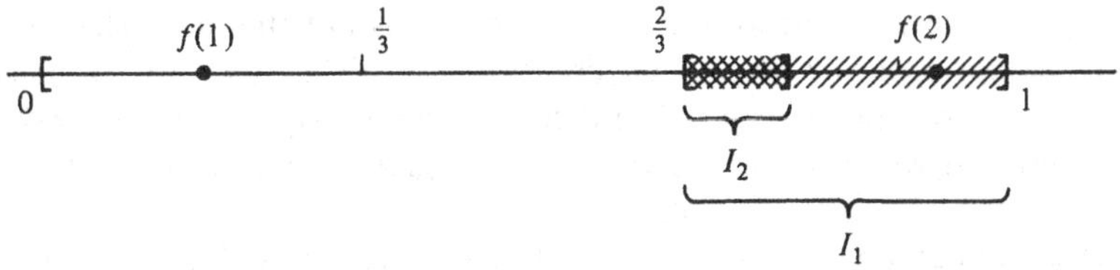


In this way we construct inductively a nested sequence $\langle I_k \rangle$ of compact intervals such that

$$f(k) \in \mathcal{C} I_k$$

for each $k \in \mathbb{N}$. It follows that

$$f(\mathbb{N}) \subset \bigcup_{k=1}^{\infty} \mathcal{C} I_k = \mathcal{C} \bigcap_{k=1}^{\infty} I_k.$$

This contradicts the assumption that $f: \mathbb{N} \to [0, 1]$ is a surjection because the intersection of the sequence $\langle I_k \rangle$ contains at least one point of $[0, 1]$ by the Chinese box theorem.

---

### 19.5 Compact sets and cluster points

The fact that compact intervals satisfy the Chinese box theorem is a far more significant fact than that they are closed and bounded sets. Given this fact, it is therefore sensible to seek to use the Chinese box theorem as the basis for the definition of a compact set in a general metric space $\mathcal{X}$. What we want is a definition of a compact set which ensures that any nested sequence $\langle K_n \rangle$ of non-empty compact sets in a metric space $\mathcal{X}$ has a non-empty intersection.

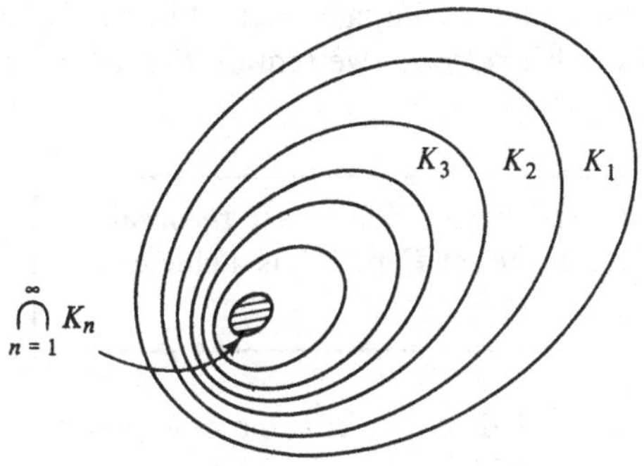


There is no shortage of different (but equivalent) ways of defining a compact set in a metric space $\mathcal{X}$. We shall give *two* definitions.

The first of these definitions is given below. It is not the more elegant of the two definitions: nor does it generalise in a satisfactory way to spaces other than metric spaces. But it is a definition whose significance is easy to understand and which is fairly easy to apply (given the material which we have covered so far in this book).

The second definition of a compact set in a metric space $\mathcal{X}$ is given in the next chapter. This should be regarded as the 'proper definition' of a compact set. The two definitions are equivalent in a metric space and most of the next chapter is concerned with proving this.

> A set $S$ in a metric space $\mathcal{X}$ is *compact* if and only if each infinite subset $E$ of $S$ has a cluster point $\xi \in S$.

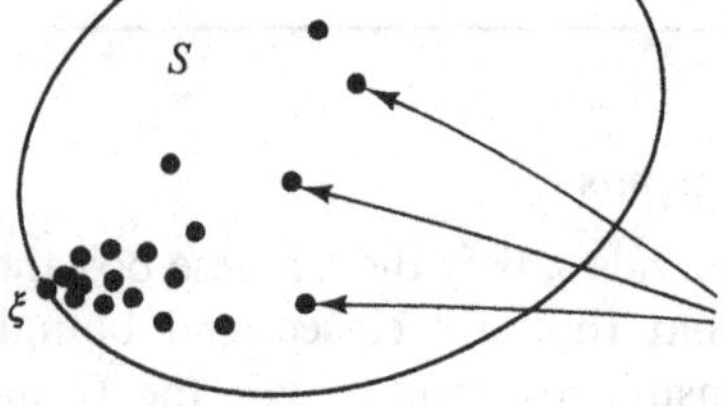


Note that this condition is satisfied 'vacuously' in the case when $S$ is finite. If expressed in formal terms, the condition is that '$\forall E \subset S$ ($E$ infinite $\Rightarrow E$ has a cluster point). But, as we know from §2.10, $P \Rightarrow Q$ is true when $P$ is false. Thus all finite sets are compact.

Our first priority is to show that a compact interval in $\mathbb{R}^1$ is compact in the sense defined above. For this purpose we require the following important theorem.

> 19.6 *Theorem* (*Bolzano–Weierstrass theorem*) Every bounded infinite set $E$ in $\mathbb{R}^n$ has a cluster point.

*Proof* Since $E$ is bounded, it lies in some closed box $S$ (exercise 13.16(5)). The box $S$ can be covered by a finite number of sub-boxes each of whose dimensions are half that of the original box $S$. Since $E$ is infinite, at least one of those sub-boxes contains an infinite subset $E_1$ of $E$. Let $S_1$ be the sub-box containing $E_1$.

We can now repeat the process with $S_1$ replacing $S$ and $E_1$ replacing $E$. Using an appropriate inductive argument we obtain a nested sequence $\langle S_k \rangle$ of closed boxes each of which contains an infinite subset of $E$.

By the Chinese box theorem, there exists a $\xi$ which belongs to each of these boxes. Let $B$ denote any open ball with centre $\xi$. Since the dimensions of $S_k$ are $2^{-k}$ times those of $S$, a closed box $S_K$ will be a subset of $B$ provided that $K$ is sufficiently large. Thus $B$ contains an infinite subset of $E$ and so $\xi$ is a cluster point of $E$ (§18.1(Iv)).

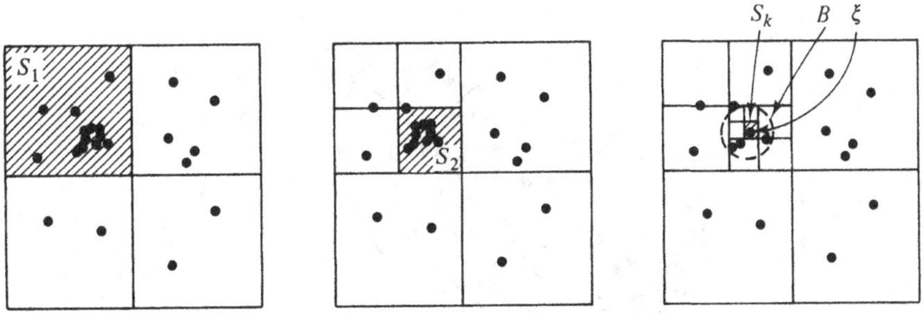


---

19.7 *Note* It is of great importance to remember that the Bolzano–Weierstrass theorem (and hence the Heine–Borel theorem which follows) are *FALSE* in a general metric space $\mathcal{X}$. This point is taken up again in chapter 20.

> 19.8 *Theorem (Heine–Borel theorem)* A set $K$ in $\mathbb{R}^n$ is compact if and only if it is closed and bounded.

*Proof* The proof that a compact set in $\mathbb{R}^n$ is closed and bounded is quite easy and, since we prove the same result for a general metric space later on (theorems 20.12 and 20.13), we shall therefore consider only the deeper part of the proof – i.e. that a closed, bounded set in $\mathbb{R}^n$ is compact.

Suppose that $K$ is a closed, bounded set in $\mathbb{R}^n$ and let $E$ be an infinite subset of $K$. From the Bolzano–Weierstrass theorem it follows that $E$ has a cluster point $\xi$. By theorem 18.7, the fact that $K$ is closed implies that $\xi \in K$. Thus $K$ is compact.

---

19.9 *Theorem (Cantor intersection theorem)* Let $\langle F_n \rangle$ be a nested sequence of non-empty closed subsets of a compact set $K$ in a metric space $\mathcal{X}$. Then

$$\bigcap_{n=1}^{\infty} F_n \neq \emptyset.$$

*Proof* If one of the sets $F_n$ is finite, the result is trivial. Otherwise we can construct an infinite subset $E$ of $K$ consisting of one point from each of the sets $F_n$.

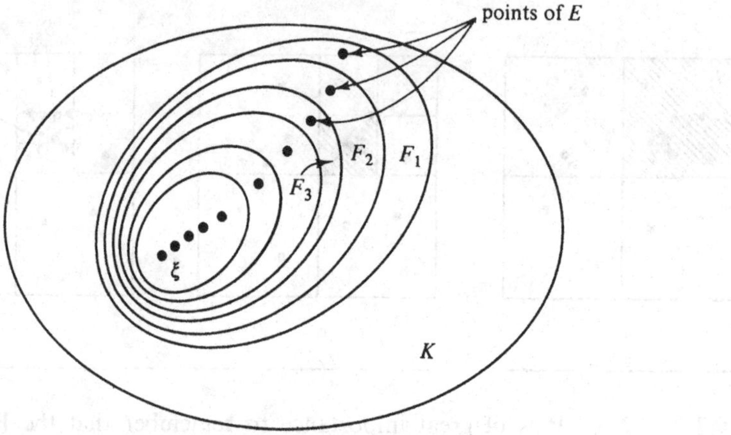


Since $E$ is an infinite subset of a compact set $K$ it has a cluster point $\xi$. Since all but a finite number of points of $E$ belong to each $F_n$, $\xi$ must be a cluster point of each of the sets $F_n$. But each $F_n$ is closed. Hence, by theorem 18.7, $\xi \in F_n$ for each $n \in \mathbb{N}$. *Thus*

$$\xi \in \bigcap_{n=1}^{\infty} F_n.$$

---

19.10 *Example* The sequence $\langle F_n \rangle$ of sets constructed in §18.9 is a nested sequence of non-empty closed subsets of the compact set [0, 1]. By the Cantor intersection theorem, its intersection $F$ is non-empty.

In fact, as is explained in §18.9, $F$ is uncountable.

---

19.11 *Exercise*

(1) Which of the sets given in exercise 18.3(1) are compact?
(2) Which of the sets given in exercise 18.3(2) are compact?

(3) Give examples of a nested sequence of non-empty closed sets in $\mathbb{R}^1$ and a nested sequence of non-empty bounded sets in $\mathbb{R}^1$ which have empty intersections.

---

## 19.12 Compactness and continuity

> **19.13** *Theorem* Let $\mathcal{X}$ and $\mathcal{Y}$ be metric spaces and let $S \subset \mathcal{X}$. If $f: S \to \mathcal{Y}$ is continuous on the set $S$, then
>
> $$S \text{ compact} \Rightarrow f(S) \text{ compact.}$$

*Proof* Suppose that $S$ is compact. Let $E$ be an infinite subset of $f(S)$. We need to show that $E$ has a cluster point in $f(S)$.

Define $g: f(S) \to S$ so that $f(g(\mathbf{y})) = \mathbf{y}$ (see example 6.9). Then $D = g(E)$ is an infinite subset of $S$. Since $S$ is compact, it follows that $D$ has a cluster point $\xi \in S$ – i.e. $d(\xi, D \setminus \{\xi\}) = 0$.

Now let $\eta = f(\xi)$. Then $\eta \in f(S)$. Also since $f$ is continuous on $S$,

$$d(\xi, D \setminus \{\xi\}) = 0 \Rightarrow d(\eta, f(D \setminus \{\xi\})) = 0$$

by theorem 16.4. But $f(D \setminus \{\xi\}) = E \setminus \{\eta\}$ and so $\eta$ is a cluster point of $E$.

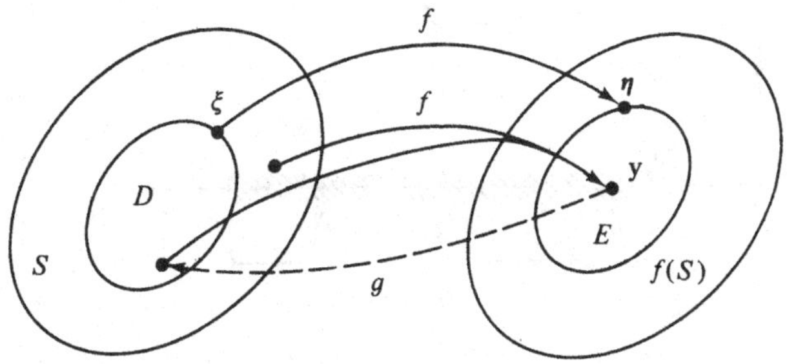


---

The next theorem is of fundamental importance in optimisation theory. It asserts that any real-valued continuous function achieves a maximum and minimum value on a *compact* set $K$. It is worth noting that this theorem need not be true if $K$ is *not* compact. Consider, for example, the function $f: \mathbb{R} \to \mathbb{R}$ defined by $f(x) = x$. This does *not* achieve a maximum nor does it achieve a minimum on the open set $(0, 1)$.

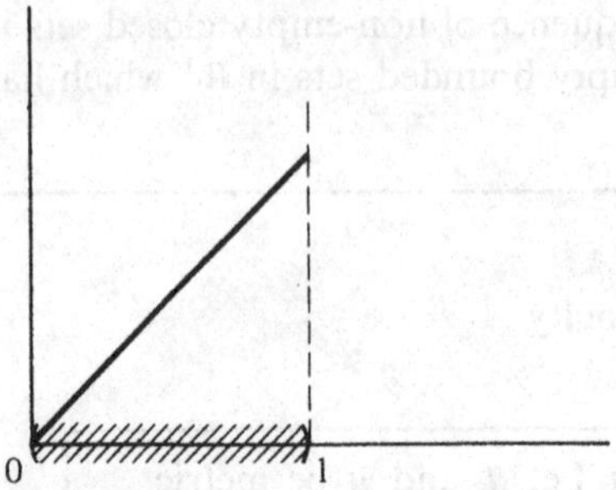


19.14 *Theorem* Suppose that $K$ is a non-empty compact set in a metric space $\mathcal{X}$ and that $f: K \to \mathbb{R}$ is continuous on the set $K$. Then $f$ achieves a maximum and a minimum value on the set $K$ – i.e. there exists $\boldsymbol{\xi} \in K$ and $\boldsymbol{\eta} \in K$ such that, for any $\mathbf{x} \in K$,

$$f(\boldsymbol{\xi}) \leqq f(\mathbf{x}) \leqq f(\boldsymbol{\eta}).$$

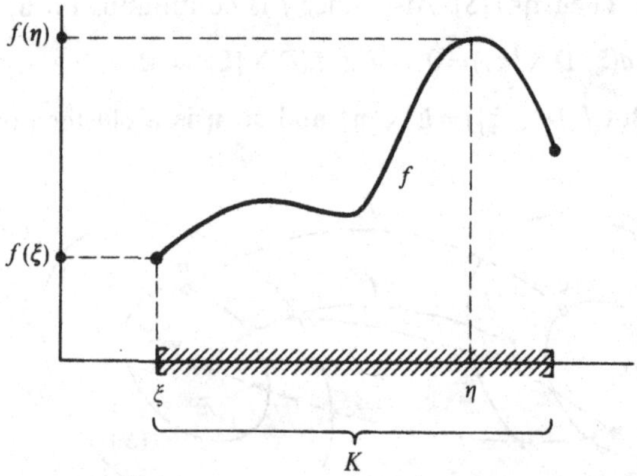


*Proof* By theorem 19.13, $f(K)$ is compact. A compact set in $\mathbb{R}^1$ is closed and bounded by the Heine–Borel theorem (19.8). Because $f(K)$ is a closed, non-empty set of real numbers which is bounded above it has a maximum (corollary 14.11). For similar reasons, $f(K)$ has a minimum.

19.15 *Corollary* Suppose that $K$ is a non-empty compact set in a metric space $\mathcal{X}$ and that $\mathbf{y} \in \mathcal{X}$. Then there exists a $\boldsymbol{\xi} \in K$ such that

$$d(\mathbf{y}, K) = d(\mathbf{y}, \boldsymbol{\xi}).$$

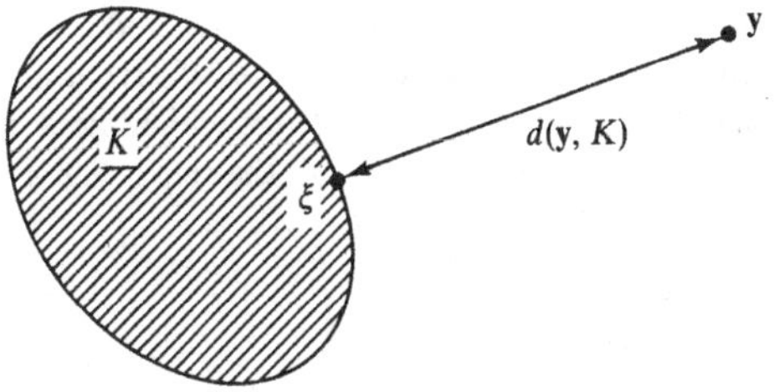


*Proof* The function $f\colon \mathscr{X} \to \mathbb{R}$ defined by $f(\mathbf{x}) = d(\mathbf{y}, \mathbf{x})$ is continuous on $\mathscr{X}$ by exercise 16.16(4). Hence it achieves a minimum on $K$ by theorem 19.14.

---

19.16 *Corollary* Suppose that $F$ is a non-empty closed set in $\mathbb{R}^n$ and that $\mathbf{y} \in \mathbb{R}^n$. Then there exists a $\boldsymbol{\xi} \in F$ such that

$$d(\mathbf{y}, F) = d(\mathbf{y}, \boldsymbol{\xi}).$$

*Proof* Let $\bar{B}$ be the *closed* ball with centre $\mathbf{y}$ and radius $d(\mathbf{y}, F) + 1$. Then $\bar{B} \cap F$ is a non-empty, closed and bounded set in $\mathbb{R}^n$. It follows from the Heine–Borel theorem (19.8) that $\bar{B} \cap F$ is compact. The result therefore follows from corollary 19.15. (Note that the result does *not* hold in a general metric space.)

---

19.17 *Exercise*

(1) The sets $A$ and $B$ in $\mathbb{R}^2$ are defined by

$$A = \{(x, y) : |x| + |y| \leqq 1\}; \quad B = \{(x, y) : 0 < x \leqq 1\}.$$

Explain why a function $f\colon \mathbb{R}^2 \to \mathbb{R}^2$ cannot be continuous on $A$ if $f(A) = B$.

(2) Let $I$ be an interval in $\mathbb{R}^1$ and let $f\colon \mathbb{R}^1 \to \mathbb{R}^1$ be continuous on $I$. Give counter-examples to each of the following (false) propositions:
   (i) $I$ closed $\Rightarrow f(I)$ closed
   (ii) $I$ closed $\Rightarrow f(I)$ bounded
   (iii) $I$ open $\Rightarrow f(I)$ open
   (iv) $I$ bounded $\Rightarrow f(I)$ bounded
   (v) $I$ open and bounded $\Rightarrow f(I)$ has no maximum.

(3) Which of the propositions above are necessarily true when $f\colon \mathbb{R}^1 \to \mathbb{R}^1$ is continuous on a *compact* interval $J$ containing $I$.

†(4) The distance $d(S, T)$ between two non-empty sets $S$ and $T$ in a metric space $\mathscr{X}$ is defined by

$$d(S, T) = \inf\{d(\mathbf{x}, \mathbf{y}) : \mathbf{x} \in S \text{ and } \mathbf{y} \in T\}.$$

Show that, if $S$ and $T$ are compact, then there exist $\mathbf{s} \in S$ and $\mathbf{t} \in T$ such that $d(S, T) = d(\mathbf{s}, \mathbf{t})$. Show that the same result holds when $\mathscr{X} = \mathbb{R}^n$ if $S$ is compact but $T$ is only closed. Give examples of closed sets $S$ and $T$ in $\mathbb{R}^2$ for which the result is *false*.

†(5) Let $S \subset \mathbb{R}$ and suppose that $f: S \to \mathbb{R}$ is continuous on the set $S$. The *graph* of $f$ is the set $T$ in $\mathbb{R}^2$ defined by

$$T = \{(x, f(x)) : x \in S\}.$$

Prove that

(i) $S$ connected $\Rightarrow$ $T$ connected

(ii) $S$ compact $\Rightarrow$ $T$ compact.

Give examples of *non-continuous* functions $f: S \to \mathbb{R}$ for which (i) and (ii) are true.

†(6) Let $f: \mathbb{R} \to \mathbb{R}$ be a bounded function whose graph is a closed set in $\mathbb{R}^2$. Prove that $f$ is continuous on $\mathbb{R}$.

# 20† COMPACT SETS (II)

## 20.1† Introduction

In this chapter we give the 'proper definition' of a compact set (see §19.5). Most of the chapter is then concerned with proving that this definition is equivalent to that of the previous chapter for a metric space $\mathcal{X}$. This is quite a lengthy piece of work and many readers will prefer to skip this chapter for the moment. The book has been written with this possibility in mind.

---

## 20.2† Open coverings

A collection $\mathcal{U}$ of sets is said to *cover* a set $E$ if and only if

$$E \subset \bigcup_{S \in \mathcal{U}} S.$$

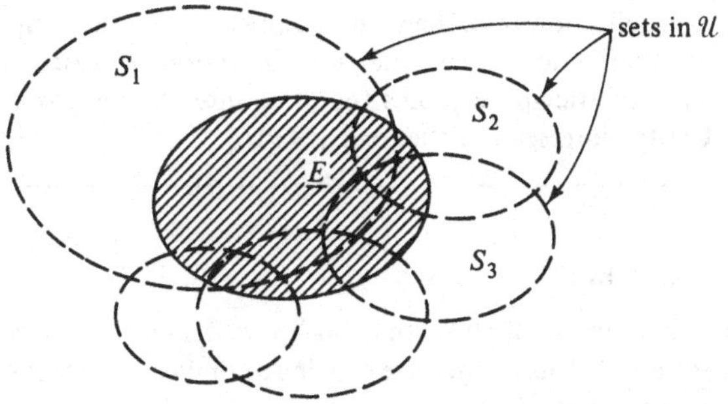


---

20.3 *Example* Let $\delta: [0, 1] \to (0, \infty)$. Then the collection

$$\mathcal{U} = \{(x - \delta(x), x + \delta(x)) : x \in [0, 1]\}$$

covers the set $E = [0, 1]$.

---

## 20.4† Compact sets

A set $K$ in a metric space $\mathcal{X}$ is said to be *compact* if any collection $\mathcal{U}$ of *open* sets which covers $K$ has a *finite* subcollection $\mathcal{F} \subset \mathcal{U}$ which covers $K$.

---

20.5 *Example* Suppose the function $\delta:[0, 1] \to (0, \infty)$ of example 20.3 is given by

$$\delta(x) = \frac{x+1}{x+2}.$$

Then a finite subcollection $\mathcal{F}$ of $\mathcal{U}$ which covers $[0, 1]$ is

$$\mathcal{F} = \{(-\tfrac{1}{2}, \tfrac{1}{2}), (\tfrac{1}{3}, \tfrac{5}{3})\}$$

– i.e. one needs only the intervals $(x-\delta(x), x+\delta(x))$ with $x=0$ and $x=1$.

---

20.6† *Exercise*

(1) A function $f: \mathbb{R} \to \mathbb{R}$ has the property that, for each $x \in [0, 1]$, there exists a $\delta > 0$ such that $f$ is either increasing or else decreasing on $(x-\delta, x+\delta)$. Prove that $f$ is either increasing or decreasing on the whole interval $[0, 1]$. (Use the definition of §20.4 but assume the Heine–Borel theorem.)

(2) Prove that any closed subset of a compact set in a metric space $\mathcal{X}$ is compact. (This may be deduced fairly easily from both of our definitions of compactness.)

(3) A collection $\mathcal{V}$ of sets has the finite intersection property if and only if the fact that each *finite* subcollection of $\mathcal{V}$ has a non-empty intersection implies that $\mathcal{V}$ has a non-empty intersection. Prove that any collection of closed subsets of a compact set $K$ in a metric space $\mathcal{X}$ has the finite intersection property.

Deduce the Cantor intersection theorem (19.9).

---

## 20.7† Compactness in $\mathbb{R}^n$

In this section we shall show that both our definitions for compactness are equivalent in the space $\mathbb{R}^n$. This will involve proving a number of interesting theorems en route. The logical scheme we shall use is illustrated on p. 79.

In this section we shall prove only those implications marked with a single line. The other implications are either obvious or have been proved already. (The fact that the Bolzano–Weierstrass property implies the Chinese box property, for example, is the substance of the proof of the Cantor intersection theorem (19.9).)

The implications indicated with firm lines will be proved for any metric space $\mathcal{X}$. The implications indicated with broken lines will be established only for $\mathbb{R}^n$. As we know, the Bolzano–Weierstrass theorem is *false* in a general metric space. This is

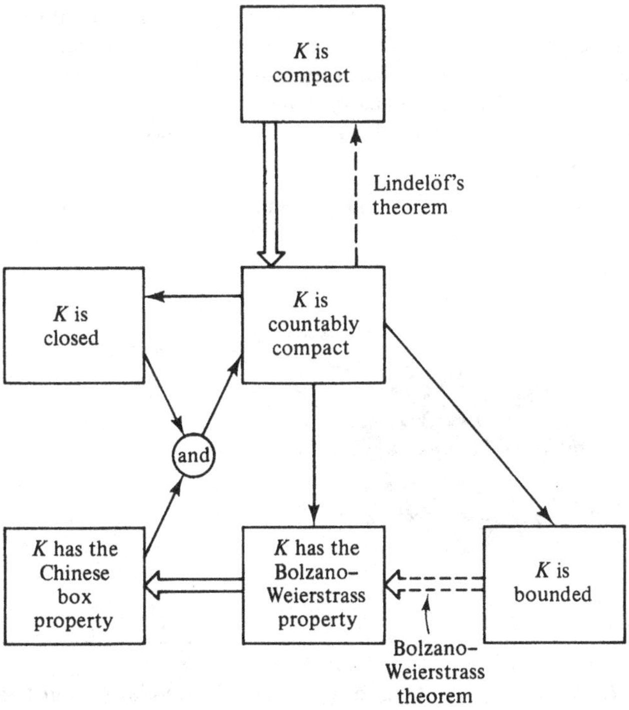


also true of Lindelöf's theorem. However, in §20.16 we shall see that the implication labelled with Lindelöf's theorem can be proved in any metric space using a somewhat more complicated method.

Three of the terms in the diagram are undefined as yet. We therefore begin with definitions of these terms.

A set $K$ in a metric space $\mathcal{X}$ has the *Bolzano–Weierstrass property* if and only if each infinite subset $E$ of $K$ has a cluster point $\xi$.

A set $K$ in a metric space $\mathcal{X}$ has the *Chinese box property* if and only if any nested sequence $\langle F_k \rangle$ of closed, non-empty subsets of $K$ has a non-empty intersection – i.e.

$$\bigcap_{k=1}^{\infty} F_k \neq \emptyset.$$

The proof that we gave of the Cantor intersection theorem (19.9) shows that any set $K$ with the Bolzano–Weierstrass property has the Chinese box property.

A set $K$ in a metric space $\mathcal{X}$ is said to be *countably compact* if and only if each *countable* collection $\mathcal{U}$ of open sets which covers $K$ has a finite subcovering $\mathcal{F}$ which covers $K$.

This definition is, of course, exactly the same as that for a compact set (§20.4) except for the insertion of the word 'countable'. It is therefore not particularly surprising that a compact set and a countably compact set are precisely the same thing in a metric space.

It is obvious that a compact set $K$ in a metric space $\mathcal{X}$ is countably compact. For the converse implication, we need it to be true that every collection $\mathcal{U}$ of open sets which covers $K$ has a *countable* subcollection which covers $K$. A set $K$ for which this is true will be said to have the *Lindelöf property*.

Let $E$ be a set in a metric space $\mathcal{X}$. We say that $D$ is *dense* in $E$ if and only if $E \subset \bar{D}$ – i.e. for each $\mathbf{x} \in E$,

$$d(\mathbf{x}, D) = 0.$$

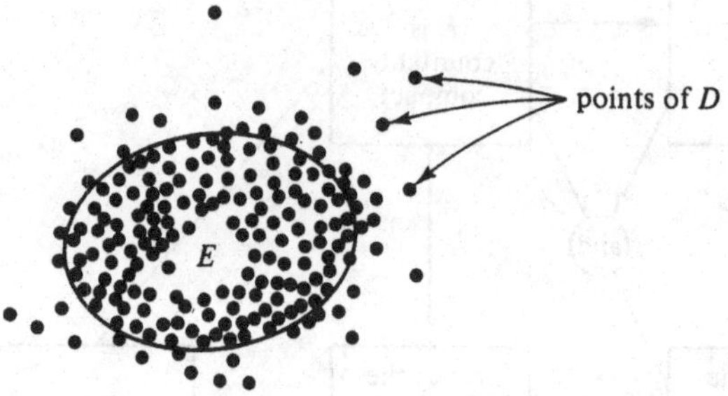


---

20.8 *Example* The set $\mathbb{Q}$ is dense in $\mathbb{R}^1$. This follows from theorem 9.20. Hence $\mathbb{Q}^n$ is dense in $\mathbb{R}^n$ by exercise 13.24(6).

---

20.9† *Theorem* Let $E$ be a set in a metric space $\mathcal{X}$. Then $E$ has the Lindelöf property provided that there exists a *countable* set $D$ which is dense in $E$.

*Proof* Let $\mathcal{U}$ be any collection of open sets which covers $E$. We must show that a *countable* subcollection covers $E$.

Let $\mathcal{B}$ denote the collection of all open balls with centres in $D$ and *rational* radii which are subsets of at least one of the sets $G \in \mathcal{U}$. Then $\mathcal{B}$ is countable. (Consider $D \times \mathbb{Q}$ and apply theorem 12.10.)

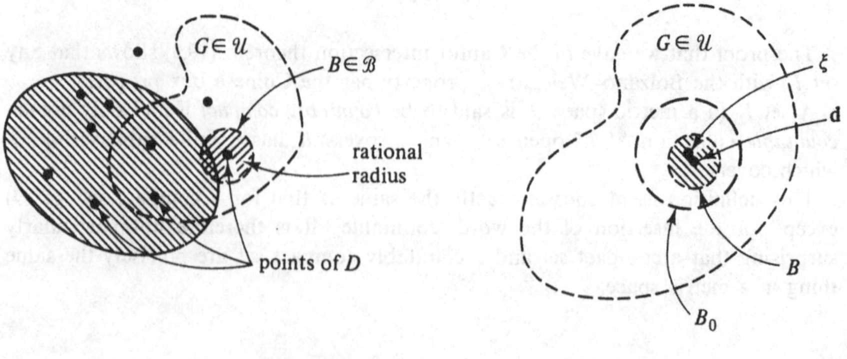

If $G \in \mathcal{U}$, then each $\xi \in G$ lies in an open ball $B \in \mathcal{B}$. Hence

$$E \subset \bigcup_{G \in \mathcal{U}} G \subset \bigcup_{B \in \mathcal{B}} B.$$

The first sentence of this paragraph is proved in the following way. Since $G$ is open, there exists an open ball $B_0$ with centre $\xi$ and radius $r > 0$ such that $B_0 \subset G$ (theorem 14.12). Since $D$ is dense in $E$, there exists a $\mathbf{d} \in D$ such that $d(\xi, \mathbf{d}) < \frac{1}{3}r$ (theorem 15.5). Choose a rational number $q$ such that $\frac{1}{3}r < q < \frac{2}{3}r$ (theorem 9.20). The open ball $B$ with centre $\mathbf{d}$ and radius $q$ then contains $\xi$ because $d(\xi, \mathbf{d}) < \frac{1}{3}r < q$ and is contained in $G$ because, if $\mathbf{y} \in B$, then

$$d(\xi, \mathbf{y}) \leqq d(\xi, \mathbf{d}) + d(\mathbf{d}, \mathbf{y}) < \tfrac{1}{3}r + q < r$$

and so $\mathbf{y} \in B_0 \subset G$.

Now let $f: \mathcal{B} \to \mathcal{U}$ be chosen so that $B \subset f(B)$. Then $f(\mathcal{B})$ is a countable subcollection of $\mathcal{U}$ and

$$E \subset \bigcup_{B \in \mathcal{B}} B \subset \bigcup_{B \in \mathcal{B}} f(B) = \bigcup_{G \in f(\mathcal{B})} G.$$

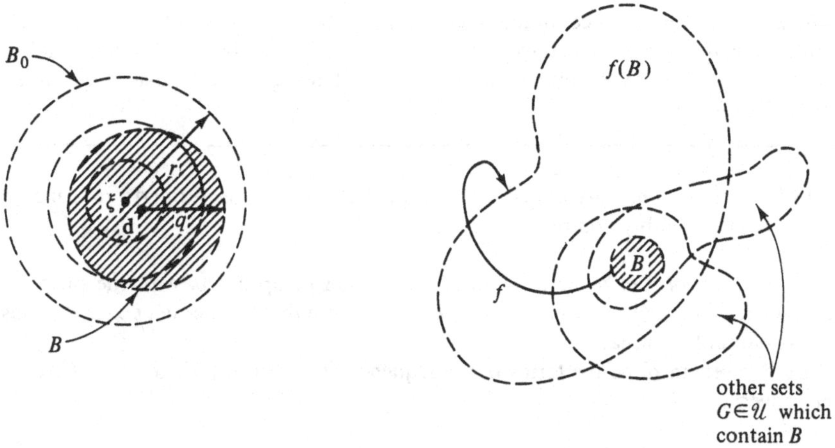


We have shown that $f(\mathcal{B})$ is a countable subcollection of $\mathcal{U}$ which covers $E$.

---

20.10† *Corollary* (*Lindelöf's theorem*) Any set $E$ in $\mathbb{R}^n$ has the Lindelöf property.

*Proof* The set $\mathbb{Q}^n$ is countable by exercise 12.13(3). Since $\mathbb{Q}^n$ is dense in $\mathbb{R}^n$ (example 20.8), $\mathbb{Q}^n$ is dense in $E$. The result therefore follows from theorem 20.9.

---

20.11† *Corollary* Any countably compact set $K$ in $\mathbb{R}^n$ is compact.

*Proof* This follows immediately from Lindelöf's theorem.

---

20.12† *Theorem* Any countably compact set $K$ in a metric space $\mathcal{X}$ is bounded.

*Proof* Let $\xi$ be any point of $\mathcal{X}$ and let $\mathcal{U}$ be the nested collection of open balls of the form

$$\{\mathbf{x} : d(\xi, \mathbf{x}) < n\}$$

where $n \in \mathbb{N}$. Then $\mathcal{U}$ is countable. Since $K$ is countably compact, a finite subcollection $\mathcal{F}$ of $\mathcal{U}$ covers $K$. Let $B$ denote the open ball of maximum radius in $\mathcal{F}$. Then $K \subset B$ and so $K$ is bounded.

---

20.13† *Theorem* Any countably compact set $K$ in a metric space $\mathcal{X}$ is closed.

*Proof* Let $\xi$ be any point *not* in $K$ and let $\mathcal{U}$ be the nested collection of all open sets of the form

$$\{\mathbf{x} : d(\xi, \mathbf{x}) > 1/n\}$$

where $n \in \mathbb{N}$. Then $\mathcal{U}$ is countable and covers $K$. Since $K$ is countably compact, a finite subcollection $\mathcal{F}$ of $\mathcal{U}$ covers $K$. Let $B = \mathcal{X} \setminus S$ be the ball of minimum radius such that $S \in \mathcal{F}$. Then $B$ contains no point of $K$. Hence $\xi$ is *not* a cluster point of $K$. Thus $K$ is closed.

---

20.14† *Theorem* Any closed set $K$ with the Chinese box property in a metric space $\mathcal{X}$ is countably compact.

*Proof* Let $\mathcal{U}$ be a countable collection of open sets with the property that *no* finite subcollection covers $K$. We shall deduce that $\mathcal{U}$ does *not* cover $K$. Thus $K$ is countably compact.

Let the sets in $\mathcal{U}$ be the terms of the sequence $\langle G_k \rangle$. Since $\{G_1, G_2, \ldots, G_k\}$ does *not* cover $K$,

$$F_k = K \cap \mathcal{C} \bigcup_{j=1}^{k} G_j = \bigcup_{j=1}^{k} (K \cap \mathcal{C} G_k)$$

is non-empty.

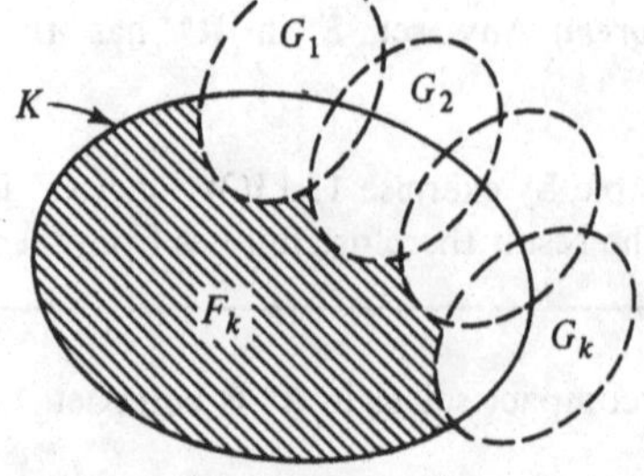

But $\langle F_k \rangle$ is a nested sequence of closed subsets (theorem 14.14) of $K$. Since $K$ has the Chinese box property, $\langle F_k \rangle$ has a non-empty intersection. Thus

$$\emptyset \neq \bigcap_{k=1}^{\infty} F_k = K \cap \mathcal{C} \bigcup_{j=1}^{\infty} G_j$$

and so $\mathcal{U}$ does *not* cover $K$.

---

This theorem completes the proof of the implications of the diagram of §20.7.

---

## 20.15† Completeness

In §19.5 we took note of the fact that the Bolzano–Weierstrass theorem is *not* true in a general metric space $\mathcal{X}$. The same is therefore true of the Heine–Borel theorem. Indeed, it can be shown that if the 'sphere' $S = \{\mathbf{x} : \|\mathbf{x}\| = 1\}$ in a normed vector space $\mathcal{X}$ is compact, then $\mathcal{X}$ is necessarily finite-dimensional. Since $S$ is always closed and bounded, this shows that the Heine–Borel theorem is always false in an infinite-dimensional normed vector space.

Why does the Bolzano–Weierstrass theorem fail in a general metric space? There are two reasons. The first is that it may be impossible to prove a suitable analogue of the Chinese box theorem. When no such analogue exists there is no point in seeking to obtain some substitute for the Bolzano–Weierstrass theorem. The deficiency is too severe. However, there is a second reason why the Bolzano–Weierstrass theorem may fail in a general metric space $\mathcal{X}$.

Observe that, in the proof of the Bolzano–Weierstrass theorem given in §19.5, it is necessary that, given any $l > 0$, it be possible to cover the bounded set $E$ with a *finite* number of closed boxes whose sides are all of length at most $l$. One can then deduce from the fact that $E$ is infinite, the conclusion that at least one of the boxes contains an infinite subset of $E$. If one needed an *infinite* collection of such boxes to cover $E$, the proof would fail.

This observation leads us to the following definition. We say that a set $S$ in a metric space $\mathcal{X}$ is *totally bounded* if and only if, given any $r > 0$, there exists a *finite* collection $\mathcal{F}_r$ of open balls of radius $r > 0$ which covers $S$.

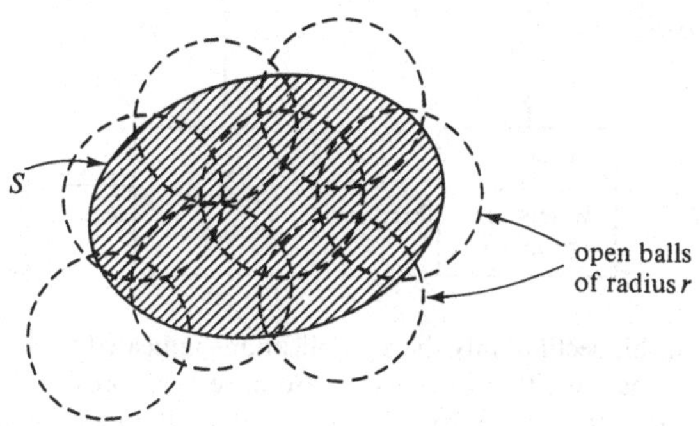


It is obvious that any totally bounded set $S$ in a metric space $\mathcal{X}$ is bounded. (If $\mathcal{F}_1$

contains $N$ elements, $S$ is a subset of any open ball $B$ with radius $2N$ provided that the centre of $B$ lies in $S$.) In $\mathbb{R}^n$ the converse is also true because, by the Heine–Borel theorem, if $S$ is bounded, then $\bar{S}$ is compact. But $\bar{S}$ is covered by the collection of *all* open balls of radius $r$. Hence $\bar{S}$ and therefore $S$ is covered by a *finite* number of open balls of radius $r$. In a general metric space $\mathfrak{X}$, however, a bounded set certainly need *not* be totally bounded (see §20.20).

The second reason we gave for the failure of the Bolzano–Weierstrass theorem in a general metric space $\mathfrak{X}$ clearly disappears if we replace the word 'bounded' in its statement by the words 'totally bounded'.

These considerations lead us to the notion of a *complete metric space*. The usual definition is given in chapter 27, but this definition is equivalent to the assertion that, in a complete metric space $\mathfrak{X}$, every *totally* bounded set has the Bolzano–Weierstrass property. Thus a complete metric space is one in which a suitably amended version of the Bolzano–Weierstrass theorem is true. A large number of important metric spaces are complete and so this is an important idea. An example is given in §20.20.

## 20.16† Compactness in general metric spaces

The diagram below summarises the results of this chapter.

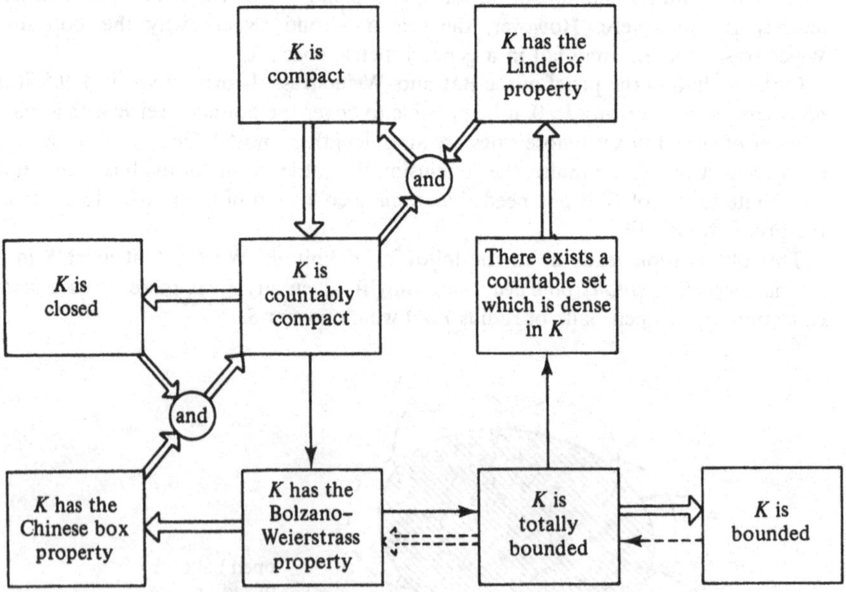


We shall prove in this section only those implications indicated with a single firm line. Other implications are either obvious or else have been proved already.

The meaning of the different styles of arrows representing implications in the diagram is explained in the following table.

| | | |
|---|---|---|
| $\longrightarrow$ | True in any metric space | Not yet proved |
| $\Longrightarrow$ | | Proved already |
| ⇢ (dashed double arrow) | True in a complete metric space | |
| ⇢ (dashed arrow) | True in $\mathbb{R}^n$ | |

---

20.17† *Theorem* A set $S$ in a metric space $\mathcal{X}$ which has the Bolzano–Weierstrass property is totally bounded.

*Proof* If $S$ is *not* totally bounded, then for some $r>0$ no finite collection of open balls of radius $r$ covers $S$. If $\mathbf{x}_1, \mathbf{x}_2, \ldots, \mathbf{x}_{n-1}$ are points of $S$ one can therefore always find a point $\mathbf{x}_n \in S$ such that $d(\mathbf{x}_n, \mathbf{x}_k) \geqq r$ $(k=1, 2, \ldots, n-1)$.

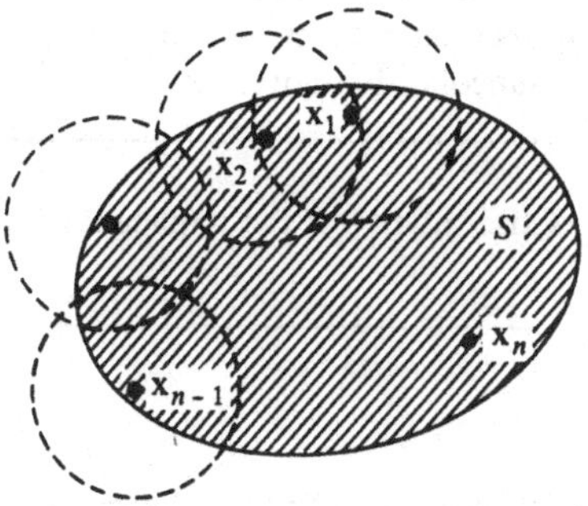


The set $S$ therefore contains an infinite subset with *no* cluster point (exercise 18.8(1).)

---

20.18† *Theorem* If $S$ is a totally bounded set in a metric space $\mathcal{X}$, there exists a countable set $D$ which is dense in $S$.

*Proof* Let $C_n$ denote the set of centres of a finite collection of open balls of radius $1/n$ which covers $S$. Then given any $\mathbf{x} \in S$, $d(\mathbf{x}, C_n) < 1/n$. Now let

$$D = \bigcup_{n=1}^{\infty} C_n.$$

Then $D$ is countable because it is the countable union of countable sets (theorem 12.12) and $D$ is dense in $S$ because, for each $\mathbf{x} \in S$,

$$d(\mathbf{x}, D) \leqq d(\mathbf{x}, C_n) < 1/n.$$

Since this is true for *all* $n\in\mathbb{N}$, $d(\mathbf{x}, D)=0$.

---

20.19† *Theorem* Any countably compact set $K$ in a metric space $\mathcal{X}$ has the Bolzano–Weierstrass property.

*Proof* Let $E$ be any infinite subset of $K$. If $E$ has no cluster point, this is also true of any countable subset $F$ of $E$. By exercise 18.8(2), $F$ is closed and all of its points are isolated.

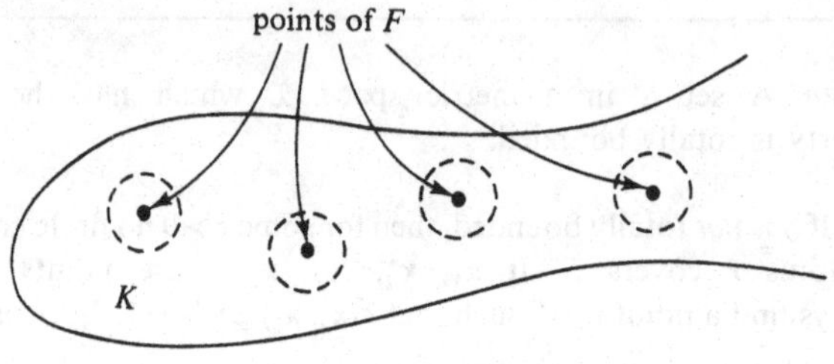


For each $\mathbf{x}\in F$ we can find an open ball which contains no point of $F$ except $\mathbf{x}$ (§18.1(IIiv)). The collection of all such open balls together with the open set $\mathcal{C}F$ is countable and covers $K$. But no finite subcollection covers $K$.

---

This theorem completes the proof of the validity of the implications indicated in the diagram of §20.16.

---

## 20.20† A spherical cube

In this section, we give one of our rare examples of a metric space other than $\mathbb{R}^n$.

We begin by observing that one can think of a sequence $\langle x_k\rangle$ of real numbers as the 'infinite-dimensional' vector

$$\mathbf{x}=(x_1, x_2, x_3, \ldots)$$

in which the co-ordinates of the vector $\mathbf{x}$ are just the terms of the sequence $\langle x_k\rangle$. Vector addition and scalar multiplication may then be defined just as we did in §13.1 for vectors in $\mathbb{R}^n$.

We shall consider the normed vector space $l^\infty$ consisting of all *bounded* sequences of real numbers. The norm on $l^\infty$ is defined by

$$\|\mathbf{x}\|=\sup_{k\in\mathbb{N}} |x_k|.$$

The distance between points $\mathbf{x}$ and $\mathbf{y}$ of $l^\infty$ is therefore given by

$$d(\mathbf{x}, \mathbf{y})=\|\mathbf{x}-\mathbf{y}\|=\sup_{k\in\mathbb{N}} |x_k-y_k|.$$

The first thing to notice is that the proof that we gave for the Chinese box theorem (19.3) works equally well for $l^\infty$. The same is therefore true of the Bolzano–Weierstrass theorem provided that 'bounded' is replaced by 'totally bounded'. Thus $l^\infty$ is a *complete* metric space.

Consider the set $S=\{\mathbf{x}:\|\mathbf{x}\|\leqq 1\}$ in $l^\infty$. This can also be written in the form

$$S=\{\mathbf{x}:|x_k|\leqq 1 \text{ for each } k\in\mathbb{N}\}.$$

The first expression would describe a ball in $\mathbb{R}^3$ and the second would describe a cube in $\mathbb{R}^3$. Hence the title of this section.

The set $S$ is clearly a closed and bounded set in $l^\infty$. But it satisfies *none* of the other properties listed in the diagram of §20.16. This is most easily shown by proving that the subset $V$ of $S$ consisting of all its 'corners' does not satisfy the Lindelöf property. The set $V$ is the set of all sequences each of whose terms is either 1 or $-1$. It is therefore an *uncountable* set (exercise 12.22(2)).

The fact that $V$ does not have the Lindelöf property simply follows from the fact that the collection $\mathcal{U}$ of all open balls with centres in $V$ and of radius 1 is uncountable and covers $V$ but has no proper subcover at all.

Note in particular that $S$ is an example of a closed, bounded set which is *not* compact.

For an example of a non-trivial compact set in a space other than $\mathbb{R}^n$ see exercise 20.21(6).

---

## 20.21† *Exercise*

(1) Prove that $l^\infty$ as described in §20.20 is a metric space.

(2) The normed vector space $l^2$ consists of all sequences $\mathbf{x}=\langle x_k\rangle$ of real numbers for which

$$\sum_{k=1}^{\infty} x_k{}^2$$

converges. An inner product is defined on $l^2$ by

$$\langle \mathbf{x}, \mathbf{y}\rangle=\sum_{k=1}^{\infty} x_k y_k.$$

Prove a version of the Cauchy–Schwarz inequality (13.6) and deduce that $l^2$ is a metric space if distance is defined in the usual way for a normed vector space (§13.18).

(3) A closed box in the set of all sequences of real numbers is a set $S$ of the form

$$S=I_1\times I_2\times I_3\times\ldots$$

where $\langle I_n\rangle$ is a sequence of compact intervals.

If $\langle S_k\rangle$ is a nested sequence of such closed boxes, prove that its intersection is not empty. Deduce from this version of the Chinese box theorem that both $l^\infty$ and $l^2$ are complete.

(4) Prove that $l^2$ contains a countable dense subset. [*Hint*: Look at the set $D$ of all points of the form $(q_1, q_2, \ldots, q_n, 0, 0, \ldots)$ where $q_1, q_2, \ldots, q_n$ are rational.] Deduce that $S=\{\mathbf{x}:\|\mathbf{x}\|\leqq 1\}$ has the Lindelöf property in $l^2$.

(5) Prove that the set $S=\{\mathbf{x}:\|\mathbf{x}\|\leq 1\}$ is closed and bounded in $l^2$ but *not* totally bounded. Deduce that $S$ is not compact in $l^2$. [*Hint*: Look at the set of all points of the form (0, 0, ..., 0, 1, 0, 0, ...).]

(6) Prove that the 'closed box' $S=\{\mathbf{x}:|x_k|\leq 1/k \text{ for each } k\in\mathbb{N}\}$ is closed and totally bounded in $l^2$. [*Hint*: Given $r>0$, choose $n$ so that

$$\left[\sum_{n+1}^{\infty} x_k^{\,2}\right]^{1/2}<\tfrac{1}{2}r$$

and then use the fact that closed boxes are totally bounded in $\mathbb{R}^n$.] Deduce that $S$ is compact in $l^2$.

# 21 TOPOLOGY

## 21.1 Topological equivalence

Suppose that $\mathcal{X}$ and $\mathcal{Y}$ are metric spaces and that $f: \mathcal{X} \to \mathcal{Y}$ is a bijection with the property that *both* $f: \mathcal{X} \to \mathcal{Y}$ and $f^{-1}: \mathcal{Y} \to \mathcal{X}$ are continuous. Such a function is called a *homeomorphism.* (This term is easily confused with the term homomorphism which means something quite different.)

If a homeomorphism $f: \mathcal{X} \to \mathcal{Y}$ exists, we say that $\mathcal{X}$ and $\mathcal{Y}$ are *topologically equivalent*. The subject of topology is concerned with studying the properties that topologically equivalent spaces have in common.

Consider, for example, a teacup made from modelling clay. This can be continuously deformed into a doughnut and the process can then be reversed.

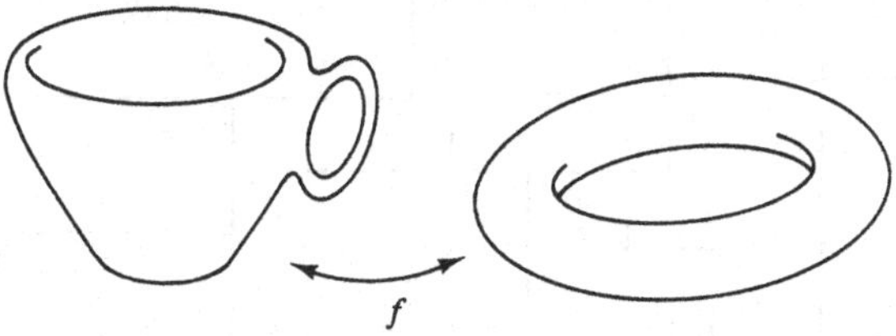


teacup doughnut

The doughnut and the teacup (regarded as metric subspaces of $\mathbb{R}^3$) are therefore topologically equivalent. The topological property that they have

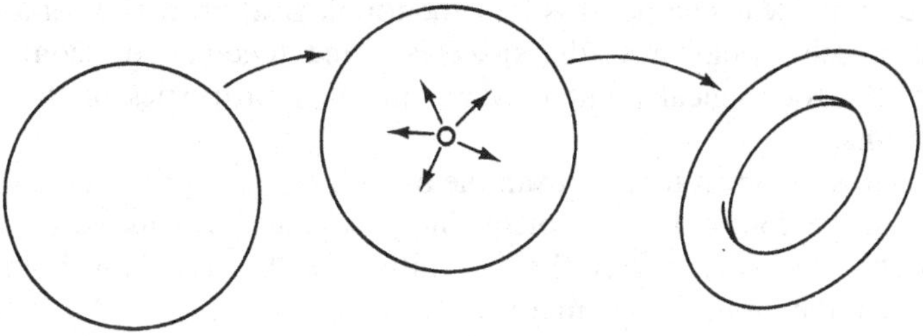

ball doughnut

in common is that they both have precisely one hole. Note that a ball is *not* topologically equivalent to a doughnut. To deform a ball into a doughnut would require making a hole at some stage and this cannot be done without pulling neighbouring pieces of modelling clay apart. But such a deformation is *not* continuous.

## 21.2 Maps

The diagram below is a topological map of the mainland of Europe. Notice, for example, that Italy looks nothing like the familiar Wellington boot. This is because quantities like distance and angle are *not* invariant under homeomorphism. However, contiguity *is* preserved. The diagram therefore indicates which nations have a common boundary and which do not. A tourist planning a trip by road might find the diagram useful in deciding what visas he will require but he would be unwise to use it in estimating how long the journey will take.

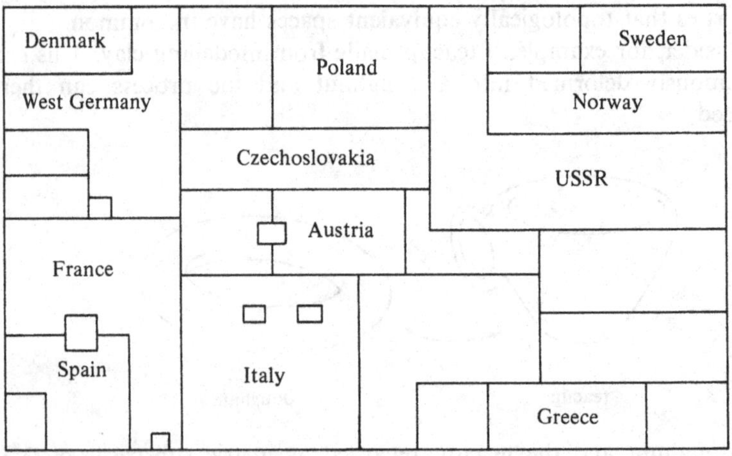


Schematic railway maps and wiring diagrams are more familiar examples of the same idea. The point is that the topological properties of a space are those which specify how the space is 'fitted together' or 'connected up'. From the topological point of view, all other properties of the space are irrelevant.

Perhaps the most famous example of a topological property is that which arises in the 'four colour problem'. Suppose that one is asked to colour the map above in such a way that no nations with a common boundary are coloured the same. How many colours will be needed? More generally, what is the minimum number of colours which will suffice for *all* such maps? (One must, of course, exclude cases where four or more nations meet at a point like the states of Arizona, Colorado, New Mexico and Utah.) The answer for maps drawn on the surface of a doughnut is *seven*. This is relatively easy to prove and has been known for a long time. Curiously, the same problem for maps drawn on a plane is very much more difficult. Only recently has it been shown that the answer is *four*. The proof (by Haken and Appel) has an extra element of interest in that it is the first computer assisted proof of a result of any substance.

---

## 21.3 Homeomorphisms between intervals

We now study the topological properties of intervals (regarded as metric subspaces of $\mathbb{R}^1$). If an interval has at least two points it falls into one of only *three* topologically distinct types. (See exercise 21.5(3).) The intervals $(-1, 1)$, $(-1, 1]$ and $[-1, 1]$ are representatives of each of the three types. In this section we shall explain why $(-1, 1)$ is *not* topologically equivalent to $(-1, 1]$ but *is* topologically equivalent to $(-1, 1]$ but *is* topologically equivalent to $\mathbb{R}$.

Suppose that $f$: $(-1, 1] \to (-1, 1)$ is a bijection. Let $f(1) = \xi$. Then $f((-1, 1)) = (-1, \xi) \cup (\xi, 1)$. Hence $f$ cannot be continuous since it maps a connected set onto a disconnected set. It follows that $(-1, 1]$ and $(-1, 1)$ are *not* topologically equivalent.

Consider the function $f$: $(-1, 1) \to \mathbb{R}$ defined by

$$f(x) = \frac{x}{1-x^2}.$$

This is a homeomorphism between $(-1, 1)$ and $\mathbb{R}$. (For a proof of this result which avoids the need for any calculation, see theorem 22.22.) It follows that $(-1, 1)$ and $\mathbb{R}$ are topologically equivalent.

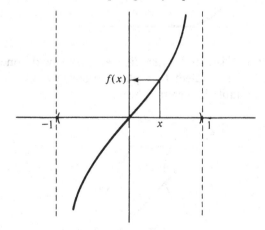

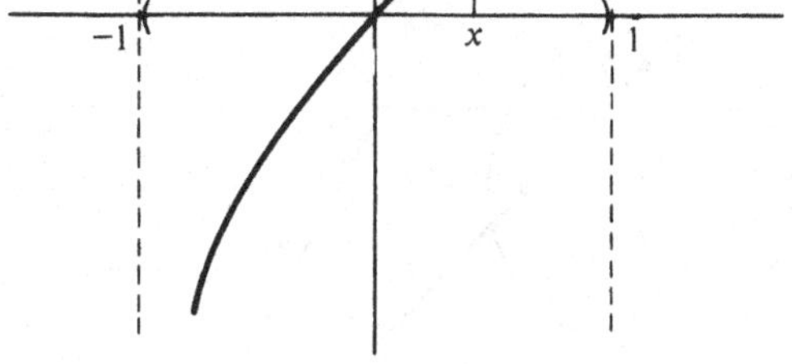


---

**21.4 Circles and spheres**

It is easy to see that a circle $C$ and a line segment $L$ (regarded as metric subspaces of $\mathbb{R}^2$) cannot be topologically equivalent. If $f: C \to L$ is a continuous bijection, then $L$ is compact because the same is true of $C$ (theorem 19.13). Hence $L$ contains its endpoints $A$ and $B$. But $f^{-1}: L \to C$ then maps the connected set $L \setminus \{A, B\}$ onto the disconnected set $C \setminus \{A', B'\}$ and hence is not continuous.

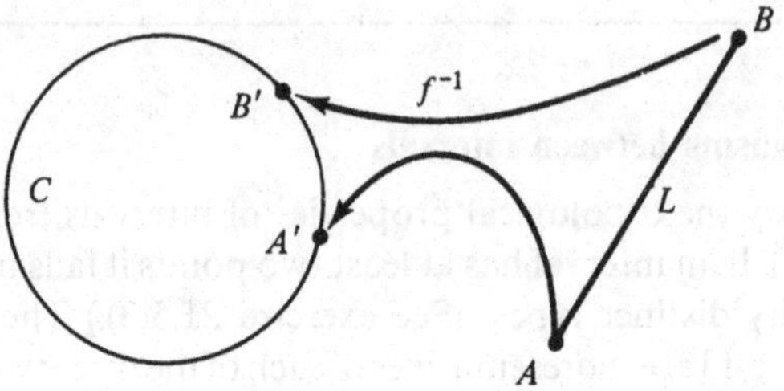


However, if just one point is removed from $C$ we obtain a set which is topologically equivalent to a line. The homeomorphism which is usually used to show this is called the *stereographic projection*. This is illustrated in the diagram below. The point deleted from $C$ (labelled $N$ in the diagram) is referred to as the 'north pole' of the projection.

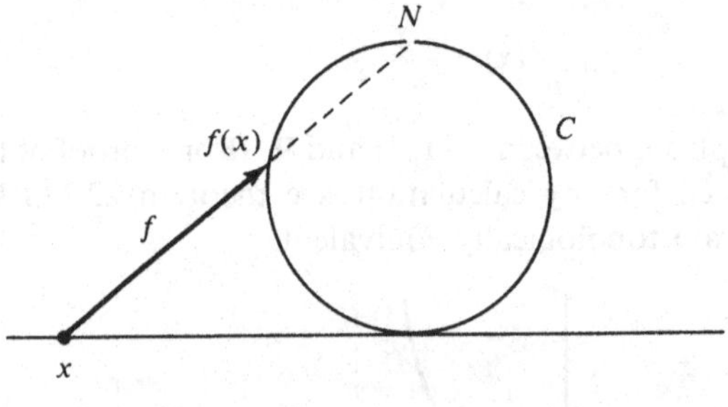


Similarly, a 'punctured sphere' is $\mathbb{R}^3$ (i.e. a sphere with one point removed) is topologically equivalent to a plane. Again, the stereographic projection provides a suitable homeomorphism.

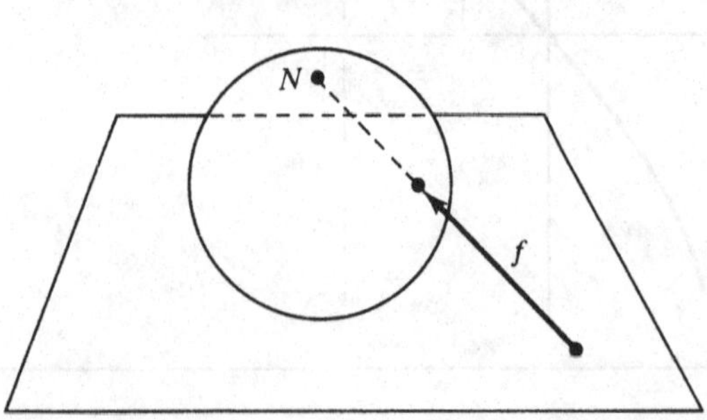

Let $S$ be the sphere in $\mathbb{R}^3$ with centre $(0, 0, \frac{1}{2})$ and radius $\frac{1}{2}$. Take as the 'north pole' the point $N=(0, 0, 1)$ and let $P$ be the plane $\{(x, y, z): z=0\}$. Then the stereographic projection $f: P \to S \setminus \{N\}$ is given by

$$f(x, y, 0)=\left(\frac{x}{1+x^2+y^2}, \frac{y}{1+x^2+y^2}, \frac{x^2+y^2}{1+x^2+y^2}\right)$$

and $f^{-1}: S \setminus \{N\} \to P$ is given by

$$f^{-1}(X, Y, Z)=\left(\frac{X}{1-Z}, \frac{Y}{1-Z}, 0\right).$$

Both functions are continuous by theorems 16.5 and 16.15.

---

21.5 *Exercise*

(1) Prove that topological equivalence as defined in §21.1 is an equivalence relation as defined in §5.5.

(2) Let $S$ be any set in $\mathbb{R}^n$ which is topologically equivalent to a closed ball $B$. Assuming Brouwer's fixed point theorem as quoted in note 17.14, show that for any function $f: S \to S$ which is continuous on $S$ there exists a $\xi \in S$ such that $f(\xi)=\xi$.

(3) Let $\mathcal{I}$ denote the collection of intervals in $\mathbb{R}^1$ which have at least two points. Let $\mathcal{U}$ denote the collection of all *open* intervals in $\mathcal{I}$, let $\mathcal{V}$ denote the collection of all *compact* intervals in $\mathcal{I}$ and let $\mathcal{W}$ denote the remaining intervals in $\mathcal{I}$. Show that $\mathcal{U}$, $\mathcal{V}$ and $\mathcal{W}$ are the equivalence classes into which $\mathcal{I}$ is split by the relation of topological equivalence.

(4) Let $L$ be a closed line segment in $\mathbb{R}^2$ and let $S$ be the closed box $[0, 1] \times [0, 1]$. Prove that the removal of any three points from $L$ produces a disconnected set but that the same is not true of $S$. Deduce that $L$ and $S$ are not topologically equivalent.

†(5) Check the formula given in §21.4 for the stereographic projection between the plane $P$ and the punctured sphere $S \setminus \{N\}$.

†(6) A function $F: \mathbb{C} \to S \setminus \{N\}$ is defined by

$$F(z)=\left(\frac{\mathcal{R}z}{1+|z|^2}, \frac{\mathcal{I}mz}{1+|z|^2}, \frac{|z|^2}{1+|z|^2}\right).$$

Prove that $F$ is a homeomorphism between the complex plane $\mathbb{C}$ and the punctured sphere $S \setminus \{N\}$ (where $S \setminus \{N\}$ is regarded as a metric subspace of $\mathbb{R}^3$). Show that

$$\|F(z_1)-F(z_2)\|=\frac{|z_1-z_2|}{\{(1+|z_1|^2)(1+|z_2|^2)\}^{1/2}}.$$

---

**21.6 Continuous functions and open sets**

21.7 *Theorem* Let $\mathcal{X}$ and $\mathcal{Y}$ be metric spaces and let $f: \mathcal{X} \to \mathcal{Y}$. Then the following statements are all equivalent.

(i) $f$ is continuous on $\mathcal{X}$.

(ii) For any $\mathbf{x} \in \mathcal{X}$ and any $S \subset \mathcal{X}$,

$$d(\mathbf{x}, S) = 0 \Rightarrow d(f(\mathbf{x}), f(S)) = 0.$$

(iii) For any $S \subset \mathcal{X}$,

$$f(\bar{S}) \subset \overline{f(S)}.$$

(iv) For any $F \subset \mathcal{Y}$,

$$F \text{ closed} \Rightarrow f^{-1}(F) \text{ closed.}$$

(v) For any $G \subset \mathcal{Y}$,

$$G \text{ open} \Rightarrow f^{-1}(G) \text{ open.}$$

*Proof* We shall show that (i)⇒(ii)⇒(iii)⇒(iv)⇒(v)⇒(i).

(i)⇒(ii). This is just theorem 16.4.

(ii) ⇒ (iii). Assume (ii) and suppose that $\mathbf{y} \in f(\bar{S})$. Then $\mathbf{y} = f(\mathbf{x})$ where $\mathbf{x} \in \bar{S}$. Since $d(\mathbf{x}, S) = 0$ it follows that $d(f(\mathbf{x}), f(S)) = 0$. Thus $\mathbf{y} = f(\mathbf{x}) \in \overline{f(S)}$.

(iii)⇒(iv). Assume (iii) and suppose that $F \subset \mathcal{Y}$ is closed. Since $f(f^{-1}(F)) \subset F$,

$$f(\overline{f^{-1}(F)}) \subset \overline{f(f^{-1}(F))} \subset \bar{F} = F.$$

Hence $\overline{f^{-1}(F)} \subset f^{-1}(F)$. Therefore $f^{-1}(F)$ is closed.

(iv)⇒(v). Assume (iv) and suppose that $G \subset \mathcal{Y}$ is open. Then $\mathcal{C}G$ is closed and hence

$$\mathcal{C}f^{-1}(G) = f^{-1}(\mathcal{C}G)$$

is closed. Thus $f^{-1}(G)$ is open.

(v)⇒(i). Assume (v) and suppose that $C$ and $D$ are separated sets in $\mathcal{Y}$. Then disjoint, open sets $G$ and $H$ exist with $C \subset G$ and $D \subset H$ by theorem 15.16. The sets $f^{-1}(G)$ and $f^{-1}(H)$ are then disjoint, open sets in $\mathcal{X}$ by (v). But then $f^{-1}(C) \subset f^{-1}(G)$ and $f^{-1}(D) \subset f^{-1}(H)$ are separated sets in $\mathcal{X}$. Thus $C$ and $D$ separated implies $f^{-1}(C)$ and $f^{-1}(D)$ separated. Hence $f$ is continuous (§16.2).

---

The most significant of the equivalences in the above theorem is that which asserts that a function $f: \mathcal{X} \to \mathcal{Y}$ is continuous if and only if, for each

$G \subset \mathcal{Y}$, $G$ open implies $f^{-1}(G)$ open. The next few sections are devoted to exploring the consequences of this result.

## 21.8 Topologies

As we know from the previous section, two metric spaces $\mathcal{X}$ and $\mathcal{Y}$ are topologically equivalent if and only if there exists a bijection $f: \mathcal{X} \to \mathcal{Y}$ such that, for each $G \subset \mathcal{Y}$,

$$G \text{ open} \Leftrightarrow f^{-1}(G) \text{ open.} \tag{1}$$

Thus, if we wish to determine whether or not $\mathcal{X}$ and $\mathcal{Y}$ are topologically equivalent, the only question we need to ask about $\mathcal{X}$ and $\mathcal{Y}$ is: what are their open sets? Any bijection $f: \mathcal{X} \to \mathcal{Y}$ can then be tested to see whether it satisfies (1). Other information about $\mathcal{X}$ and $\mathcal{Y}$ may be helpful or interesting but it is not strictly necessary. Thus we know all that there is to know about the topological structure of a space if we have a list of all its open sets. For this reason the collection of all open sets in a space is called its *topology*.

## 21.9† Relative topologies

Notice that theorem 21.7 refers to a function $f: \mathcal{X} \to \mathcal{Y}$ where $\mathcal{X}$ and $\mathcal{Y}$ are metric spaces. Various equivalent conditions are then given for $f$ to be continuous on the whole space $\mathcal{X}$. In contrast, the theorems of chapter 16 always involved a *subset* $\mathcal{Z}$ of $\mathcal{X}$ and were concerned with the continuity of a function $f: \mathcal{Z} \to \mathcal{Y}$ *on the set* $\mathcal{Z}$.

If $\mathcal{Z}$ is a subset of $\mathcal{X}$, then it is, of course, true that $\mathcal{Z}$ is itself a metric space provided that we use the same definition for distance in $\mathcal{Z}$ as is used in $\mathcal{X}$. We say that $\mathcal{Z}$ is a *metric subspace* of $\mathcal{X}$.

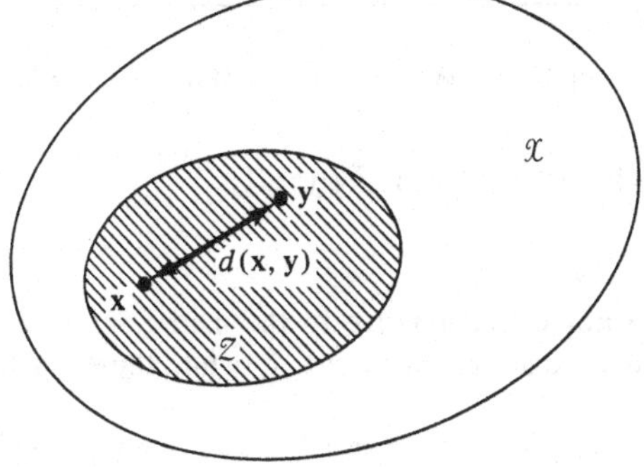


In chapter 16, we carefully chose a definition for a function $f: \mathcal{Z} \to \mathcal{Y}$ to be continuous on the set $\mathcal{Z}$ which makes it quite irrelevant whether we regard our underlying metric space to be $\mathcal{X}$ or whether we simply throw out all the points of $\mathcal{X}$ which are not in $\mathcal{Z}$ and take our underlying metric space to be $\mathcal{Z}$. This is because two subsets $A$ and $B$ of $\mathcal{Z}$ are contiguous in the metric space $\mathcal{X}$ if and only if they are contiguous in the metric space $\mathcal{Z}$.

It follows that theorem 21.7 applies equally well in respect of a function $f: \mathcal{Z} \to \mathcal{Y}$ which is continuous on a *subset* $\mathcal{Z}$ of the metric space $\mathcal{X}$. One simply takes note of the fact that $\mathcal{Z}$ is a metric subspace of $\mathcal{X}$ and replaces $\mathcal{X}$ at each occurrence by $\mathcal{Z}$. This leaves the meaning of item (i) of theorem 21.7 unaltered and the same is true of item (ii).

However, very considerable care is necessary with items (iii), (iv) and (v) of theorem 21.7 when the underlying metric space is switched from $\mathcal{X}$ to $\mathcal{Z}$. Such a switch affects the meaning of the words 'open', 'closed', 'closure' etc. This is because these ideas all depend on the notion of a boundary point. Recall that $\xi \in \mathcal{X}$ is a boundary point of $S$ if and only if $d(\xi, S)=0$ and $d(\xi, \mathcal{C}S)=0$. But, if $\xi \notin \mathcal{Z}$, it ceases to be eligible as a boundary point when one switches the underlying space from $\mathcal{X}$ to $\mathcal{Z}$. Even if $\xi \in \mathcal{Z}$ it may *still* cease to be a boundary point when we switch from $\mathcal{X}$ to $\mathcal{Z}$ because the nature of $\mathcal{C}S$ depends on whether we use $\mathcal{X}$ or $\mathcal{Z}$ as the universal set. If $\mathcal{Z}$ replaces $\mathcal{X}$ as the universal set, $\mathcal{C}S$ will in general become a smaller set and hence $d(\xi, \mathcal{C}S)$ may cease to be zero.

Since a subset $S$ of $\mathcal{Z}$ has fewer boundary points relative to the metric space $\mathcal{Z}$ than it does relative to the larger metric space $\mathcal{X}$, it is easier for $S$ to satisfy the criteria for being open or closed relative to $\mathcal{Z}$ than it is to satisfy the criteria for being open or closed relative to $\mathcal{X}$. Thus the subsets of $\mathcal{Z}$ which are *not* open or closed relative to $\mathcal{X}$ may very well become open or closed when the underlying metric space $\mathcal{X}$ is replaced by the metric subspace $\mathcal{Z}$. Returning to items (iii), (iv) and (v) of theorem 21.7, this observation simply reflects the fact that it is easier for a function $f$ to be continuous on a subset $\mathcal{Z}$ of $\mathcal{X}$ than it is for $f$ to be continuous on the whole of $\mathcal{X}$.

If $\mathcal{Z}$ is a metric subspace of the metric space $\mathcal{X}$, we define the *relative topology* of $\mathcal{Z}$ to be the collection of subsets of $\mathcal{Z}$ which are open *relative to* $\mathcal{Z}$. As explained above, the relative topology of $\mathcal{Z}$ is *NOT* just the collection of all subsets of $\mathcal{Z}$ which are open *relative to* $\mathcal{X}$. It is a *larger* collection than this.

---

21.10 *Example* Suppose that $\mathcal{X}=\mathbb{R}^2$. Let $\mathcal{Z}_1=\{(x, y): x>1\}$ and let $\mathcal{Z}_2=\mathcal{Z}_1 \cup \{(1, 0)\}$. Take

$$S_1=\{(x, y): x>1 \text{ and } x^2+y^2 \leqq 1\}.$$

Note that $S_1 \subset \mathcal{Z}_1 \subset \mathcal{Z}_2$.

The set $S_1$ is neither open nor closed relative to the metric space $\mathcal{X}=\mathbb{R}^2$. The boundary points on its straight edge belong to $\mathcal{X} \setminus S_1$ while those on the curved edge belong to $S_1$.

The set $S_1$ is *closed* relative to the metric space $\mathcal{Z}_1$. The boundary points of $S_1$ relative to the metric space $\mathcal{Z}_1$ are those on the curved edge of $S_1$ and these all belong to $S_1$. Notice that the points on the straight edge of $S_1$ do *not* belong to $\mathcal{Z}_1$ and hence are *not* eligible as boundary points relative to the metric space $\mathcal{Z}_1$.

The set $S_1$ is *not* closed relative to the metric space $\mathcal{Z}_2$. We have that (1, 0) is a boundary point of $S_1$ relative to the metric space $\mathcal{Z}_2$ which does not belong to $S_1$.

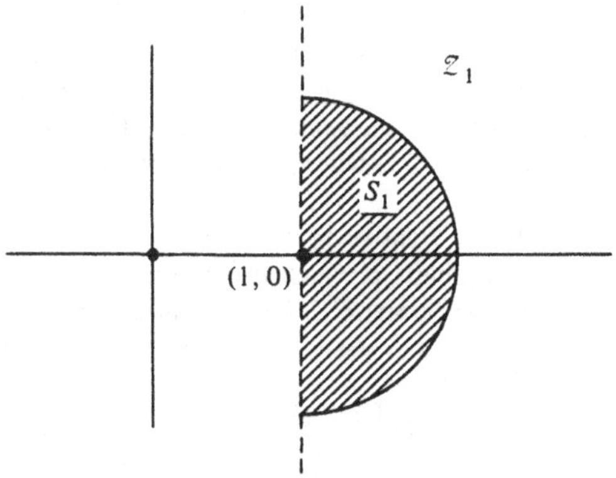


The set $S_2 = S_1 \cup \{(1, 0)\}$ is also interesting. This is *not* closed relative to the metric space $\mathcal{X}$. It is not eligible as a closed set relative to the metric space $\mathcal{Z}_1$ since it is not a subset of $\mathcal{Z}_1$. It *is* closed relative to the metric space $\mathcal{Z}_2$. Observe that although $(1, 0) \in \mathcal{Z}_2$ and is a boundary point of $S_2$ relative to the metric space $\mathcal{X}$, it is *not* a boundary point of $S_2$ relative to the metric space $\mathcal{Z}_2$. We have that $d((1, 0), \mathcal{X} \setminus S_2) = 0$ *but* $d((1, 0), \mathcal{Z}_2 \setminus S_2) = 1$.

Of course $\mathcal{Z}_1$ is both open and closed relative to the metric space $\mathcal{Z}_1$ (theorem 14.14). Similarly, $\mathcal{Z}_2$ is both open and closed relative to the metric space $\mathcal{Z}_2$. Observe finally that $\mathcal{Z}_1$ is an *open* set relative to the metric space $\mathcal{Z}_2$. Its only boundary point relative to the metric space $\mathcal{Z}_2$ is (1, 0).

---

The next theorem makes it quite easy to find the relative topology of a metric subspace $\mathcal{Z}$ of a metric space $\mathcal{X}$ provided that one knows the topology of $\mathcal{X}$ to begin with.

---

21.11† *Theorem* Let $\mathcal{Z}$ be a metric subspace of a metric space $\mathcal{X}$. Then $H$ is an open set relative to the metric space $\mathcal{Z}$ if and only if

$$H = G \cap \mathcal{Z}$$

for some set $G$ which is open relative to the metric space $\mathcal{X}$.

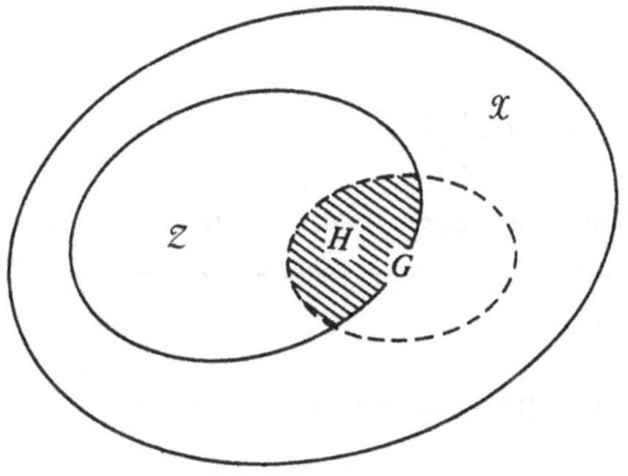

*Proof* Recall that an open ball $B$ with centre $\xi \in \mathcal{Z}$ and radius $r>0$ in the metric space $\mathcal{Z}$ is defined by $B=\{\mathbf{z}: d(\mathbf{z}, \xi)<r\}$. In this expression, of course, the variable $\mathbf{z}$ ranges over the universal set $\mathcal{Z}$. Given such an open ball $B$ in the metric space $\mathcal{Z}$, let $B'$ denote the open ball with the *same* centre $\xi$ and radius $r>0$ in the metric space $\mathcal{X}$. Thus $B'=\{\mathbf{x}: d(\mathbf{x}, \xi)<r\}$, where the variable $\mathbf{x}$ ranges over the universal set $\mathcal{X}$. Obviously $B=B'\cap \mathcal{X}$.

(i) Suppose that $H$ is an open set relative to the metric space $\mathcal{Z}$. By exercise 14.17(11), $H$ is the union of all open balls $B$ which it contains. Here $B$ denotes, of course, an open ball relative to the metric space $\mathcal{Z}$. Let $\mathcal{W}$ denote the collection of all such open balls $B$ contained in $H$. Then

$$H=\bigcup_{B\in\mathcal{W}} B.$$

Let

$$G=\bigcup_{B\in\mathcal{W}} B'$$

Then $G$ is open relative to the metric space $\mathcal{X}$ because it is the union of sets which are open relative to the metric space $\mathcal{X}$ (theorem 14.14). Also $H=G\cap \mathcal{Z}$.

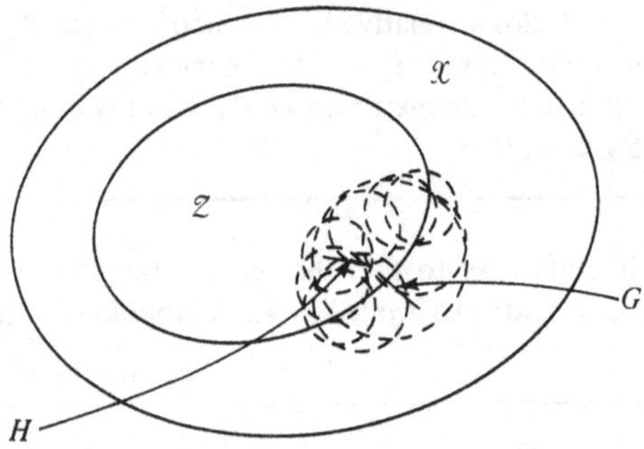


(ii) Suppose that $H=G\cap \mathcal{Z}$ where $G$ is open relative to the metric space $\mathcal{X}$. Let $\mathcal{W}'$ denote the collection of all open balls $B'$ (relative to the metric space $\mathcal{X}$) which are contained in $G$. Then

$$G=\bigcup_{B'\in\mathcal{W}'} B'.$$

But then

$$H=\bigcup_{B'\in\mathcal{W}'} B$$

and hence is the union of sets which are open relative to the metric space $\mathcal{Z}$.

---

21.12† *Corollary* Let $\mathcal{Z}$ be a metric subspace of a metric space $\mathcal{X}$. Then $H$ is a closed set relative to the metric space $\mathscr{Z}$ if and only if

$$H=F\cap\mathscr{Z}$$

for some set $F$ which is closed relative to the metric space $\mathscr{X}$.

*Proof* We have that $\mathscr{Z}\setminus H$ is open relative to the metric space $\mathscr{Z}$ if and only if $\mathscr{Z}\setminus H=G\cap\mathscr{Z}$ where $G$ is open relative to the metric space $\mathscr{X}$. But $\mathscr{Z}\setminus H=G\cap\mathscr{Z}$ is equivalent to $H=\mathscr{C}G\cap\mathscr{Z}$ and the result follows.

---

21.13 *Example* The set $[0, 1)$ is neither open nor closed relative to the metric space $\mathbb{R}^1$. However, $[0, 1)$ is *open* relative to the metric subspace $[0, 2]$ because $[0, 1)=(-1, 1)\cap[0, 2]$ and $(-1, 1)$ is an open set relative to the metric space $\mathbb{R}^1$. The set $[0, 1)$ is *closed* relative to the metric subspace $(-1, 1)$ because $[0, 1)=[0, 2]\cap(-1, 1)$ and $[0, 2]$ is a closed set relative to the metric space $\mathbb{R}^1$.

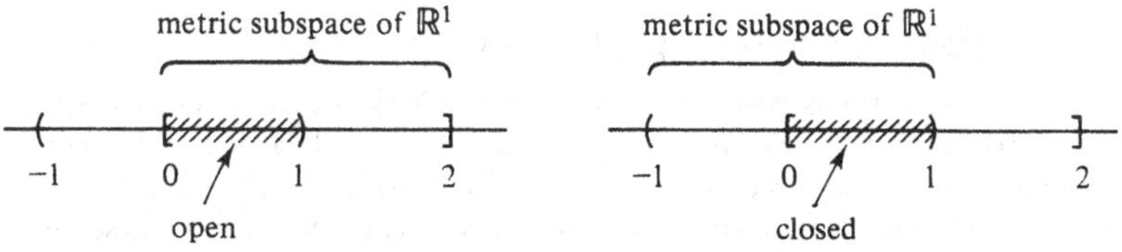


---

21.14† *Exercise*

(1) Let $A$, $B$, $C$ and $D$ be the sets in $\mathbb{R}^2$ given by

$A=\{(x, y): 1\leqq x^2+y^2\leqq 2\}$,
$B=\{(x, y): y^2<2\}$,
$C=\{(x, y): y^2\geqq 2\}$,
$D=\{(x, y): y^2>2\}$,

Explain why the function $f$: $\mathbb{R}^2\to\mathbb{R}^2$ cannot be continuous on $\mathbb{R}^2$ if any of the following properties hold:

(i) $f^{-1}(B)=C$ (ii) $f^{-1}(A)=D$
(iii) $f(A)=C$ (iv) $f(B)=D$.

(2) Let $\mathscr{X}$ be a *compact* metric space and let $f:\mathscr{X}\to\mathscr{Y}$ be a bijection which is continuous on $\mathscr{X}$. Prove that $f$ is a homeomorphism.
[*Hint*: Use theorem 21.7iv and exercise 20.6(2).]

(3) Let $S$ be a set in $\mathbb{R}^2$ defined by

$$S=\{(x, y): 1<x\leqq 2\}.$$

Decide whether or not $S$ is open or closed relative to the three metric spaces considered in example 21.10.

(4) Characterise the subsets of $[0, 1]$ which are open relative to $[0, 1]$ regarded as a metric subspace of $\mathbb{R}^1$. [*Hint*: Use theorem 17.27.]

(5) Prove that every subset of $\mathbb{N}$ is both open and closed relative to $\mathbb{N}$ regarded as a metric subspace of $\mathbb{R}^1$. (The relative topology of $\mathbb{N}$ is said to be *discrete*.)
(6) Let $A$ and $B$ be subsets of a metric subspace $\mathcal{Z}$ of a metric space $\mathcal{X}$. Show that $A$ and $B$ are contiguous relative to the metric space $\mathcal{Z}$ if and only if they are contiguous relative to the metric space $\mathcal{X}$.
(7) Let $\mathcal{Z}$ be a metric subspace of a metric space $\mathcal{X}$. Show that a subset $C$ of $\mathcal{Z}$ is connected relative to $\mathcal{Z}$ if and only if it is connected relative to $\mathcal{X}$. Give an example of a connected set $D$ in $\mathbb{R}^2$ such that $D \cap \mathcal{Z}$ is *not* connected relative to $\mathcal{Z} = \{(x, y): x^2 + y^2 \leqq 1\}$.
(8) Let $\mathcal{Z}$ be a metric subspace of a metric space $\mathcal{X}$. Prove that a subset $K$ of $\mathcal{Z}$ is compact relative to $\mathcal{Z}$ if and only if it is compact relative to $\mathcal{X}$. Give an example of a compact set $L$ in $\mathbb{R}^1$ such that $L \cap \mathcal{Z}$ is *not* compact relative to $\mathcal{Z} = (0, 1)$.
(9) Let $\mathcal{Z}$ be a metric subspace of $\mathcal{X}$. Prove that $\mathcal{Z}$ is a connected subset of $\mathcal{X}$ if and only if the only subsets of $\mathcal{Z}$ which are both open and closed *relative to* $\mathcal{Z}$ are $\emptyset$ and $\mathcal{Z}$.

---

## 21.15† Introduction to topological spaces

We sometimes wish to discuss the topological properties of a space on which a metric is *not* given. From §21.18, it is clear that this will be no problem provided that we know what the open sets of the space are. This leads us to define a *topological space* to be a non-empty set $\mathcal{X}$ and a collection $\mathcal{T}$ of sets in $\mathcal{X}$. The collection $\mathcal{T}$ will serve as the topology of $\mathcal{X}$. In order that it be sensible to regard the sets in $\mathcal{T}$ as the 'open sets' of $\mathcal{X}$ we require that $\mathcal{T}$ satisfy the conclusions of theorem 14.13 – i.e.

(i) $\emptyset \in \mathcal{T}$, $\mathcal{X} \in \mathcal{T}$.
(ii) Any union of sets in $\mathcal{T}$ is in $\mathcal{T}$.
(iii) Any finite intersection of sets in $\mathcal{T}$ is in $\mathcal{T}$.

Of course, a topological space is a more abstract notion than a metric space just as a metric space is a more abstract notion than $\mathbb{R}^n$. But it would be naïve to suppose that this makes a topological space mathematically more difficult to deal with. On the contrary, the more abstract a space is, the less structure it has and therefore the fewer theorems that can be true of it. The usual pattern in passing from a concrete space to a more abstract space is that many results which were *theorems* in the concrete space become *definitions* in the abstract space. They therefore do not need to be proved. For example, the triangle inequality for distance in $\mathbb{R}^n$ which we proved as a *theorem* (theorem 13.7) with the help of the Cauchy–Schwarz inequality becomes part of the *definition* of distance in a metric space.

These remarks do not mean, of course, that it is not useful to pass from a concrete space to a more abstract space. In the first place, one obtains a theory that is more widely applicable. In the case of topological spaces, this fact is particularly important. In the second place, one often gains considerable insight into the structure of the original concrete space since the process of abstraction forces one to focus on those properties of the original space which *really* matter for the proof of a particular theorem.

Returning to the idea of topological space, the immediate question is: how many of the properties of a metric space are valid in a topological space?

We begin by defining a *closed set* in a topological space to be the complement of an open set. Thus theorem 14.9 becomes a *definition* for a topological space. Theorem 14.13 is part of the *definition* of an open set in a topological space. Theorem 14.14 follows from theorems 14.9 and 14.13 and is therefore true in any topological space.

We define the *closure* $\bar{E}$ of a set $E$ in a topological space to be the intersection of all closed sets which contain $E$. Similarly, the *interior* $\mathring{E}$ of $E$ is defined to be the union of all open sets contained in $E$. Thus theorems 15.7 and 15.9 become *definitions* in a topological space. The following properties of the closure and interior of sets remain valid and are worth remembering:

(i) $A\cup B=\bar{A}\cup\bar{B}$; $(A\cap B)^{\circ}=\mathring{A}\cap\mathring{B}$
(ii) $A\subset B\Rightarrow\mathring{A}\subset\mathring{B}$; $A\subset B\Rightarrow\mathring{A}\subset\mathring{B}$
(iii) $A\cap B\subset\mathring{A}\cap\mathring{B}$; $(A\cup B)^{\circ}\subset\mathring{A}\cup\mathring{B}$.

(See exercise 15.10(5).) We can now *define* the boundary $\partial E$ of a set $E$ in a topological space by

$$\partial E=\bar{E}\backslash\mathring{E}.$$

If $\mathcal{X}$ and $\mathcal{Y}$ are topological spaces, we use theorem 21.7(v) to provide a definition for the continuity of a function $f:\mathcal{X}\rightarrow\mathcal{Y}$. We say that $f:\mathcal{X}\rightarrow\mathcal{Y}$ is *continuous* (on $\mathcal{X}$) if and only if, for each $G\subset\mathcal{Y}$,

$$G \text{ open} \Rightarrow f^{-1}(G) \text{ open}.$$

The proof of theorem 21.7 shows that this condition implies the more intuitively satisfying criterion given in §16.2.

If $\mathcal{X}$ and $\mathcal{Y}$ are topological spaces, a homeomorphism $f:\mathcal{X}\rightarrow\mathcal{Y}$ is a continuous bijection with a continuous inverse $f^{-1}:\mathcal{Y}\rightarrow\mathcal{X}$. If a homeomorphism $f:\mathcal{X}\rightarrow\mathcal{Y}$ exists, we say that the topological spaces $\mathcal{X}$ and $\mathcal{Y}$ are topologically equivalent (or *homeomorphic*). From the topological point of view, two homeomorphic spaces are essentially the same (see the discussion of §9.21).

Let $\mathcal{Z}$ be a subset of a topological space $\mathcal{X}$ with topology $\mathcal{T}$. Then

$$\mathcal{V}=\{\mathcal{Z}\cap G: G\in\mathcal{T}\}$$

is a collection of subsets of $\mathcal{Z}$ which satisfies the conditions for a topology on $\mathcal{Z}$. We call $\mathcal{V}$ the topology on $\mathcal{Z}$ *relative to* $\mathcal{X}$. Thus theorem 21.11 for a metric space becomes a *definition* for a topological space. The set $\mathcal{Z}$ with the topology $\mathcal{V}$ is called a *topological subspace* of $\mathcal{X}$.

The introduction of the idea of a relative topology means that we do not need a separate definition for continuity on a subset $\mathcal{Z}$ of a topological space $\mathcal{X}$. We use the definition given above but with $\mathcal{Z}$ (regarded as a topological subspace of $\mathcal{X}$) replacing $\mathcal{X}$.

Similarly, we need only provide definitions of a connected topological *space* and a compact topological *space*.

A *connected* topological space $\mathcal{X}$ is one in which the only sets which are *both* open *and* closed are $\emptyset$ and $\mathcal{X}$. Theorem 17.3 for a metric space therefore becomes a definition for a topological space. It is important that theorem 17.9 remains true in a topological space. With our new definitions, the proof is even easier.

---

**21.16†** *Theorem* Let $\mathcal{X}$ and $\mathcal{Y}$ be topological spaces and let $f:\mathcal{X}\to\mathcal{Y}$ be a continuous surjection. Then

$$\mathcal{X} \text{ connected} \Rightarrow \mathcal{Y} \text{ connected}.$$

*Proof* Suppose that $E$ is a set in $\mathcal{Y}$ which is both open and closed. Since $f:\mathcal{X}\to\mathcal{Y}$ is continuous, $f^{-1}(E)$ is both open and closed in $\mathcal{X}$. But $\mathcal{X}$ is connected. Thus, $f^{-1}(E)=\emptyset$ or $f^{-1}(E)=\mathcal{X}$. Because $f$ is a surjection, it follows that $E=\emptyset$ or $E=\mathcal{X}$. Thus $\mathcal{Y}$ is connected.

---

A *compact* topological space $\mathcal{X}$ is one with the property that any collection $\mathcal{U}$ of open sets which covers $\mathcal{X}$ has a finite subcollection which covers $\mathcal{X}$. This definition is identical with that of §20.4. It is important that theorem 19.13 remains true in a topological space and, again, with our new definitions, the proof is even easier.

---

**21.17†** *Theorem* Let $\mathcal{X}$ and $\mathcal{Y}$ be topological spaces and let $f:\mathcal{X}\to\mathcal{Y}$ be a continuous surjection. Then

$$\mathcal{X} \text{ compact} \Rightarrow \mathcal{Y} \text{ compact}.$$

*Proof* Let $\mathcal{V}$ be a collection of open sets which covers $\mathcal{Y}$. Then

$$\mathcal{U}=\{f^{-1}(H): H\in\mathcal{V}\}$$

is a collection of open sets which covers $\mathcal{X}$. Since $\mathcal{X}$ is compact, a finite subcollection $\mathcal{E}$ covers $\mathcal{X}$. Let

$$\mathcal{F}=\{f(G): G\in\mathcal{E}\}.$$

Then $\mathcal{F}$ is a finite subcollection of $\mathcal{V}$ which covers $\mathcal{Y}$. Hence $\mathcal{Y}$ is compact.

---

## 21.18† Product topologies

In the space $\mathbb{R}^2$ the Euclidean metric $d$: $\mathbb{R}^2\to\mathbb{R}$ is defined by

$$d(\mathbf{x}, \mathbf{y})=\{(x_1-y_1)^2+(x_2-y_2)^2\}^{1/2}.$$

We use this metric because it corresponds to the notion of the distance between two points as understood in Euclidean geometry. However, the Euclidean metric is not the only possible metric which can be used in $\mathbb{R}^2$. The function $m$: $\mathbb{R}^2\to\mathbb{R}$ defined by

$$m(\mathbf{x}, \mathbf{y})=\max\{|x_1-y_1|, |x_2-y_2|\}$$

is an example of an alternative metric. So is the function $l:\mathbb{R}^2\to\mathbb{R}$ defined by

$$l(\mathbf{x}, \mathbf{y})=|x_1-y_1|+|x_2-y_2|.$$

To verify that these functions satisfy the requirements for a metric given in §13.1 is very easy.

Of course, an open ball with respect to one of the metrics $l$ and $m$ looks very different from a Euclidean open ball as the diagrams below illustrate.

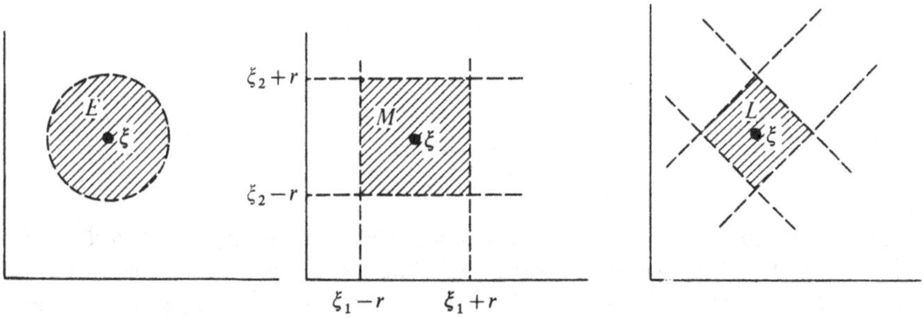


$E=\{\mathbf{x}: d(\xi, \mathbf{x})<r\}$ $M=\{\mathbf{x}: m(\xi, \mathbf{x})<r\}$ $L=\{\mathbf{x}: l(\xi, \mathbf{x})<r\}$

However, the geometrical shape of these open balls is irrelevant from the topological point of view. We know that any open set in a metric space $\mathcal{X}$ is the union of all open balls which it contains (exercise 14.17(11)). But it is evident that, given any one of the open balls illustrated above, we can find an open ball of either of the other two types with the same centre which fits inside the original open ball. Thus a set $G$ is open with respect to one of the metrics if and only if it is open with respect to them all – i.e. the three metrics all generate the *same* open sets in $\mathbb{R}^2$.

Now suppose that $\mathcal{X}_1$ and $\mathcal{X}_2$ are topological spaces. What is the appropriate topology for $\mathcal{X}_1 \times \mathcal{X}_2$?

If $\mathcal{X}_1$ and $\mathcal{X}_2$ are metric spaces with metrics $d_1$: $\mathcal{X}_1 \to \mathbb{R}$ and $d_2$: $\mathcal{X}_2 \to \mathbb{R}$, we can proceed as with the space $\mathbb{R}^2 = \mathbb{R} \times \mathbb{R}$. In particular, the functions $d$: $\mathcal{X}_1 \times \mathcal{X}_2 \to \mathbb{R}$ and $m$: $\mathcal{X}_1 \times \mathcal{X}_2 \to \mathbb{R}$ defined by

$$d(\mathbf{x}, \mathbf{y}) = \{(d_1(\mathbf{x}_1, \mathbf{y}_1))^2 + (d_2(\mathbf{x}_2, \mathbf{y}_2))^2\}^{1/2} \tag{1}$$

and

$$m(\mathbf{x}, \mathbf{y}) = \max\{d_1(\mathbf{x}_1, \mathbf{y}_1), d_2(\mathbf{x}_2, \mathbf{y}_2)\} \tag{2}$$

are metrics on $\mathcal{X}_1 \times \mathcal{X}_2$ which generate the *same* topology on $\mathcal{X}_1 \times \mathcal{X}_2$. Why should this topology be more useful than the various other topologies which one might impose on $\mathcal{X}_1 \times \mathcal{X}_2$?

The significant fact about the metrics $d$ and $m$ is that they generate a topology on $\mathcal{X}_1 \times \mathcal{X}_2$ which makes the projection functions $P_1: \mathcal{X}_1 \times \mathcal{X}_2 \to \mathcal{X}_1$ and $P_2$: $\mathcal{X}_1 \times \mathcal{X}_2 \to \mathcal{X}_2$ *continuous*. For example, if we use the metric $d$ in $\mathcal{X}_1 \times \mathcal{X}_2$, then

$$d_1(P_1(\mathbf{x}), P_1(\mathbf{y})) = d_1(\mathbf{x}_1, \mathbf{y}_1) \leqq d(\mathbf{x}, \mathbf{y})$$

and hence $P_1$ is continuous on $\mathcal{X}_1 \times \mathcal{X}_2$ by lemma 16.8.

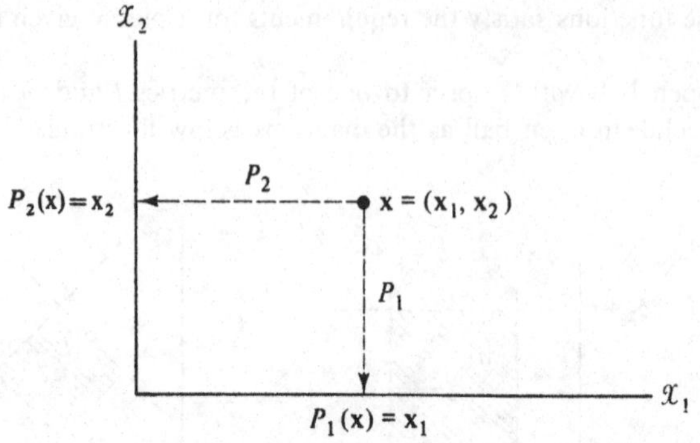


This observation makes it natural to define the *product topology* on $\mathscr{X}_1 \times \mathscr{X}_2$ to be the weakest topology (i.e. the topology with the fewest open sets) with respect to which the projection functions $P_1 : \mathscr{X}_1 \times \mathscr{X}_2 \to \mathscr{X}_1$ and $P_2 : \mathscr{X}_1 \times \mathscr{X}_2 \to \mathscr{X}_2$ are continuous.

It follows that, if $G_1$ is an open set in $\mathscr{X}_1$, then $P_1^{-1}(G_1)$ must be an open set in the product topology of $\mathscr{X}_1 \times \mathscr{X}_2$. Similarly, if $G_2$ is an open set in $\mathscr{X}_2$, then $P_2^{-1}(G_2)$ must be an open set in the product topology of $\mathscr{X}_1 \times \mathscr{X}_2$.

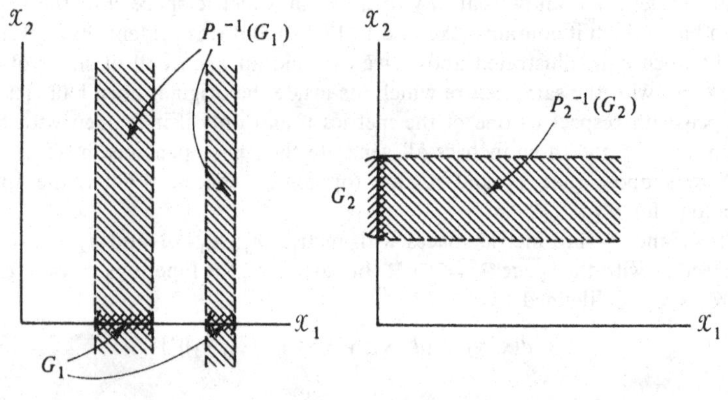


Since the intersection of a *finite* collection of open sets must be open, it follows that

$$G_1 \times G_2 = P_1^{-1}(G_1) \cap P_2^{-1}(G_2)$$

must be an open set in the product topology of $\mathscr{X}_1 \times \mathscr{X}_2$. Finally, the union of *any* collection of open sets must be open. Thus, if $\mathscr{W}$ is any collection of sets of the form $G_1 \times G_2$ (where $G_1$ is open in $\mathscr{X}_1$ and $G_2$ is open in $\mathscr{X}_2$), then

$$S = \bigcup_{G_1 \times G_2 \in \mathscr{W}} G_1 \times G_2 \tag{3}$$

must be an open set in the product topology of $\mathscr{X}_1 \times \mathscr{X}_2$.

The collection $\mathscr{T}$ of *all* sets $S$ of the form (3) satisfies the requirements for a topology on $\mathscr{X}_1 \times \mathscr{X}_2$ given in §21.15. It follows that $\mathscr{T}$ is the product topology on $\mathscr{X}_1 \times \mathscr{X}_2$.

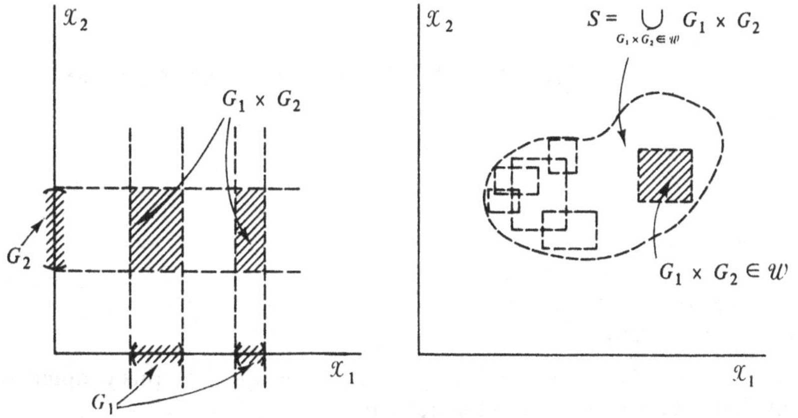


Now return to the case in which $\mathscr{X}_1$ and $\mathscr{X}_2$ are metric spaces with metrics $d_1$ and $d_2$ respectively. If we use the metric $m$: $\mathscr{X}_1 \times \mathscr{X}_2 \to \mathbb{R}$ defined by (2), then the open balls $B$ in $\mathscr{X}_1 \times \mathscr{X}_2$ are of the form $B = B_1 \times B_2$ where $B_1$ is an open ball in $\mathscr{X}_1$ and $B_2$ is an open ball in $\mathscr{X}_2$. It is therefore apparent from the preceding discussion, that the open sets in $\mathscr{X}_1 \times \mathscr{X}_2$ generated by the metric $m$ are precisely those which lie in the product topology $\mathscr{T}$. The same is therefore true of the metric $d$: $\mathscr{X}_1 \times \mathscr{X}_2 \to \mathbb{R}$ defined by (1) and of numerous other metrics.

From the topological point of view it does not matter which of these metrics we choose to use in $\mathscr{X}_1 \times \mathscr{X}_2$ since they all generate the same topology. When discussing topological matters, we therefore work with the metric in $\mathscr{X}_1 \times \mathscr{X}_2$ which happens to be most convenient for the problem in hand. In $\mathbb{R}^2$ (or $\mathbb{R}^n$) this is usually the Euclidean metric. When $\mathscr{X}_1$ and $\mathscr{X}_2$ are more general metric spaces, however, the metric $m$ is often much less cumbersome.

# 22 LIMITS AND CONTINUITY (I)

## 22.1 Introduction

Let $f:\mathcal{X}\to\mathcal{Y}$ and suppose that $\boldsymbol{\xi}\in\mathcal{X}$ and $\boldsymbol{\eta}\in\mathcal{Y}$. In this chapter, we shall study the meaning of the statement

$$f(\mathbf{x})\to\boldsymbol{\eta} \quad \text{as} \quad \mathbf{x}\to\boldsymbol{\xi}.$$

Sometimes this is written in the equivalent form

$$\lim_{\mathbf{x}\to\boldsymbol{\xi}} f(\mathbf{x})=\boldsymbol{\eta}.$$

We say that '$f(\mathbf{x})$ tends to $\boldsymbol{\eta}$ as $\mathbf{x}$ tends to $\boldsymbol{\xi}$' or that '$f(\mathbf{x})$ converges to the *limit* $\boldsymbol{\eta}$ as $\mathbf{x}$ approaches $\boldsymbol{\xi}$'.

The diagrams below illustrate the idea we are trying to capture. The first diagram is of a function $f:\mathbb{R}^2\to\mathbb{R}^2$ for which $f(\mathbf{x})\to\eta$ as $\mathbf{x}\to\xi$. It shows $\mathbf{x}$ approaching $\boldsymbol{\xi}$ along a path. As $\mathbf{x}$ describes this path, $f(\mathbf{x})$ approaches $\boldsymbol{\eta}$. We shall of course want the same to be true *however* $\mathbf{x}$ approaches $\boldsymbol{\xi}$.

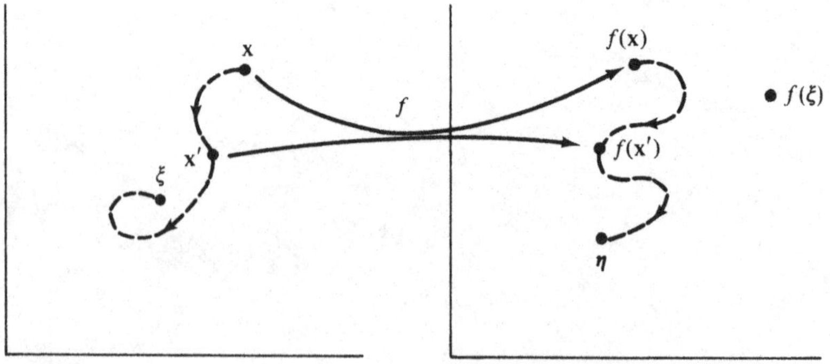


The next diagram shows the graph of a function $f:\mathbb{R}\to\mathbb{R}$ for which $f(x)\to\eta$ as $x\to\xi$.

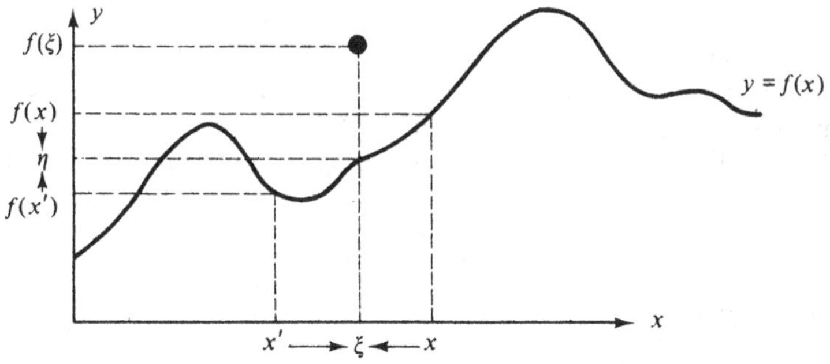


Note that in both diagrams it is *false* that $\boldsymbol{\eta}=f(\boldsymbol{\xi})$. It is *not* the case that one can always find the value of

$$\lim_{x\to\xi} f(\mathbf{x})$$

by replacing $\mathbf{x}$ in the formula for $f(\mathbf{x})$ by $\boldsymbol{\xi}$. This point is of some importance in calculus. The derivative of a function $f\colon \mathbb{R}\to\mathbb{R}$ at the point $\xi$ is defined by

$$f'(\xi)=\lim_{x\to\xi}\frac{f(x)-f(\xi)}{x-\xi}$$

(provided that the limit exists). Observe, however, that if we replace $x$ on the right-hand side by $\xi$ we obtain a meaningless expression because the right-hand side is not defined when $x=\xi$. In computing the limit, we are interested in what happens as $x$ *approaches* $\xi$ not in what happens when $x$ *equals* $\xi$. Our definition of a limit will take this into account by explicitly excluding consideration of what happens when $\mathbf{x}=\boldsymbol{\xi}$.

Sometimes, of course, it *will* be true that

$$f(\mathbf{x})\to f(\xi) \quad \text{as} \quad \mathbf{x}\to\xi.$$

We then say that $f$ is *continuous at the point* $\boldsymbol{\xi}$. The diagram below illustrates a function $f\colon \mathbb{R}\to\mathbb{R}$ which is continuous at $\xi$.

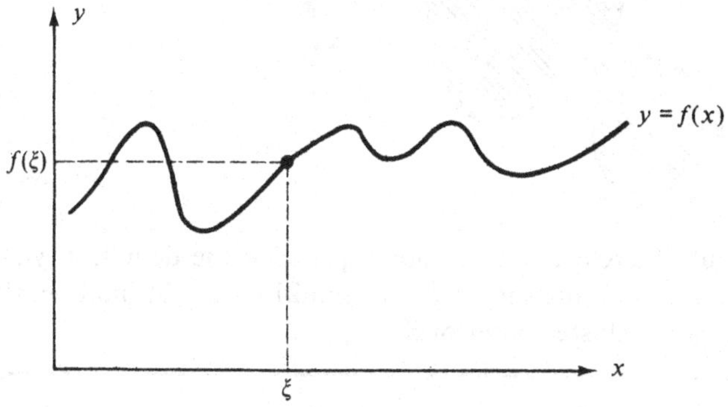

Observe that, in drawing the graph of $f$, one does not have to lift the pencil from the paper when passing through the point where $x = \xi$. This provides one explanation for the use of the word 'continuous'. A more adequate explanation is provided by theorem 22.5.

We shall also be interested in providing a definition of what it means to say that

$$f(\mathbf{x}) \to \eta \quad \text{as} \quad \mathbf{x} \to \xi$$

*through the set S.*

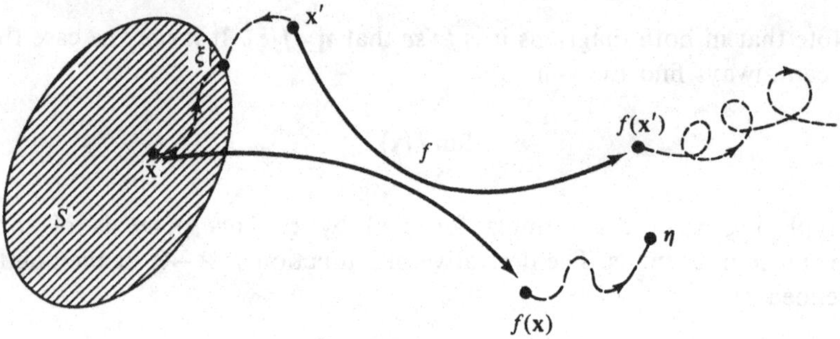


When considering a limit through the set $S$, we are interested *only* in what happens as $\mathbf{x}$ approaches $\xi$ through the set $S$. What happens as $\mathbf{x}$ approaches $\xi$ from outside the set $S$ is irrelevant and our definition will specifically exclude consideration of such points.

The definition we shall give works for any set $S$ and any point $\xi$. Note, however, that it does not make very much intuitive sense to talk about '$\mathbf{x}$ approaching $\xi$ through the set $S$' if $\xi$ is not a cluster point of $S$. There is then no way for '$\mathbf{x}$ to approach $\xi$' without going outside the set $S$. The diagram below illustrates this point.

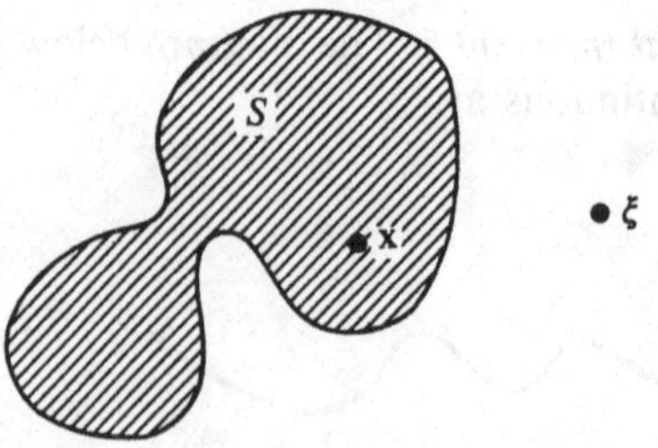


We should therefore not be too surprised if the definition yields results which are not in accordance with our intuitions about limits in those cases when $\xi$ is not a cluster point of $S$.

## 22.2 Open sets and the word 'near'

We are seeking a precise mathematical definition of the statement

$$f(\mathbf{x}) \to \boldsymbol{\eta} \quad \text{as} \quad \mathbf{x} \to \boldsymbol{\xi}.$$

The intuitive notion which we wish to capture is that $f(\mathbf{x})$ gets 'nearer and nearer' to $\boldsymbol{\eta}$ as $\mathbf{x}$ gets 'nearer and nearer' to $\boldsymbol{\xi}$. This is a rather woolly statement which could mean various things. We choose to interpret it as meaning that

'If $\mathbf{x}$ is sufficiently near to $\boldsymbol{\xi}$, then $f(\mathbf{x})$ is near to $\boldsymbol{\eta}$.'

This formulation allows us to concentrate on the crucial issue which is: what do we mean in mathematical terms by the word 'near'.

Recall that an open set $G$ in a metric space $\mathscr{X}$ is a set which contains none of its boundary points. Thus, if $\boldsymbol{\xi} \in G$, it cannot lie on the 'edge' of $G$. Hence $G$ must contain all points of $\mathscr{X}$ which are 'sufficiently near' to $\boldsymbol{\xi}$.

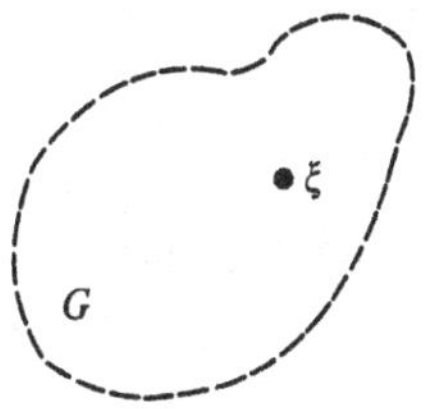


It is therefore natural to frame the definition of a limit in terms of the idea of an open set. Such a definition has the additional advantage that it still makes sense in a topological space – i.e. a space in which one knows what the open sets are although one may not be provided with a metric.

---

## 22.3 Limits

Let $\mathscr{X}$ and $\mathscr{Y}$ be metric (or topological) spaces and let $f: \mathscr{X} \to \mathscr{Y}$ where $S$ is a set in $\mathscr{X}$. Suppose that $\boldsymbol{\xi} \in \mathscr{X}$ and $\boldsymbol{\eta} \in \mathscr{Y}$. Then we say that

$$f(\mathbf{x}) \to \boldsymbol{\eta} \quad \text{as} \quad \mathbf{x} \to \boldsymbol{\xi}$$

through the set $S$ if and only if

> For any open set $G$ containing $\boldsymbol{\eta}$, there exists an open set $H$ containing $\boldsymbol{\xi}$ such that
>
> $$\mathbf{x} \in H \cap S \Rightarrow f(\mathbf{x}) \in G$$
>
> provided $\mathbf{x} \neq \boldsymbol{\xi}$.

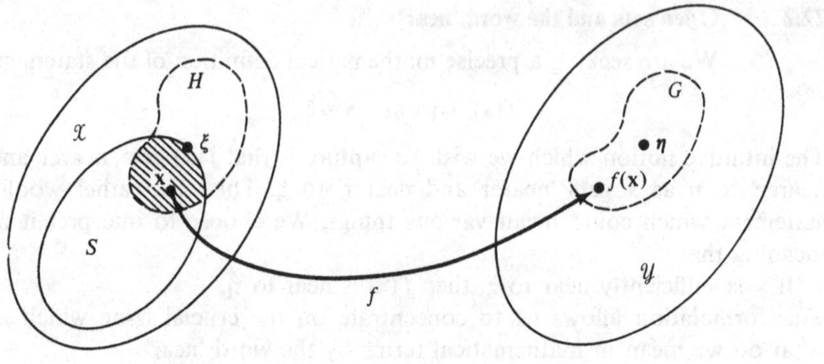


One thinks of the choice of $G$ as specifying how 'near' we want $f(\mathbf{x})$ to be to $\boldsymbol{\eta}$. Since we are interested only in the limit through the set $S$, we restrict attention to $\mathbf{x} \in S$. We also exclude the case $\mathbf{x} = \boldsymbol{\xi}$. With these provisos, the definition says that, if we take $\mathbf{x}$ 'sufficiently near' to $\boldsymbol{\xi}$ – i.e. $\mathbf{x} \in H$ – then $f(\mathbf{x})$ will be as 'near' to $\boldsymbol{\eta}$ as we specified – i.e. $f(\mathbf{x}) \in G$.

We now define the statement

$$f(\mathbf{x}) \to \boldsymbol{\eta} \quad \text{as} \quad \mathbf{x} \to \boldsymbol{\xi}.$$

This should mean that $f(\mathbf{x})$ approaches $\boldsymbol{\eta}$ *however* $\mathbf{x}$ approaches $\boldsymbol{\xi}$. The appropriate definition is therefore that:

> For any open set $G$ containing $\boldsymbol{\eta}$, there exists an open set $H$ containing $\boldsymbol{\xi}$ such that
>
> $$\mathbf{x} \in H \Rightarrow f(\mathbf{x}) \in G$$
>
> provided $\mathbf{x} \neq \boldsymbol{\xi}$.

In order that this definition makes sense, it is necessary that $f$ be defined on some open set $S$ containing $\boldsymbol{\xi}$ (except possibly at $\boldsymbol{\xi}$ itself). Such an open set $S$ contains all points of $\mathcal{X}$ which are 'sufficiently near' to $\boldsymbol{\xi}$. Indeed, our definition of the statement '$f(\mathbf{x}) \to \boldsymbol{\eta}$ as $\mathbf{x} \to \boldsymbol{\xi}$' is equivalent to the assertion that $f(\mathbf{x}) \to \boldsymbol{\eta}$ as $\mathbf{x} \to \boldsymbol{\xi}$ through the set $S$ for some open set $S$ containing $\boldsymbol{\xi}$.

---

**22.4 Limits and continuity**

If $f(\mathbf{x}) \to f(\boldsymbol{\xi})$ as $\mathbf{x} \to \boldsymbol{\xi}$, we say that $f$ is *continuous at the point* $\boldsymbol{\xi}$. The next theorem relates this terminology to our previous work on continuity.

---

22.5 *Theorem* Let $\mathcal{X}$ and $\mathcal{Y}$ be metric (or topological) spaces and let

$f: \mathcal{X} \to \mathcal{Y}$ where $S$ is a set in $\mathcal{X}$. Then $f$ is continuous on the set $S$ if and only if, for each $\boldsymbol{\xi} \in S$,

$$f(\mathbf{x}) \to f(\boldsymbol{\xi}) \quad \text{as} \quad \mathbf{x} \to \boldsymbol{\xi}$$

through the set $S$.

*Proof* We give the simplest proof. This depends on the results of chapter 21. An alternative proof using only the ideas of chapter 16 is suggested in exercise 22.24(4).

(i) Suppose that $f$ is continuous on the set $S$. Let $\boldsymbol{\xi}$ be any point of $S$ and let $G$ be an open set containing $f(\boldsymbol{\xi})$. Then $f^{-1}(G)$ is an open set relative to $S$ by theorem 21.7. From theorem 21.11, it follows that there exists an open set $H$ in $\mathcal{X}$ such that $f^{-1}(G) = H \cap S$. We then have that $\mathbf{x} \in H \cap S \Rightarrow f(\mathbf{x}) \in G$. This shows that $f(\mathbf{x}) \to f(\boldsymbol{\xi})$ as $\mathbf{x} \to \boldsymbol{\xi}$ through the set $S$.

(ii) Suppose that $f$ is not continuous on the set $S$. Then there exists an open set $G$ in $\mathcal{Y}$ such that $f^{-1}(G)$ is not open relative to $S$ (theorem 21.7). Let $\boldsymbol{\xi} \in f^{-1}(G)$ be a boundary point of $f^{-1}(G)$ relative to $S$. Then any open set $H$ in $\mathcal{X}$ which contains $\boldsymbol{\xi}$ has the property that $H \cap S$ contains a point $\mathbf{z} \in \mathcal{C} f^{-1}(G)$.

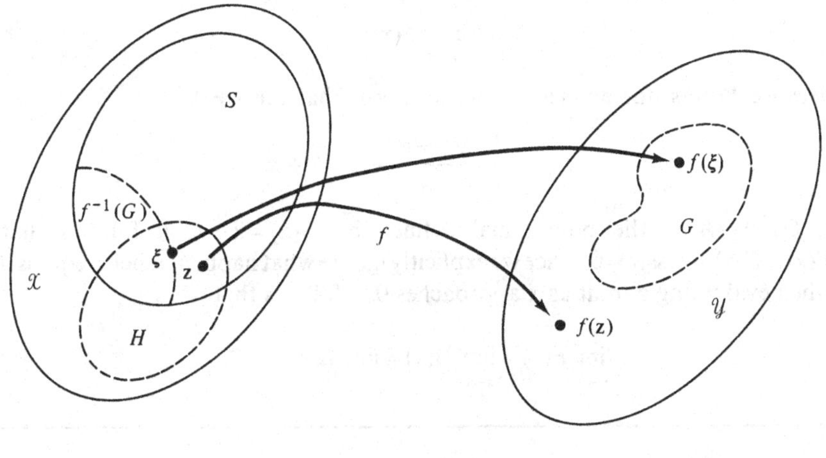


Since $\mathbf{z} \in H \cap S$ but $f(\mathbf{z}) \notin G$, it follows that it is false that $f(\mathbf{x}) \to f(\xi)$ as $\mathbf{x} \to \xi$ through the set $S$.

---

22.6 *Corollary* Let $P: \mathbb{R}^n \to \mathbb{R}$ and $Q: \mathbb{R}^n \to \mathbb{R}$ be polynomials and suppose that $Q(\boldsymbol{\xi}) \neq 0$. Then

$$\frac{P(\mathbf{x})}{Q(\mathbf{x})} \to \frac{P(\boldsymbol{\xi})}{Q(\boldsymbol{\xi})} \quad \text{as} \quad \mathbf{x} \to \boldsymbol{\xi}.$$

*Proof* This follows immediately from theorems 16.15 and 22.5.

---

Note that the same result holds with $\mathbb{R}$ replaced throughout by $\mathbb{C}$.

---

22.7 *Examples* (i) Let $S=\mathbb{R}^2\backslash\{(0, 0)\}$ and consider the rational function $f: S\to\mathbb{R}$ defined by

$$f(x, y)=\frac{x^3y+2xy^2+3}{x^2+y^2}.$$

By corollary 22.6,

$$f(x, y)\to 3 \quad\text{as}\quad (x, y)\to(1, 1).$$

(ii) Let $S=\mathbb{R}\backslash\{0\}$ and consider the rational function $F: S\to\mathbb{R}$ defined by

$$F(x)=\frac{(1+x)^2-1}{x}.$$

We know from corollary 22.6 that $F(x)\to 1$ as $x\to -1$ but corollary 22.6 does not help immediately in evaluating

$$\lim_{x\to 0} F(x)$$

because $F(0)$ is not defined. Note, however, that for $x\neq 0$,

$$F(x)=\frac{1+2x+x^2-1}{x}=2+x.$$

If $G$: $\mathbb{R}\to\mathbb{R}$ is the polynomial defined by $G(x)=2+x$, it follows that $F(x)=G(x)$ unless $x=0$. Since we explicitly ignore what happens when $x$ equals 0 when evaluating a limit as $x$ approaches 0, it follows that

$$\lim_{x\to 0} F(x)=\lim_{x\to 0} G(x)=\lim_{x\to 0} (2+x)=2.$$

---

22.8 *Exercise*
(1) Evaluate the following limits

(i) $\displaystyle\lim_{(x, y)\to(1, 1)} \frac{y^2-x^2}{y^2+x^2}$ (ii) $\displaystyle\lim_{(x, y)\to(1, -1)} \frac{y^2-x^2}{y+x}$.

(2) Let $K=\{(x, y): 0\leqq x\leqq 1 \text{ and } 0\leqq y\leqq 1\}$ and let $f$: $\mathbb{R}^2\to\mathbb{R}$ be defined by

$$f(x, y)=\begin{cases}1 & (x, y)\in K\\ 0 & (x, y)\notin K.\end{cases}$$

Prove that $f(x, y) \to 1$ as $(x, y) \to (1, 1)$ through the set $K$ and $f(x, y) \to 0$ as $(x, y) \to (1, 1)$ through the set $\mathcal{C}K$.

(3) Suppose that $\boldsymbol{\xi}$ is not a cluster point of $S$. Explain why it is true that $f(\mathbf{x}) \to \boldsymbol{\eta}$ as $\mathbf{x} \to \boldsymbol{\xi}$ through the set $S$ for *all* $\boldsymbol{\eta}$. If $\boldsymbol{\xi}$ is a cluster point of $S$, prove that there exists at most one $\boldsymbol{\eta}$ such that $f(\mathbf{x}) \to \boldsymbol{\eta}$ as $\mathbf{x} \to \xi$ through the set $S$ provided that, for each pair of distinct points $\boldsymbol{\eta}_1$ and $\boldsymbol{\eta}_2$ in $\mathcal{Y}$, there exist disjoint open sets $G_1$ and $G_2$ such that $\boldsymbol{\eta}_1 \in G_1$ and $\boldsymbol{\eta}_2 \in G_2$. Show that such disjoint open sets always exist when $\mathcal{Y}$ is a metric space.

(4) Suppose that $S_1 \cup S_2 = S$. Prove that $f(\mathbf{x}) \to \boldsymbol{\eta}$ as $\mathbf{x} \to \boldsymbol{\xi}$ through $S$ if and only if $f(\mathbf{x}) \to \boldsymbol{\eta}$ as $\mathbf{x} \to \boldsymbol{\xi}$ through $S_1$ *and* through $S_2$.
Deduce that, if $f$ is the function of question 2, then there is no $\boldsymbol{\eta} \in \mathbb{R}$ for which $f(x, y) \to \boldsymbol{\eta}$ as $(x, y) \to (1, 1)$.

(5) Suppose that $T$ is a subset of $S$. If $f(\mathbf{x}) \to \boldsymbol{\eta}$ as $\mathbf{x} \to \boldsymbol{\xi}$ through the set $S$, prove that $f(\mathbf{x}) \to \boldsymbol{\eta}$ as $\mathbf{x} \to \boldsymbol{\xi}$ through the set $T$.

(6) Suppose that $f\colon \mathbb{R} \to \mathbb{R}$ and $g\colon \mathbb{R} \to \mathbb{R}$ are continuous on the set $\mathbb{R}$ and $f(x) = g(x)$ for each $x \in \mathbb{Q}$. Prove that $f = g$. [*Hint*: Let $\xi$ be irrational. Then $\xi$ is a cluster point of $\mathbb{Q}$ (why?). Also $f(x) \to f(\xi)$ as $x \to \xi$ through $\mathbb{Q}$ and $g(x) \to g(\xi)$ as $x \to \xi$ through $\mathbb{Q}$.]

---

## 22.9 Limits and distance

In a metric space $\mathcal{X}$ a set $G$ is open if and only if each $\mathbf{x} \in G$ is the centre of an open ball $B$ which is entirely contained in $G$. This fact means that, if $\mathcal{X}$ and $\mathcal{Y}$ are metric spaces, then we can rewrite the definition of a limit given in §22.3 in terms of open balls rather than open sets.

Let $\mathcal{X}$ and $\mathcal{Y}$ be metric spaces and let $f\colon S \to \mathcal{Y}$ where $S$ is a set in $\mathcal{X}$. Then $f(\mathbf{x}) \to \boldsymbol{\eta}$ as $\mathbf{x} \to \boldsymbol{\xi}$ through the set $S$ if and only if, for each open ball $E$ with centre $\boldsymbol{\eta}$, there exists an open ball $\Delta$ with centre $\boldsymbol{\xi}$ such that

$$\mathbf{x} \in \Delta \cap S \Rightarrow f(\mathbf{x}) \in E$$

provided that $\mathbf{x} \neq \boldsymbol{\xi}$.

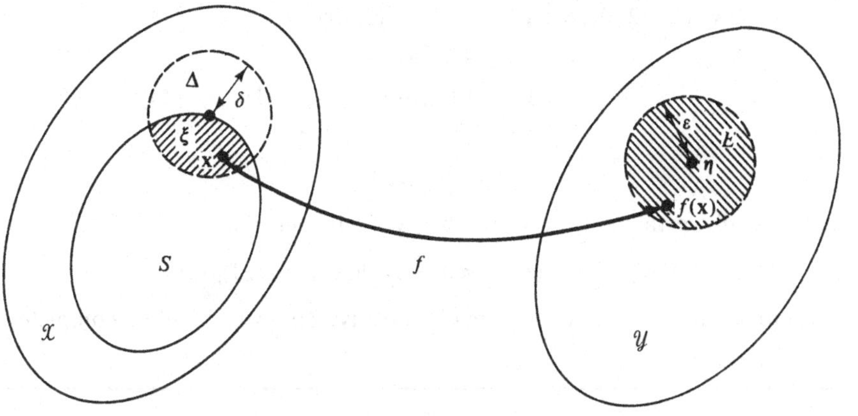

If we take the radius of the open ball $E$ to be $\varepsilon$ and that of $\Delta$ to be $\delta$, the definition assumes the following familiar form:

> For any $\varepsilon>0$, there exists a $\delta>0$ such that for each $\mathbf{x}\in S$,
> $$0<d(\boldsymbol{\xi}, \mathbf{x})<\delta \Rightarrow d(\boldsymbol{\eta}, f(\mathbf{x}))<\varepsilon.$$

It is sometimes helpful to think of $f(\mathbf{x})$ as an approximation to $\boldsymbol{\eta}$. The quantity $d(\boldsymbol{\eta}, f(\mathbf{x}))$ is then the error in approximating to $\boldsymbol{\eta}$ by $f(\mathbf{x})$. The definition then asserts that this error can be made as small as we choose by taking $\mathbf{x}$ sufficiently close to $\boldsymbol{\xi}$. (Here, of course, it is understood that values of $\mathbf{x}$ outside $S\setminus\{\boldsymbol{\xi}\}$ are to be ignored.)

---

22.10 *Example* Consider the function $f: \mathbb{R}^2\to\mathbb{R}^2$ defined by $(u, v)=f(x, y)$ where

$$\left.\begin{aligned} u&=x+y\\ v&=x-y.\end{aligned}\right\}$$

The fact that $f(x, y)\to(2, 0)$ as $(x, y)\to(1, 1)$ follows from theorem 22.5 because $f$ is continuous on $\mathbb{R}^2$. In this example we give an alternative proof using the criterion for convergence given in §22.9.

*Proof* Given any $\varepsilon>0$, we shall demonstrate the existence of a $\delta>0$ such that

$$0<\|(x, y)-(1, 1)\|<\delta \Rightarrow \|f(x, y)-(2, 0)\|<\varepsilon.$$

We begin by observing that

$$\|(x, y)-(1, 1)\|=\{(x-1)^2+(y-1)^2\}^{1/2}$$

and

$$\begin{aligned}\|f(x, y)-(2, 0)\| &= \|(x+y, x-y)-(2, 0)\|\\ &=\{(x+y-2)^2+(x-y)^2\}^{1/2}\\ &=\{[(x-1)+(y-1)]^2+[(x-1)-(y-1)]^2\}^{1/2}\\ &=\{2(x-1)^2+2(y-1)^2\}^{1/2}\\ &=\sqrt{2}\,\|(x, y)-(1, 1)\|.\end{aligned}$$

Given $\varepsilon>0$ we therefore have to choose $\delta>0$ such that

$$0<\|(x, y)-(1, 1)\|<\delta \Rightarrow \sqrt{2}\|(x, y)-(1, 1)\|<\varepsilon.$$

The choice of $\delta=\varepsilon/\sqrt{2}$ (or any smaller positive number) therefore completes the proof.

---

22.11 *Example* Consider the function $f\colon \mathbb{R}^2 \to \mathbb{R}^1$ defined by

$$f(x, y) = \begin{cases} \dfrac{y^2 - x^2}{y^2 + x^2} & (x, y) \neq (0, 0) \\ 0 & (x, y) = (0, 0). \end{cases}$$

Let $L_\alpha$ be the line $L_\alpha = \{(x, y): y = \alpha x\}$. For points $(x, y) \in L_\alpha$, we have that

$$f(x, y) = f(x, \alpha x) = \frac{\alpha^2 x^2 - x^2}{\alpha^2 x^2 + x^2} = \frac{\alpha^2 - 1}{\alpha^2 + 1} \quad (x \neq 0).$$

Hence

$$f(x, y) \to \frac{\alpha^2 - 1}{\alpha^2 + 1} \quad \text{as} \quad (x, y) \to (0, 0)$$

*along the line* $y = \alpha x$ – i.e. through the set $L_\alpha$.

It follows that $f(x, y)$ cannot tend to a limit as $(x, y) \to (0, 0)$ through $\mathbb{R}^2$. If it did, it would have to tend to the *same* limit through each of the sets $L_\alpha$ (see exercise 22.8(4)).

The diagram indicates how the function tends to different limits along different straight lines through (0, 0).

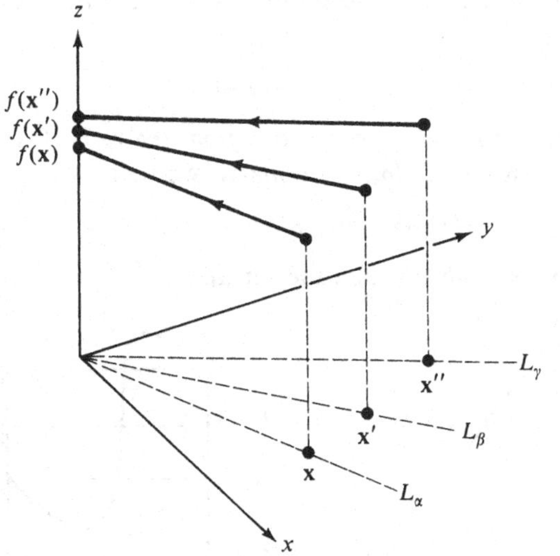


The part of the surface $z = f(x, y)$ drawn looks somewhat like a piece of spiral staircase.

## 22.12 Right and left hand limits

For a function $f: \mathbb{R} \to \mathbb{R}$ we have that $f(x) \to \eta$ as $x \to \xi$ if and only if for any $\varepsilon > 0$, there exists a $\delta > 0$ such that

$$0 < |x - \xi| < \delta \Rightarrow |f(x) - \eta| < \varepsilon.$$

This definition can be illustrated graphically as below.

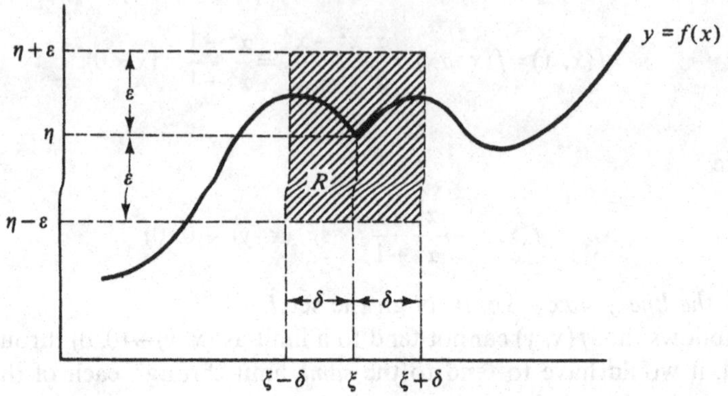


In this diagram $R$ is a rectangle with centre $(\xi, \eta)$ of height $2\varepsilon$ and width $2\delta$. The definition asserts that, however small we choose to make the height of $R$, its width can be chosen so as to ensure that the part of the graph of $f$ above $(\xi - \delta, \xi + \delta)$ lies within $R$ (except possibly for the point $(\xi, f(\xi))$).

Now suppose that $I = (a, b)$ is an open interval in $\mathbb{R}$ and that $f: (a, b) \to \mathbb{R}$. We write

$$f(x) \to \eta \quad \text{as} \quad x \to a+$$

and say that $f(x)$ tends to $\eta$ as $x$ tends to *a from the right* if and only if $f(x) \to \eta$ as $x \to a$ through the set $(a, b)$. Similarly, we write

$$f(x) \to \eta \quad \text{as} \quad x \to b-$$

if and only if $f(x) \to \eta$ as $x \to b$ through the set $(a, b)$.

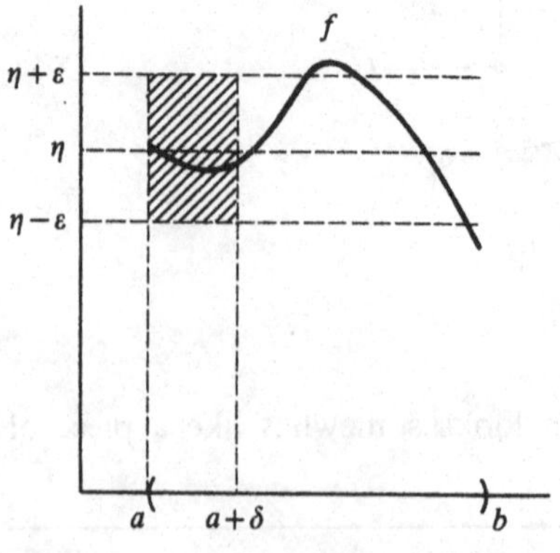


$f(x) \to \eta$ as $x \to a+$

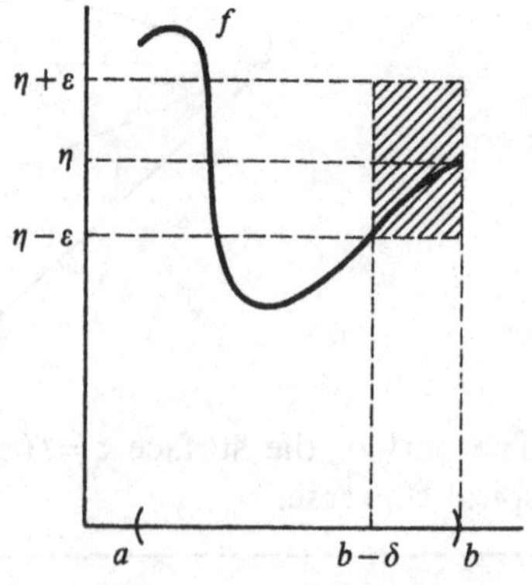


$f(x) \to \eta$ as $x \to b-$

It is useful to note that $f(x)\to\eta$ as $x\to a+$ if and only if, for any $\varepsilon>0$, there exists a $\delta>0$ such that

$$a<x<a+\delta \Rightarrow |f(x)-\eta|<\varepsilon.$$

Also $f(x)\to\eta$ as $x\to b-$ if and only if, for any $\varepsilon>0$, there exists a $\delta>0$ such that

$$b-\delta<x<b \Rightarrow |f(x)-\eta|<\varepsilon.$$

---

22.13 *Theorem* Let $I$ be an open interval containing $\xi$ and suppose that $f: I\backslash\{\xi\}\to\mathbb{R}$. Then

$$f(x)\to\eta \quad \text{as} \quad x\to\xi$$

if and only if $f(x)\to\eta$ as $x\to\xi-$ *and* $f(x)\to\eta$ as $x\to\xi+$.

*Proof* This follows immediately from exercise 22.8(4).

---

22.14 *Example* Let $f\colon \mathbb{R}\to\mathbb{R}$ be defined by

$$f(x)=\begin{cases}1-x & (x\leqq 1)\\ 2x & (x>1).\end{cases}$$

Then $f(x)\to 0$ as $x\to 1-$ and $f(x)\to 2$ as $x\to 1+$. It follows from theorem 22.13 that

$$\lim_{x\to 1} f(x)$$

does not exist. In particular, $f$ is not continuous at the point 1.

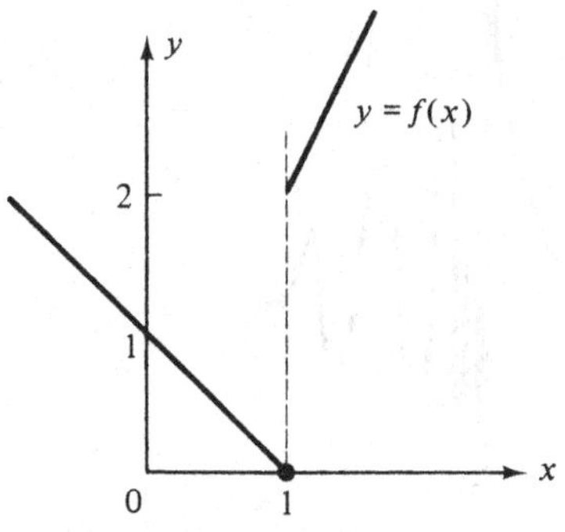


Note that it is not necessary to use an argument involving $\varepsilon$ and $\delta$ to show that $f(x)\to 2$ as $x\to 1+$. The function $g\colon \mathbb{R}\to\mathbb{R}$ defined by $g(x)=2x$ is a polynomial and hence is continuous everywhere. Also $f(x)=g(x)$ when $x>1$.

In calculating a limit as $x \to 1+$ we ignore what happens when $x \leq 1$. Hence

$$\lim_{x\to 1+} f(x) = \lim_{x\to 1+} g(x) = \lim_{x\to 1+} 2x = 2.$$

Similarly for $f(x) \to 0$ as $x \to 1-$.

---

## 22.15 Some notation

When the limits exist, the notation

$$f(\xi-) = \lim_{x\to\xi} f(x); \quad f(\xi+) = \lim_{x\to\xi+} f(x)$$

can often be useful. However, it is important not to confuse $f(\xi-)$ or $f(\xi+)$ with $f(\xi)$. These three quantities are equal only when $f$ is continuous at the point $\xi$.

If $f(\xi-) = f(\xi)$ we say that $f$ is *continuous on the left* at $\xi$. If $f(\xi+) = f(\xi)$, we say that $f$ is *continuous on the right* at $\xi$. The diagrams below illustrate some of the possibilities.

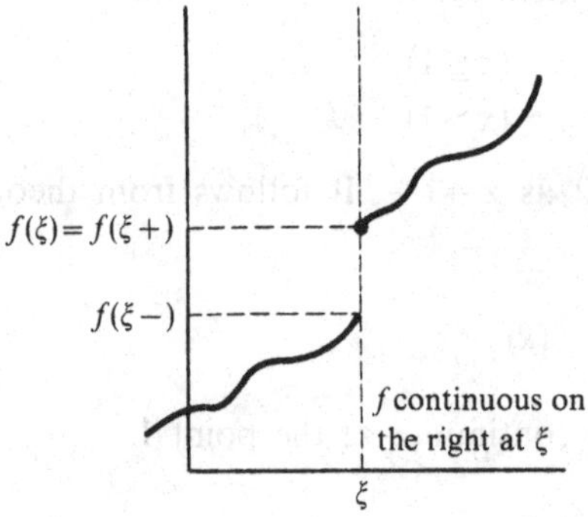


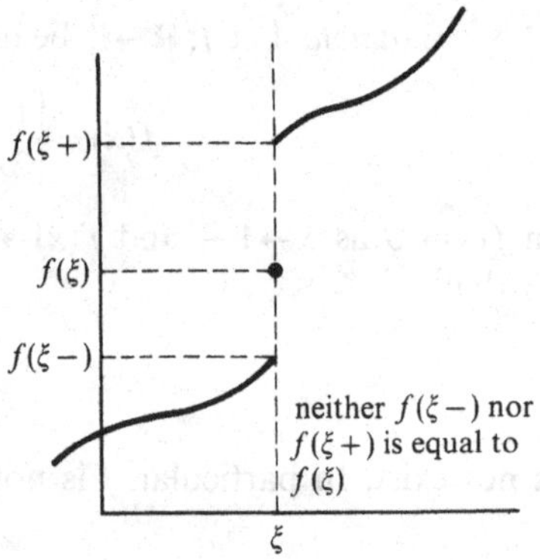


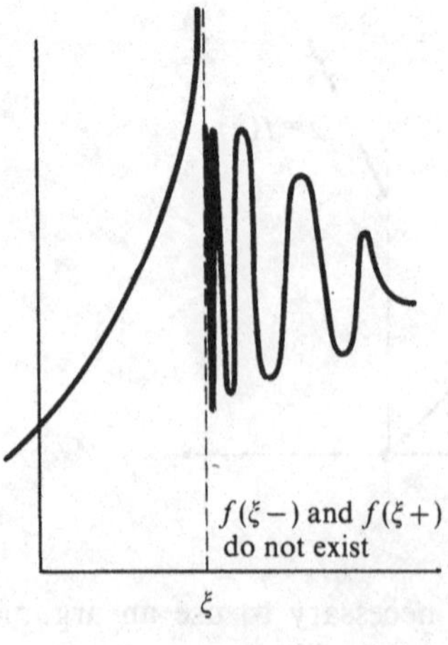

## 22.16 Monotone functions

Suppose that $S$ is a non-empty subset of $\mathbb{R}$. We say that $f: S \to \mathbb{R}$ *increases* on $S$ if and only if

$$x < y \Rightarrow f(x) \leqq f(y)$$

for each $x$ and $y$ in $S$. Similarly, $f: S \to \mathbb{R}$ *decreases* on $S$ if and only if

$$x < y \Rightarrow f(x) \geqq f(y)$$

for each $x$ and $y$ in $S$. We say that $f$ is *strictly* increasing on $S$ if and only if

$$x < y \Rightarrow f(x) < f(y)$$

for each $x$ and $y$ in $S$. Finally, $f$ is *strictly* decreasing on $S$ if and only if

$$x < y \Rightarrow f(x) > f(y).$$

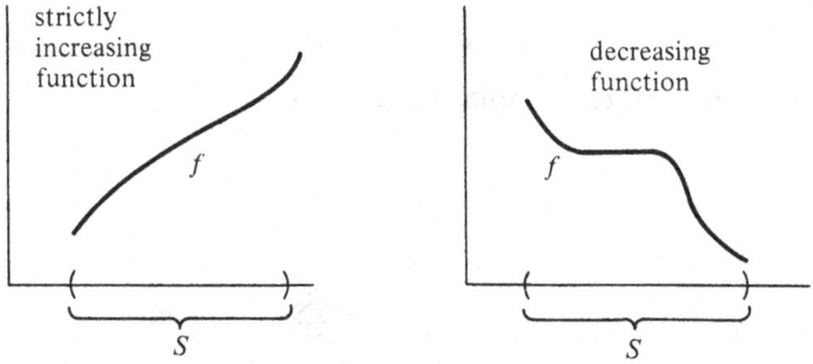


A function which is either increasing or else decreasing on $S$ is said to be *monotone* on $S$. (A function can be *both* increasing *and* decreasing on $S$. But then it must be constant.) A function which is either strictly increasing or strictly decreasing is said to be *strictly monotone.*

---

22.17 *Theorem* Suppose that $I = (a, b)$ is an open interval in $\mathbb{R}$ and that $f: I \to \mathbb{R}$ is increasing on $I$.

(i) If $f$ is bounded above with supremum $L$ on $I$, then $f(x) \to L$ as $x \to b-$.

(ii) If $f$ is bounded below with infimum $l$ on $I$, then $f(x) \to l$ as $x \to a+$.

*Proof* We prove only (i). From §22.12, we know that, given any $\varepsilon > 0$, we have to demonstrate the existence of a $\delta > 0$ such that

$$b - \delta < x < b \Rightarrow |f(x) - L| < \varepsilon.$$

But $|f(x) - L| < \varepsilon \Leftrightarrow L - \varepsilon < f(x) < L + \varepsilon$ (exercise 13.13(3)). The inequality

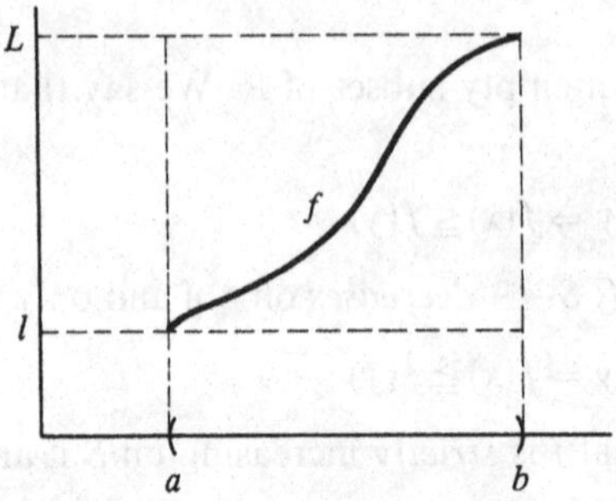


$f(x) < L + \varepsilon$ is automatically satisfied because $L$ is an upper bound for $f$ on $(a, b)$. Since $L - \varepsilon$ is *not* an upper bound (because $L$ is the *smallest* upper bound), there exists a $y \in (a, b)$ such that $f(y) > L - \varepsilon$. But $f$ increases on $(a, b)$. Therefore, for any $x$ satisfying $y < x < b$,

$$L - \varepsilon < f(y) \leqq f(x).$$

The choice $\delta = b - y$ then completes the proof.

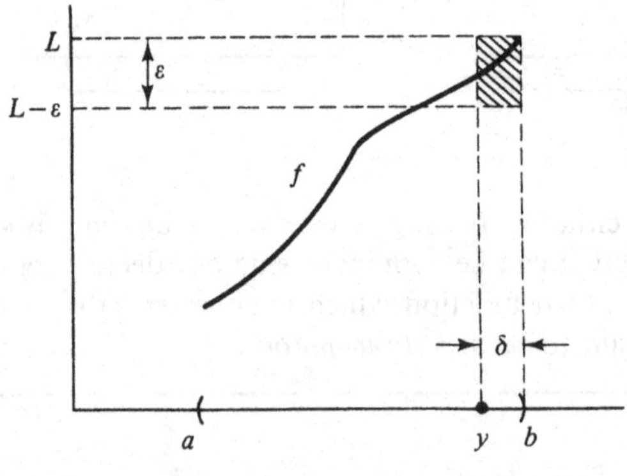


---

22.18 *Corollary* Suppose that $I = (a, b)$ is an open interval in $\mathbb{R}$ and that $f: I \to \mathbb{R}$ is increasing on $I$. If $\xi \in (a, b)$, then $f(\xi -)$ and $f(\xi +)$ both exist and

$$f(x) \leqq f(\xi -) \leqq f(\xi) \leqq f(\xi +) \leqq f(y)$$

provided that $a < x < \xi < y < b$.

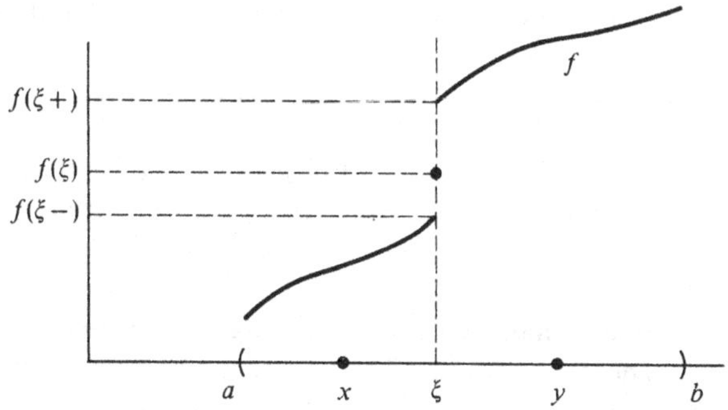


*Proof* The function $f$ is bounded above on the interval $(a, \xi)$ by $f(\xi)$. By theorem 22.17, the *smallest* upper bound of $f$ on $(a, \xi)$ is $f(\xi-)$. It follows that, for any $x \in (a, \xi)$,

$$f(x) \leqq f(\xi-) \leqq f(\xi).$$

A similar argument for the interval $(\xi, b)$ yields the other inequalities.

---

## 22.19 Inverse functions

Recall that a function $f$: $S \to T$ admits an inverse function $f^{-1}$: $T \to S$ if and only if $f$ is bijective. A function is bijective if and only if it is both surjective and injective. Surjective means that $f(S) = T$ and injective means that $f(x_1) = f(x_2) \Leftrightarrow x_1 = x_2$.

In this section, we are interested in the case when $S$ and $T$ are intervals in $\mathbb{R}$ and $f$ is continuous on $S$. It then seems 'intuitively obvious' that $f$ is bijective if and only if $T = f(S)$ and $f$ is strictly increasing or strictly decreasing on $S$. Furthermore, it seems equally clear that the inverse function $f^{-1}$: $T \to S$ will be continuous on $T$.

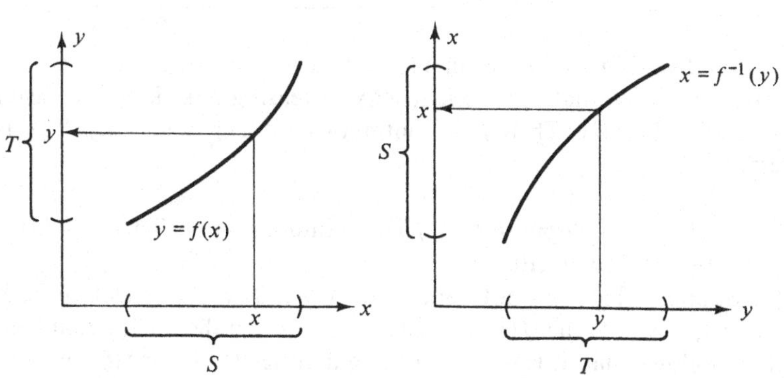

But, like so many 'intuitively obvious' theorems, the proof requires some deep results.

---

22.20 *Proposition* Let $I$ be an interval in $\mathbb{R}$ and suppose that $f: I \to \mathbb{R}$ is continuous on $I$. Then $f$ is injective if and only if $f$ is either strictly increasing or strictly decreasing on $I$.

*Proof* The fact that a strictly increasing or decreasing function is injective is trivial. The proof of the other half of the theorem is the content of exercise 17.28(3). It requires a simple application of the intermediate value theorem (17.11).

The diagram illustrates a function $f: I \to \mathbb{R}$ which is continuous on $I$ but neither increasing nor decreasing on $I$. If $a<b<c$ and $f(a)<f(c)<f(b)$, the intermediate value theorem implies that there exists a $\xi \in [a, b]$ such that $f(\xi)=f(c)$. Hence $f$ is not injective.

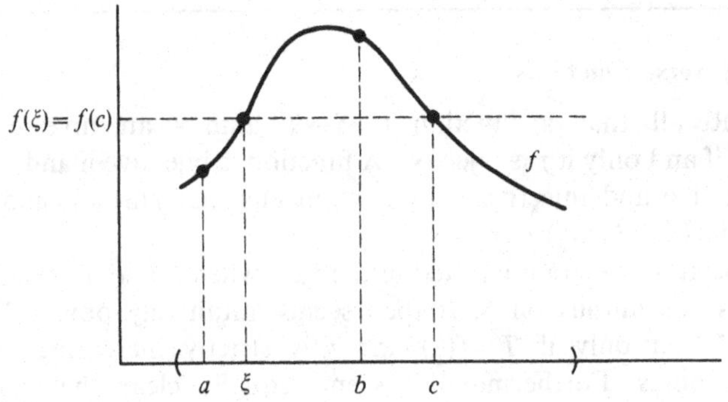


---

22.21 *Theorem* Let $I$ be an interval in $\mathbb{R}$ and let $J$ be a set of real numbers. Suppose that $f: I \to J$ is strictly increasing or strictly decreasing on $I$ and that $J=f(I)$. Then $f$ is continuous on $I$ if and only if $J$ is an interval.

*Proof* (i) Suppose that $f$ is continuous on $I$. Then $J=f(I)$ is an interval by theorem 17.10.

(ii) Suppose that $f$ is not continuous on $I$. Then, by theorem 22.5, for some $\xi \in I$, it is false that $f(x) \to f(\xi)$ as $x \to \xi$ through the set $I$. From theorem 22.17 it follows that at least one of the quantities $f(\xi-)$ or $f(\xi+)$ exists but is not equal to $f(\xi)$. But corollary 22.18 then implies than $J$ is not an interval.

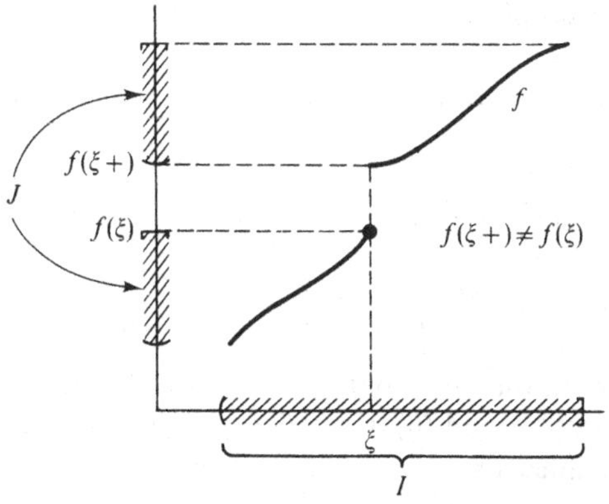


22.22 *Theorem* Let $I$ and $J$ be intervals in $\mathbb{R}$ and suppose that $f: I \to J$ is surjective, continuous and strictly increasing or decreasing on $I$. Then $f: I \to J$ is a homeomorphism – i.e. $f$ is bijective and $f^{-1}: J \to I$ is continuous on $J$.

*Proof* The fact that $f$ is bijective follows from the two preceding results. Since $f^{-1}: J \to I$ is surjective and strictly increasing or decreasing, then it is continuous on $J$ by theorem 22.21.

**22.23 Roots**

In §9.13 we proved that, if $y \geqq 0$, then there exists a unique $x \geqq 0$ such that $y = x^n$. We call $x$ the *nth root* of $y$ and write

$$x = y^{1/n}.$$

A more elegant proof of this result uses the theorems of the preceding section. Let $n \in \mathbb{N}$ and define $f: [0, \infty) \to [0, \infty)$ by

$$f(x) = x^n.$$

The function $f$ is strictly increasing and continuous on $[0, \infty)$. To use theorem 22.22 we need to prove that it is surjective. We know from theorem 17.10 that $J = f([0, \infty))$ is an interval. But $f(0) = 0$ and $J$ is clearly unbounded above. It follows that $J = [0, \infty)$ and so $f$ is surjective. By theorem 22.22, $f$ admits an inverse function $f^{-1}: [0, \infty) \to [0, \infty)$. We have that

$$y = x^n \Leftrightarrow x = f^{-1}(y)$$

provided $x \geqq 0$ and $y \geqq 0$. Thus $f^{-1}: [0, \infty) \to [0, \infty)$ is the $n$th root function – i.e.

$$f^{-1}(y) = y^{1/n} \quad (y \geqq 0).$$

Note that theorem 22.22 supplies the extra information that this function is continuous on $[0, \infty)$.

---

22.24 *Exercise*

(1) Let $f: \mathbb{R}^2 \to \mathbb{R}^2$ be defined by $f(x, y) = (u, v)$ where

$$u = \begin{cases} 1+x & (y \geqq x) \\ y & (y < x) \end{cases} \qquad v = \begin{cases} 1+y & (y \geqq x) \\ x & (y < x). \end{cases}$$

What can be said about the limits of $f$ as $(x, y) \to (0, 0)$ through (i) the set $S = \{(x, y): x < 0 \text{ and } y > 0\}$ and (ii) the set $T = \{(x, y): x > 0 \text{ and } y < 0\}$?

(2) Let $g: \mathbb{R}^2 \to \mathbb{R}^1$ be defined by

$$g(x, y) = \begin{cases} \dfrac{xy^2}{x^2+y^4} & ((x, y) \neq (0, 0)) \\ 0 & ((x, y) = (0, 0)). \end{cases}$$

Prove that $g(x, y) \to 0$ as $(x, y) \to (0, 0)$ along each of the lines $y = \alpha x$. But show that $g(x, y) \not\to 0$ as $(x, y) \to (0, 0)$ along the parabola $x = \alpha y^2$.

(3) A function $f: \mathbb{R}^2 \setminus \{(0, 0)\} \to \mathbb{R}$ is defined by

$$f(x, y) = \frac{xy}{x^2+y^2}.$$

If $0 < \varepsilon < \frac{1}{2}$, show that $|f(x, y)| < \varepsilon$ if and only if

$$\left|\frac{y}{x}\right| < 2\varepsilon\{1+\sqrt{(1-4\varepsilon^2)}\}^{-1} \quad \text{or} \quad \left|\frac{x}{y}\right| < 2\varepsilon\{1+\sqrt{(1-4\varepsilon^2)}\}^{-1}.$$

If $g(x, y) \to 0$ as $(x, y) \to (0, 0)$, prove that $f(x, y) \to 0$ as $(x, y) \to (0, 0)$ through the set

$$S = \left\{(x, y): \left|\frac{y}{x}\right| < g(x, y) \quad \text{or} \quad \left|\frac{x}{y}\right| < g(x, y)\right\}.$$

(4) Let $\mathscr{X}$ and $\mathscr{Y}$ be metric spaces and let $f: S \to \mathscr{Y}$ where $S$ is a subset of $\mathscr{X}$. If $\boldsymbol{\xi} \in \mathscr{X}$ and $\boldsymbol{\eta} \in \mathscr{Y}$, prove that $f(\mathbf{x}) \to \boldsymbol{\eta}$ as $\mathbf{x} \to \boldsymbol{\xi}$ through the set $S$ if and only if, for each subset $E$ of $S$ not containing $\boldsymbol{\xi}$,

$$d(\boldsymbol{\xi}, E) = 0 \Rightarrow d(\boldsymbol{\eta}, f(E)) = 0.$$

Use this result to deduce theorem 22.5 from theorem 16.4.

(5) Let $I$ be an interval in $\mathbb{R}$ and let $f: I \to \mathbb{R}$ be increasing on $I$. Prove that the set of points in $I$ at which $f$ is *not* continuous is countable. [*Hint*: Recall the proof of theorem 17.27.]

(6) Let $f: \mathbb{R} \to \mathbb{R}$ be defined by $f(x) = 1 + x + x^3$. Show that $f$ has an inverse function $f^{-1}: \mathbb{R} \to \mathbb{R}$ and that this is continuous on $\mathbb{R}$.

---

**22.25 Combining limits**

22.26 *Theorem* Suppose that $\mathcal{X}$, $\mathcal{Y}$ and $\mathcal{Z}$ are metric (or topological) spaces and that $f: T \to \mathcal{Z}$ and $g: S \to T$ where $S \subset \mathcal{X}$ and $T \subset \mathcal{Y}$. Let $\boldsymbol{\xi} \in \mathcal{X}$ and let $\boldsymbol{\eta} \in T$.

If $f$ is continuous on the set $T$ and $g(\mathbf{x}) \to \boldsymbol{\eta}$ as $\mathbf{x} \to \boldsymbol{\xi}$ through the set $S$, then

$$f(g(\mathbf{x})) \to f(\boldsymbol{\eta}) \quad \text{as} \quad \mathbf{x} \to \boldsymbol{\xi}$$

through the set $S$.

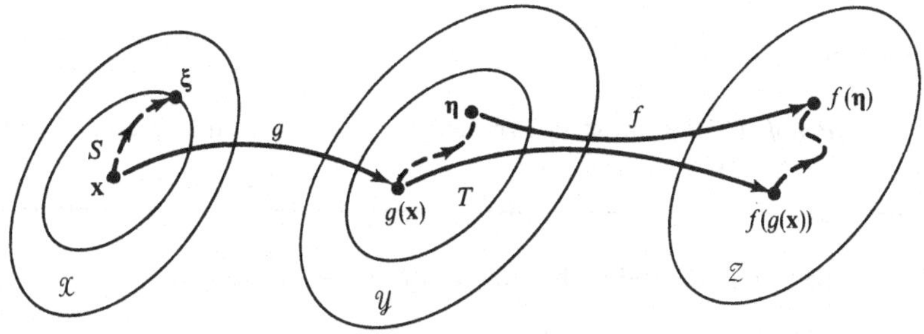


*Proof* Let $G$ be any open set in $\mathcal{Z}$ which contains $f(\boldsymbol{\eta})$. Since $f$ is continuous on the set $T$, it follows from theorem 22.5 that there exists an open set $H$ in $\mathcal{Y}$ containing $\boldsymbol{\eta}$ such that

$$\mathbf{y} \in H \cap T \Rightarrow f(\mathbf{y}) \in G. \tag{1}$$

(Note that we do not need to exclude the possibility that $\mathbf{y} = \boldsymbol{\eta}$ because $f(\boldsymbol{\eta}) \in G$ by assumption.)

Also, since $H$ is an open set containing $\boldsymbol{\eta}$, there exists an open set $J$ containing $\boldsymbol{\xi}$ such that

$$\mathbf{x} \in J \cap S \Rightarrow g(\mathbf{x}) \in H \cap T \tag{2}$$

provided $\mathbf{x} \neq \boldsymbol{\xi}$. (Note that $g(S) \subset T$ and so $\mathbf{x} \in S \Rightarrow g(\mathbf{x}) \in T$.) Combining (1) and (2), we obtain that

$$\mathbf{x} \in J \cap S \Rightarrow f(g(\mathbf{x})) \in G$$

provided $\mathbf{x} \neq \boldsymbol{\xi}$. It follows that

$$f(g(\mathbf{x})) \to f(\boldsymbol{\eta}) \quad \text{as} \quad \mathbf{x} \to \boldsymbol{\xi}$$

through the set $S$.

---

22.27 *Corollary* Suppose that $\mathcal{X}$ is a metric (or topological) space and that $g: S \to [0, \infty)$. If $g(\mathbf{x}) \to \eta$ as $\mathbf{x} \to \boldsymbol{\xi}$ through the set $S$, then

$$\{g(\mathbf{x})\}^{1/n} \to \eta^{1/n} \quad \text{as} \quad \mathbf{x} \to \boldsymbol{\xi}$$

through the set $S$ for any $n \in \mathbb{N}$.

*Proof* This follows immediately from §22.23 and theorem 22.26.

---

22.28 *Theorem* Suppose that $\mathscr{X}$ is a metric (or topological) space and that $f\colon S\to\mathbb{R}^n$ where $S$ is a set in $\mathscr{X}$. Let $\boldsymbol{\xi}\in\mathscr{X}$ and write

$$f=(f_1, f_2, \ldots, f_n).$$

Then $f(\mathbf{x})\to\boldsymbol{\eta}$ as $\mathbf{x}\to\boldsymbol{\xi}$ through the set $S$ if and only if, for each $k=1, 2, \ldots, n$,

$$f_k(\mathbf{x})\to\eta_k \quad \text{as} \quad \mathbf{x}\to\boldsymbol{\xi}$$

through the set $S$.

*Proof* This is most conveniently deduced from exercise 22.24(4) and exercise 13.24(6).

---

22.29 *Example* Consider the function $f\colon \mathbb{R}^2\to\mathbb{R}^2$ defined by

$$f(x, y)=(g(x, y), h(x, y))$$

where $g\colon\mathbb{R}^2\to\mathbb{R}^1$ and $h\colon\mathbb{R}^2\to\mathbb{R}^1$ are the functions of exercise 22.24(2) and (3). We have that $g(x, y)\to 0$ as $(x, y)\to 0$ along the line $x=y$. Similarly, $h(x, y)\to 0$ as $(x, y)\to 0$ along the line $x=y$. From theorem 22.28 it follows that

$$f(x, y)\to(0, 0) \quad \text{as} \quad (x, y)\to(0, 0)$$

along the line $x=y$.

On the other hand, $g(x, y) \not\to 0$ as $(x, y)\to(0, 0)$ along the parabola $x=y^2$ and hence

$$f(x, y)\not\to(0, 0) \quad \text{as} \quad (x, y)\to(0, 0)$$

along $x=y^2$.

---

22.30 *Theorem* Suppose that $\mathscr{X}$ is a metric (or topological) space and that $f_1\colon S\to\mathbb{R}^n$ and $f_2\colon S\to\mathbb{R}^n$ where $S$ is a set in $\mathscr{X}$. Let $\boldsymbol{\xi}\in\mathscr{X}$ and suppose that

$$f_1(\mathbf{x})\to\boldsymbol{\eta}_1 \quad \text{as} \quad \mathbf{x}\to\boldsymbol{\xi}$$

and

$$f_2(\mathbf{x})\to\boldsymbol{\eta}_2 \quad \text{as} \quad \mathbf{x}\to\boldsymbol{\xi}$$

through the set $S$. Then for any real numbers $a$ and $b$,

$$af_1(\mathbf{x})+bf_2(\mathbf{x})\to a\boldsymbol{\eta}_1+b\boldsymbol{\eta}_2 \quad \text{as} \quad \mathbf{x}\to\boldsymbol{\xi}$$

through the set $S$.

*Proof* Define $F: S \to \mathbb{R}^n$ by $f(\mathbf{x}) = af_1(\mathbf{x}) + bf_2(\mathbf{x})$. From theorem 22.28, we have that

$$(f_1(\mathbf{x}), f_2(\mathbf{x})) \to (\boldsymbol{\eta}_1, \boldsymbol{\eta}_2) \quad \text{as} \quad \mathbf{x} \to \boldsymbol{\xi}$$

through the set $S$. But, with the notation of theorem 16.9,

$$F = L \circ (f_1, f_2).$$

Since $L$ is continuous on $\mathbb{R}^n \times \mathbb{R}^n$ (theorem 16.9), it follows from theorem 22.26 that

$$f(\mathbf{x}) \to L(\boldsymbol{\eta}_1, \boldsymbol{\eta}_2) \quad \text{as} \quad \mathbf{x} \to \boldsymbol{\xi}$$

through the set $S$.

---

22.31 *Theorem* Suppose that $\mathcal{X}$ is a metric (or topological) space and that $f_1: S \to \mathbb{R}$ and $f_2: S \to \mathbb{R}$ where $S$ is a set in $\mathcal{X}$. Let $\boldsymbol{\xi} \in \mathcal{X}$ and suppose that

$$f_1(\mathbf{x}) \to \eta_1 \quad \text{as} \quad \mathbf{x} \to \boldsymbol{\xi}$$

and

$$f_2(\mathbf{x}) \to \eta_2 \quad \text{as} \quad \mathbf{x} \to \boldsymbol{\xi}$$

through the set $S$. Then

$$f_1(\mathbf{x}) f_2(\mathbf{x}) \to \eta_1 \eta_2 \quad \text{as} \quad \mathbf{x} \to \boldsymbol{\xi}$$

through the set $S$.

*Proof* Define $F: S \to \mathbb{R}$ by $F(\mathbf{x}) = f_1(\mathbf{x}) f_2(\mathbf{x})$. From theorem 22.28, we have that

$$(f_1(\mathbf{x}), f_2(\mathbf{x})) \to (\eta_1, \eta_2) \quad \text{as} \quad \mathbf{x} \to \boldsymbol{\xi}$$

through the set $S$. But, with the notation of theorem 16.9, $F = M \circ (f_1, f_2)$. Since $M$ is continuous on $\mathbb{R} \times \mathbb{R}$ (theorem 16.9), it follows from theorem 22.26 that $F(\mathbf{x}) \to \eta_1 \eta_2$ as $\mathbf{x} \to \boldsymbol{\xi}$ through the set $S$.

---

22.32 *Theorem* With the same hypotheses as in the previous theorem,

$$\frac{f_1(\mathbf{x})}{f_2(\mathbf{x})} \to \frac{\eta_1}{\eta_2} \quad \text{as} \quad \mathbf{x} \to \boldsymbol{\xi}$$

through the set $S$ provided that $\eta_2 \neq 0$.

*Proof* The proof is the same as that of theorem 22.31 except that, in the notation of theorem 16.9, $D$ replaces $M$.

(Note that, if $f(\mathbf{x}) \to \eta_2$ as $\mathbf{x} \to \boldsymbol{\xi}$ through the set $S$ and $\eta_2 \neq 0$, then there exists an open set $J$ containing $\boldsymbol{\xi}$ such that $f(\mathbf{x}) \neq 0$ for any $\mathbf{x} \in J \cap S$.)

---

22.33 *Exercise*

(1) Suppose that $g\colon \mathbb{R}^2 \to \mathbb{R}^1$ and $h\colon \mathbb{R}^2 \to \mathbb{R}^1$ have the property that $g(x, y) \to l$ as $(x, y) \to (0, 0)$ and $h(x, y) \to m$ as $(x, y) \to (0, 0)$. Find

$$\text{(i)} \quad \lim_{(x, y)\to(0, 0)} f_1(x, y) \quad \text{(ii)} \quad \lim_{(x, y)\to(0, 0)} f_2(x, y)$$

when $f_1\colon \mathbb{R}^2 \to \mathbb{R}^1$ and $f_2\colon \mathbb{R}^2 \to \mathbb{R}^1$ are defined by

$$f_1(x, y) = \{g(x, y)h(x, y)\}^{1/2}$$
$$f_2(x, y) = [1 + \{g(x, y)\}^2 + \{h(x, y)\}^2]^{-2}.$$

Also find

$$\lim_{(x, y)\to(0, 0)} f(x, y)$$

where $f\colon \mathbb{R}^2 \to \mathbb{R}^2$ is given by $f = (f_1, f_2)$.

(2) Let $f\colon \mathbb{R} \to \mathbb{R}$ and $g\colon \mathbb{R} \to \mathbb{R}$ be defined by

$$f(y) = \begin{cases} 1 & (y = 3) \\ 2 & (y \neq 3) \end{cases} \qquad g(x) = 3.$$

Prove that $f(y) \to 2$ as $y \to 3$ and $g(x) \to 3$ as $x \to 4$ but show that it is *false* that $f(g(x)) \to 2$ as $x \to 4$.

(3) Suppose that $S \subset \mathscr{X}$ and that $g\colon S \to \mathscr{Y}$ has the property that $g(\mathbf{x}) \to \boldsymbol{\eta}$ as $\mathbf{x} \to \boldsymbol{\xi}$ through the set $S$ while $f\colon \mathscr{Y} \to \mathscr{Z}$ has the property that $f(\mathbf{y}) \to \boldsymbol{\zeta}$ as $\mathbf{y} \to \boldsymbol{\eta}$. Prove that

$$f(g(\mathbf{x})) \to \boldsymbol{\zeta} \quad \text{as} \quad \mathbf{x} \to \boldsymbol{\xi}$$

through the set $S$ provided that either of the conditions

(i) $f$ is continuous at the point $\boldsymbol{\eta}$

or (ii) $\boldsymbol{\eta} \notin g(S)$

is satisfied.

---

## 22.34† Complex functions

Let $S$ be a set of complex numbers (§ 10.20) and consider a function $f\colon S \to \mathbb{C}$. If $\omega$ and $\zeta$ are complex numbers, we are interested in the statement

$$f(z) \to \omega \quad \text{as} \quad z \to \zeta \tag{1}$$

through the set $S$.

As we know from § 13.9, the complex number $z = x + iy$ can be identified with the point $(x, y)$ in $\mathbb{R}^2$ in which case $\|z\| = \{x^2 + y^2\}^{1/2} = \|(x, y)\|$. There is therefore no difficulty in interpreting (1) which simply means that:

$$\forall \varepsilon > 0\, \exists \delta > 0\, \forall z \in S,$$
$$0 < |z - \zeta| < \delta \Rightarrow |f(z) - \omega| < \varepsilon.$$

It is useful to note that theorems 22.30, 22.31 and 22.32 apply equally well if $\mathbb{R}$ is replaced throughout by $\mathbb{C}$.

---

22.35 *Example* Consider the function $f: \mathbb{C} \to \mathbb{C}$ defined by $f(z) = z^3$. We shall prove that

$$z^3 \to -i \quad \text{as} \quad z \to i.$$

Observe that $i^3 = i^2 i = -i$. Hence

$$|z^3 - (-i)| = |z^3 - i^3| = |z - i| \cdot |z^2 + zi + i^2|$$

(see exercise 10.24(5)). We have that $|i| = 1$ and, if $|z - i| < 1$, then $|z| \leqq |i| + 1 = 2$. Hence, if $|z - i| < 1$,

$$|z^3 - (-i)| < |z - i|(4 + 2 + 1) = 7|z - i|.$$

Now suppose that $\varepsilon > 0$ is given and choose $\delta = \min\{\varepsilon/7, 1\}$. Then, if $|z - i| < \delta$,

$$|z^3 - (-i)| < 7|z - i| < 7\delta \leqq \varepsilon$$

and the result follows.

This calculation, of course, can be avoided by observing that $f$ is a polynomial and hence is continuous at every point.

---

# 23† LIMITS AND CONTINUITY (II)

## 23.1† Double limits

We begin with some notation. Let $\mathscr{X}$, $\mathscr{Y}$ and $\mathscr{Z}$ be metric spaces and let $\xi\in\mathscr{X}$, $\eta\in\mathscr{Y}$ and $\zeta\in\mathscr{Z}$. Suppose that $A$ is a set in $\mathscr{X}$ and $B$ is a set in $\mathscr{Y}$ and write $S=A\times B$. We shall consider a function

$$f\colon S\backslash\{(\xi,\eta)\}\to\mathscr{Z}.$$

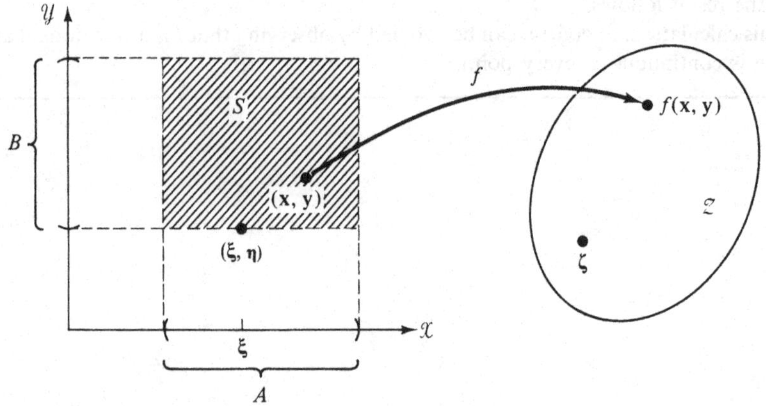


In mathematical applications one often comes across statements of the type '$f(\mathbf{x},\mathbf{y})\to\zeta$ as $\mathbf{x}\to\xi$ through $A$ and $\mathbf{y}\to\eta$ through $B$'. A limit specified in this manner will be called a *double limit*. It is important to be aware of the fact that statements of this sort about double limits can be highly ambiguous and that it is therefore always necessary to examine the context carefully in order to determine precisely what the author means.

The most straightforward case is that in which $\mathbf{x}$ and $\mathbf{y}$ are intended to be 'independent variables'. In this case the double limit statement simply means that

$$f(\mathbf{x},\mathbf{y})\to\zeta \quad \text{as} \quad (\mathbf{x},\mathbf{y})\to(\xi,\eta)$$

through the set $S=A\times B$. However, in order to use the definition of a limit given in §22.3 we need to know what sets are open in $\mathscr{X}\times\mathscr{Y}$. We therefore augment our definition of a double limit in this case by observing that the open sets of $\mathscr{X}\times\mathscr{Y}$ are to be those in the product topology of $\mathscr{X}\times\mathscr{Y}$ (§21.18).

As explained in §21.18, when $\mathscr{X}$ and $\mathscr{Y}$ are metric spaces with metrics $d_1$ and $d_2$ respectively, there are various metrics which we can introduce into $\mathscr{X}\times\mathscr{Y}$ which generate the product topology of $\mathscr{X}\times\mathscr{Y}$ and the choice of which of these metrics to use in $\mathscr{X}\times\mathscr{Y}$ is largely a matter of convenience. When $\mathscr{X}=\mathscr{Y}=\mathbb{R}$ and so $\mathscr{X}\times\mathscr{Y}=\mathbb{R}^2$, we usually use the Euclidean metric but, in many cases, it is easier to use the metric $m$: $\mathscr{X}\times\mathscr{Y}\to\mathbb{R}$ defined by

$$m((\mathbf{x}_1, \mathbf{y}_1), (\mathbf{x}_2, \mathbf{y}_2))=\max\{d_1(\mathbf{x}_1, \mathbf{x}_2), d_2(\mathbf{y}_1, \mathbf{y}_2)\}.$$

The use of this metric in the formulation of the definition of a limit given in §22.9 yields the following criterion for the existence of a double limit:

For any $\varepsilon>0$, there exists a $\delta>0$ such that for each $(\mathbf{x}, \mathbf{y})\in S=A\times B$,

$$0<m((\boldsymbol{\xi}, \boldsymbol{\eta}), (\mathbf{x}, \mathbf{y}))<\delta \Rightarrow d(\boldsymbol{\zeta}, f(\mathbf{x}, \mathbf{y}))<\varepsilon.$$

This can in turn be written in the following less clumsy form:

For any $\varepsilon>0$, there exists a $\delta>0$ such that for any $\mathbf{x}\in A$ and any $\mathbf{y}\in B$,

$$d_1(\boldsymbol{\xi}, \mathbf{x})<\delta \text{ and } d_2(\boldsymbol{\eta}, \mathbf{y})<\delta \Rightarrow d(\boldsymbol{\zeta}, f(\mathbf{x}, \mathbf{y}))<\varepsilon$$

provided that $(\mathbf{x}, \mathbf{y})\neq(\boldsymbol{\xi}, \boldsymbol{\eta})$.

---

23.2 *Example* Suppose that $f$: $\mathbb{R}^2\to\mathbb{R}^1$ and that $f(x, y)\to\zeta$ as $(x, y)\to(\xi, \eta)$. If we use the Euclidean metric in $\mathbb{R}^2$, the definition of this statement can be expressed in the form:

For any $\varepsilon>0$, there exists a $\delta>0$ such that

$$0<\{(x-\xi)^2+(y-\eta)^2\}^{1/2}<\delta \Rightarrow |f(x, y)-\zeta|<\varepsilon. \qquad (1)$$

Alternatively, we can use the metric $m$ in $\mathbb{R}^2$ and obtain the definition in the form:

For any $\varepsilon>0$, there exists a $\Delta>0$ such that

$$|x-\xi|<\Delta \text{ and } |y-\eta|<\Delta \Rightarrow |f(x, y)-\zeta|<\varepsilon \qquad (2)$$

provided $(x, y)\neq(\xi, \eta)$.

The first form of the definition says that $|f(x, y)-\zeta|<\varepsilon$ provided that $(x, y)\neq(\xi, \eta)$ is in a sufficiently small *disc* with centre $(\xi, \eta)$. The second says that $|f(x, y)-\zeta|<\varepsilon$ provided that $(x, y)\neq(\xi, \eta)$ is in a sufficiently small *box* with centre $(\xi, \eta)$. Both definitions are plainly equivalent because, if $\delta=\delta_1$ works in (1), then $\Delta=\delta_1/\sqrt{2}$ works in (2). On the other hand, if $\Delta$ works in (2), then $\delta=\Delta$ works in (1).

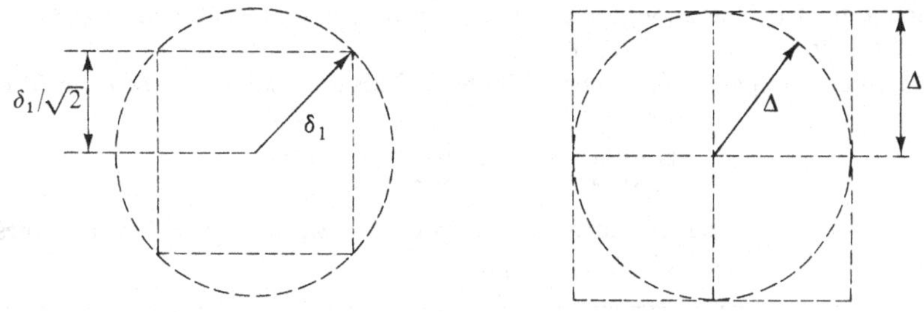


---

## 23.3† Double limits (continued)

We have not yet finished with the statement '$f(\mathbf{x}, \mathbf{y}) \to \zeta$ as $\mathbf{x} \to \xi$ through $A$ and $\mathbf{y} \to \eta$ through $B$'. In the discussion above we considered its meaning when $\mathbf{x}$ and $\mathbf{y}$ were to be understood as 'independent variables'. However, the statement is often used in circumstances when $\mathbf{x}$ and $\mathbf{y}$ are *not* 'independent variables' – i.e. some relation between $\mathbf{x}$ and $\mathbf{y}$ must be satisfied. Sometimes one is told explicitly what this relation is. On other occasions one has to guess the relation from the context.

The existence of such a relation means that there is a set $R$ in $\mathscr{X} \times \mathscr{Y}$ with the property that $(\mathbf{x}, \mathbf{y})$ not in $R$ are to be disregarded. Our double limit statement then means that

$$f(\mathbf{x}, \mathbf{y}) \to \xi \quad \text{as} \quad (\mathbf{x}, \mathbf{y}) \to (\xi, \eta)$$

through the set $R \cap S$ (where $S = A \times B$ and it is understood that the product topology is used in $\mathscr{X} \times \mathscr{Y}$).

---

23.4 *Example* Consider the function $f: \mathbb{R}^2 \to \mathbb{R}$ of exercise 22.8(2). If one is asked to consider the double limit $\zeta$ defined by the statement $f(x, y) \to \zeta$ as $x \to 1$ and $y \to 1$ where $x + y > 1$, then one should interpret this as meaning that

$$f(x, y) \to \zeta \quad \text{as} \quad (x, y) \to (1, 1)$$

through the set $J = \{(x, y) : x + y > 1\}$. Since $J \subset \mathscr{C}K$, it follows from exercise 22.8(2) that $\zeta = 0$.

---

## 23.5† Repeated limits

The notation

$$\lim_{\substack{\mathbf{x} \to \xi \\ \mathbf{y} \to \eta}} f(\mathbf{x}, \mathbf{y})$$

is available for a *double limit* as discussed in §23.1. We shall however prefer the less clumsy notation

$$\lim_{(\mathbf{x}, \mathbf{y}) \to (\xi, \eta)} f(\mathbf{x}, \mathbf{y})$$

although the latter notation is somewhat less precise in that it takes for granted the use of the product topology in $\mathscr{X} \times \mathscr{Y}$.

It is important *not* to confuse either of these pieces of notation with *repeated limits*

$$\lim_{\mathbf{x} \to \xi} \left( \lim_{\mathbf{y} \to \eta} f(\mathbf{x}, \mathbf{y}) \right); \quad \lim_{\mathbf{y} \to \eta} \left( \lim_{\mathbf{x} \to \xi} f(\mathbf{x}, \mathbf{y}) \right).$$

The reasons why one needs to be careful in dealing with repeated limits are best explained with the help of some examples.

---

23.6 *Example* Observe that

$$\lim_{y\to 0}\frac{x^2-y^2}{x^2+y^2}=1$$

and hence

$$\lim_{x\to 0}\left(\lim_{y\to 0}\frac{x^2-y^2}{x^2+y^2}\right)=1.$$

On the other hand, a similar argument shows that

$$\lim_{y\to 0}\left(\lim_{x\to 0}\frac{x^2-y^2}{x^2+y^2}\right)=-1.$$

The order in which the limits appear cannot therefore be reversed in general without changing the result.

---

23.7 *Example* Observe that, for each $x\in\mathbb{R}$,

$$\lim_{y\to 0}\frac{xy}{x^2+y^2}=0.$$

It follows that

$$\lim_{x\to 0}\left(\lim_{y\to 0}\frac{xy}{x^2+y^2}\right)=0.$$

Similarly,

$$\lim_{y\to 0}\left(\lim_{x\to 0}\frac{xy}{x^2+y^2}\right)=0.$$

However, the double limit

$$\lim_{(x,\,y)\to(0,\,0)}\frac{xy}{x^2+y^2}$$

does *not* exist (for the same reason that the similar limit in example 22.11 does not exist; see also example 23.15).

---

23.8 *Example* Consider the function $h\colon \mathbb{R}^2\to\mathbb{R}$ defined by

$$h(x,y)=\begin{cases} x & (y\geqq 0)\\ -x & (y<0.)\end{cases}$$

We have that

$$|h(x,y)|=|x|\leqq\|(x,y)\|$$

and so it is easily shown that

$$\lim_{(x, y)\to(0, 0)} h(x, y)=0.$$

On the other hand, we have that $h(x, y)\to x$ as $y\to 0+$ and $h(x, y)\to -x$ as $y\to 0-$. It follows that

$$\lim_{y\to 0} h(x, y) \tag{1}$$

does *not* exist unless $x=0$ and hence that the repeated limit

$$\lim_{x\to 0}\left(\lim_{y\to 0} h(x, y)\right)$$

does not exist.

---

The results of examples 23.6 and 23.7 are not particularly surprising. We have already seen a number of examples in which $f(x, y)$ tends to different limits as $(x, y)\to(0, 0)$ along different paths and, as the diagrams below indicate, one can think of taking a repeated limit as another special way of allowing $(x, y)$ to approach $(0, 0)$.

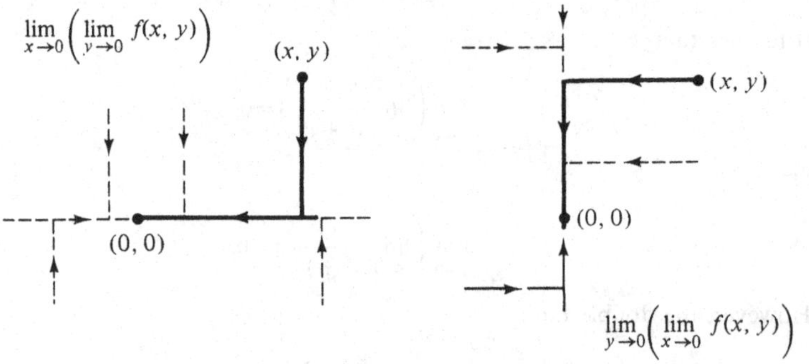


Example 23.8 requires a little more thought. The problem is that the existence of the double limit as $(x, y)\to(0, 0)$ does not guarantee that $h(x, y)$ tends to a limit as $(x, y)\to(X, 0)$ along the line $x=X$ (unless $X=0$). As the diagrams below indicate, there is no reason why matters should be otherwise.

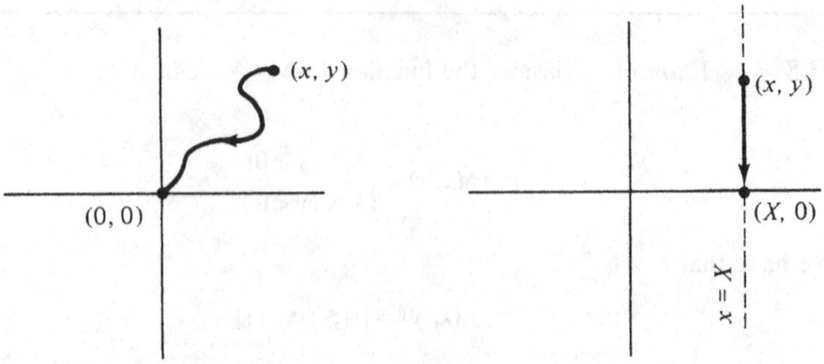

However, if we remove this pitfall by restricting our attention to those functions for which the limit (1) *does* exist, then the existence of the double limit implies the existence of the repeated limit and the two are equal. This result is the content of the next theorem.

---

23.9† *Theorem* Let $\mathcal{X}$, $\mathcal{Y}$ and $\mathcal{Z}$ be metric spaces and let $\boldsymbol{\xi}\in\mathcal{X}$, $\boldsymbol{\eta}\in\mathcal{Y}$ and $\boldsymbol{\zeta}\in\mathcal{Z}$. Suppose that $A$ is a set in $\mathcal{X}$ with cluster point $\boldsymbol{\xi}$ and that $B$ is a set in $\mathcal{Y}$. Let $f: S\backslash(\boldsymbol{\xi}, \boldsymbol{\eta})\to\mathcal{Z}$ where $S=A\times B$.

Suppose that

$$f(\mathbf{x}, \mathbf{y})\to\boldsymbol{\zeta} \quad \text{as} \quad (\mathbf{x}, \mathbf{y})\to(\boldsymbol{\xi}, \boldsymbol{\eta}) \tag{2}$$

through the set $S$ and that, for each $\mathbf{y}\in B$,

$$f(\mathbf{x}, \mathbf{y})\to l(\mathbf{y}) \quad \text{as} \quad \mathbf{x}\to\boldsymbol{\xi} \tag{3}$$

through the set $A$ where $l: B\to\mathcal{Z}$. Then

$$l(\mathbf{y})\to\boldsymbol{\zeta} \quad \text{as} \quad \mathbf{y}\to\boldsymbol{\eta} \tag{4}$$

through the set $B$.

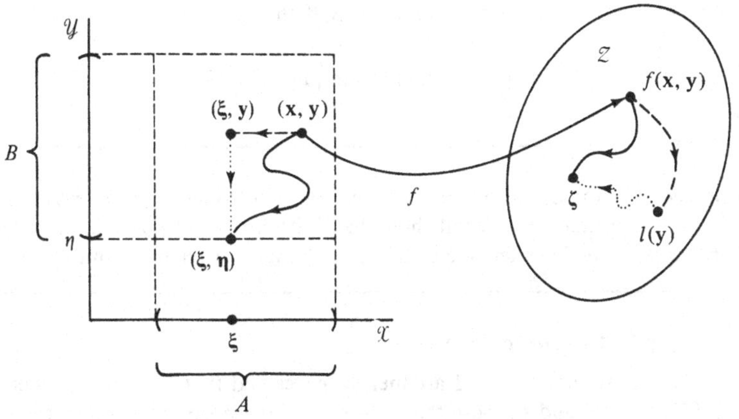


*Proof* Let $\varepsilon>0$ be given. By (2) there exists a $\delta>0$ such that for each $\mathbf{x}\in A$ and each $\mathbf{y}\in B$

$$d_1(\boldsymbol{\xi}, \mathbf{x})<\delta \text{ and } d_2(\boldsymbol{\eta}, \mathbf{y})<\delta \Rightarrow d(\boldsymbol{\zeta}, f(\mathbf{x}, \mathbf{y}))<\varepsilon/2$$

provided $(\mathbf{x}, \mathbf{y})\neq(\boldsymbol{\xi}, \boldsymbol{\eta})$. By (3), for each $\mathbf{y}\in B$ there exists a $\Delta_{\mathbf{y}}>0$ such that, for each $\mathbf{x}\in A$,

$$0<d_1(\boldsymbol{\xi}, \mathbf{x})<\Delta_{\mathbf{y}} \Rightarrow d(l(\mathbf{y}), f(\mathbf{x}, \mathbf{y}))<\varepsilon/2.$$

For each $\mathbf{y} \in B$, choose $\mathbf{x}_{\mathbf{y}} \in A$ so that $d_1(\xi, \mathbf{x}_{\mathbf{y}}) < \delta$ and $d_1(\xi, \mathbf{x}_{\mathbf{y}}) < \Delta_{\mathbf{y}}$. Then, if $0 < d_2(\eta, \mathbf{y}) < \delta$,

$$d(\zeta, l(\mathbf{y})) \leq d(\zeta, f(\mathbf{x}_{\mathbf{y}}, \mathbf{y})) + d(f(\mathbf{x}_{\mathbf{y}}, \mathbf{y}), l(\mathbf{y}))$$
$$< \varepsilon/2 + \varepsilon/2 = \varepsilon$$

and hence (4) holds.

---

It is sometimes important to know whether or not it is true that

$$\lim_{\mathbf{x}\to\xi}\left(\lim_{\mathbf{y}\to\eta} f(\mathbf{x}, \mathbf{y})\right) = \lim_{\mathbf{y}\to\eta}\left(\lim_{\mathbf{x}\to\xi} f(\mathbf{x}, \mathbf{y})\right). \tag{5}$$

If the equation holds, the two limiting operations are said to 'commute'. Applied mathematicians often write in a manner which would lead one to suppose that (5) is always true. But this is very definitely *not* the case even when both sides of the equation exist. Example 23.6 is a case in point.

However, equation (5) is true sufficiently often for it to be appropriate to show some measure of surprise should it turn out to be false in a particular case. In particular, we have the following result. The proof is trivial.

---

23.10† *Theorem* Let $\mathcal{X}$, $\mathcal{Y}$ and $\mathcal{Z}$ be metric spaces and let $S$ be an open set in $\mathcal{X} \times \mathcal{Y}$ containing $(\xi, \eta)$. If $f: S \to \mathcal{Z}$ is continuous on $S$, then

$$\lim_{\mathbf{x}\to\xi}\left(\lim_{\mathbf{y}\to\eta} f(\mathbf{x}, \mathbf{y})\right) = \lim_{\mathbf{y}\to\eta}\left(\lim_{\mathbf{x}\to\xi} f(\mathbf{x}, \mathbf{y})\right).$$

---

To proceed any further with this topic we need to introduce the subject of *uniform convergence*. In particular, we shall show that, if both sides of (5) exist, then the two sides are equal provided that one of the inner limits is a *uniform* limit.

---

## 23.11† Uniform convergence

Suppose that $\mathcal{X}$ and $\mathcal{Z}$ are metric spaces and that $f: A \times B \to \mathcal{Z}$ where $A$ is a set in $\mathcal{X}$. Let $\xi \in \mathcal{X}$ and suppose that $l: B \to \mathcal{Z}$ has the property that, for each $\mathbf{y} \in B$,

$$f(\mathbf{x}, \mathbf{y}) \to l(\mathbf{y}) \quad \text{as} \quad \mathbf{x} \to \xi$$

through the set $A$.

Expressed in full detail, this assertion means that

$$\forall \mathbf{y} \in B \ \forall \varepsilon > 0 \ \exists \delta > 0 \ \forall \mathbf{x} \in A,$$
$$0 < d(\xi, \mathbf{x}) < \delta \Rightarrow d(l(\mathbf{y}), f(\mathbf{x}, \mathbf{y})) < \varepsilon. \tag{1}$$

We know from chapter 3 that the order in which the quantifiers $\forall$ and $\exists$ occur in a statement is significant to its meaning. In particular, if $P(u, v)$ is a predicate, it is in general *false* that

$$\forall u\ \exists v\ P(u, v) \Leftrightarrow \exists v\ \forall u\ P(u, v).$$

It follows that, if we move the term '$\forall \mathbf{y} \in B$' from the beginning of our list of quantifiers to the end, we obtain a statement which means something *different* from the original statement – i.e. the statement

$$\begin{gathered}\forall \varepsilon>0\ \exists \delta>0\ \forall \mathbf{x} \in A\ \forall \mathbf{y} \in B, \\ 0<d(\xi, \mathbf{x})<\delta \Rightarrow d(l(\mathbf{y}), f(\mathbf{x}, \mathbf{y}))<\varepsilon\end{gathered} \tag{2}$$

does *not* mean the same as (1).

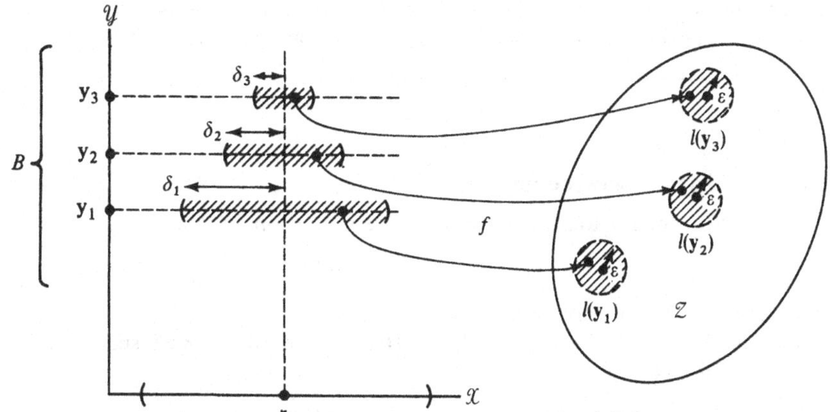


pointwise convergence

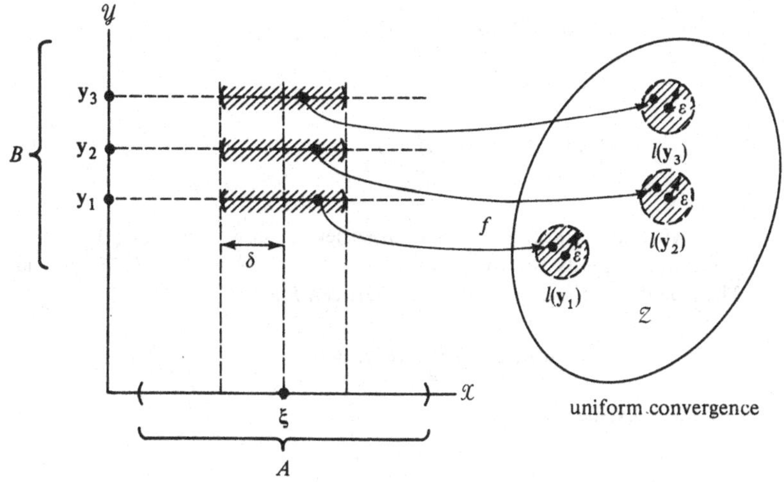


uniform convergence

The difference in the meaning of the two statements is that, in statement (1), the value of $\delta$ which is asserted to exist may depend both on the value of $\varepsilon$ *and* on the value of $\mathbf{y}$. In statement (2), the value of $\delta$ depends only on the value of $\varepsilon$ – i.e. the *same* value of $\delta$ works *uniformly* for all values of $\mathbf{y} \in B$.

If (2) holds, we say that

$$f(\mathbf{x}, \mathbf{y}) \to l(\mathbf{y}) \quad \text{as} \quad \mathbf{x} \to \xi$$

through the set $A$ *uniformly* for $\mathbf{y} \in B$.

It is sometimes necessary to emphasise that this statement is stronger than (1). When (1) holds, we therefore say

$$f(\mathbf{x}, \mathbf{y}) \to l(\mathbf{y}) \quad \text{as} \quad \mathbf{x} \to \xi$$

through the set $A$ *pointwise* for $\mathbf{y} \in B$. Note that it is obvious that uniform convergence implies pointwise convergence. But the converse is very definitely *false.*

---

## 23.12† Distance between functions

With the notation of the previous section, suppose that

$$f(\mathbf{x}, \mathbf{y}) \to l(\mathbf{y}) \quad \text{as} \quad \mathbf{x} \to \xi \tag{1}$$

through the set $A$ *uniformly* for $\mathbf{y} \in B$.

Suppose that $\varepsilon > 0$ is given. Then $\varepsilon/2 > 0$. Hence there exists a $\delta > 0$ such that for each $\mathbf{x} \in A$ and $\mathbf{y} \in B$

$$0 < d(\xi, \mathbf{x}) < \delta \Rightarrow d(l(\mathbf{y}), f(\mathbf{x}, \mathbf{y})) < \varepsilon/2.$$

Since the final inequality holds for *all* $\mathbf{y} \in B$ we may conclude that, for any $\varepsilon > 0$ there exists a $\delta > 0$ such that, for each $\mathbf{x} \in A$,

$$0 < d(\xi, \mathbf{x}) < \delta \Rightarrow \sup_{\mathbf{y} \in B} d(l(\mathbf{y}), f(\mathbf{x}, \mathbf{y})) < \varepsilon$$

i.e.

$$\sup_{\mathbf{y} \in B} d(l(\mathbf{y}), f(\mathbf{x}, \mathbf{y})) \to 0 \quad \text{as} \quad \mathbf{x} \to \xi \tag{2}$$

through the set $A$.

We have shown that (1)⇒(2) and it is even easier to show that (2)⇒(1).

This observation suggests the following definition. Suppose that $F: B \to \mathcal{Z}$ and $G: B \to \mathcal{Z}$. Then we define the *uniform distance* between $F$ and $G$ by

$$u(F, G) = \sup_{\mathbf{y} \in B} d(F(\mathbf{y}), G(\mathbf{y})) \tag{3}$$

where this quantity exists.

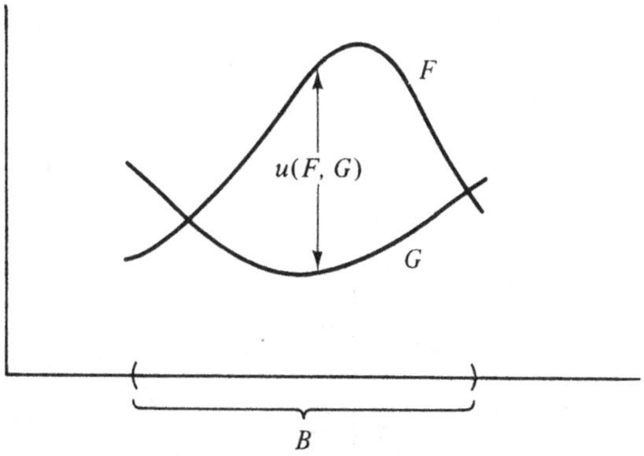


23.13 *Example* Consider the function $F:[0, 1]\to\mathbb{R}$ defined by $F(y)=8y^2+1$ and the function $G:[0, 1]\to\mathbb{R}$ defined by $G(y)=6y$. The uniform distance between $F$ and $G$ is given by

$$u(F, G)=\sup_{0\leqq y\leqq 1}|F(y)-G(y)|$$

$$=\max_{0\leqq y\leqq 1}|8y^2-6y+1|.$$

The maximum of $H(y)=|8y^2-6y+1|$ is either attained at $y=0$ or $y=1$ or else at a point $\eta\in(0, 1)$. In the latter case, differentiation yields that

$$16\eta-6=0$$

$$\eta=\tfrac{3}{8}.$$

Since $H(0)=1$, $H(1)=3$ and $H(\tfrac{3}{8})=\tfrac{1}{8}$, it follows that

$$u(F, G)=3.$$

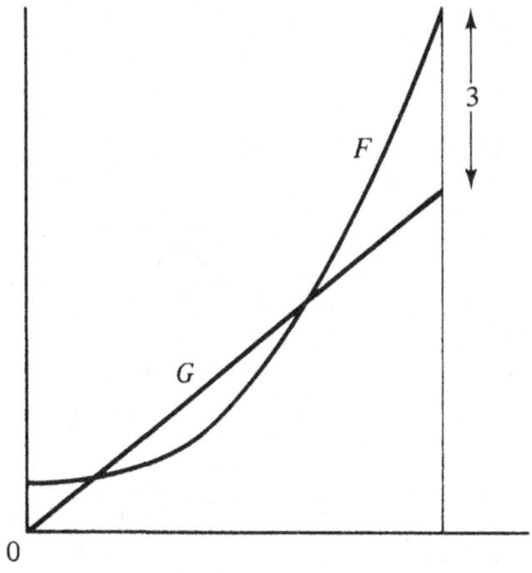

In terms of the uniform distance notation, statement (2) assumes the less clumsy form,

$$u(l, f(\mathbf{x},\bullet)) \to 0 \quad \text{as} \quad \mathbf{x} \to \boldsymbol{\xi} \tag{4}$$

through the set $A$, where $f(\mathbf{x},\bullet): B \to \mathcal{Z}$ is the function whose value at $\mathbf{y} \in B$ is $f(\mathbf{x}, \mathbf{y})$.

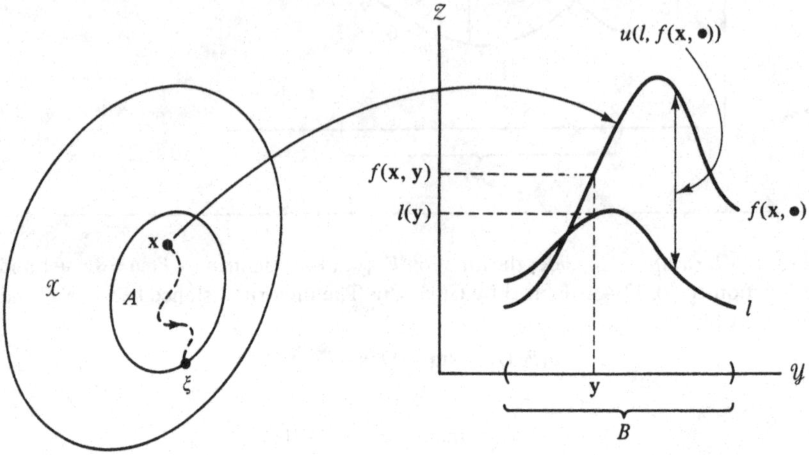


If $F$ and $G$ are *bounded* on $B$ then $u(F, G)$ defined by (3) always exists and the set of all bounded functions on $B$ is a *metric space* with $u$ as metric. (See §13.18.) When dealing with bounded functions we may therefore rewrite (4) in the simple form:

$$f(\mathbf{x},\bullet) \to l \quad \text{as} \quad \mathbf{x} \to \boldsymbol{\xi} \tag{5}$$

through the set $A$.

It is often helpful to 'forget' that $f(\mathbf{x},\bullet)$ and $l$ are functions and to think of them instead as 'points' in the metric space of bounded functions on $B$.

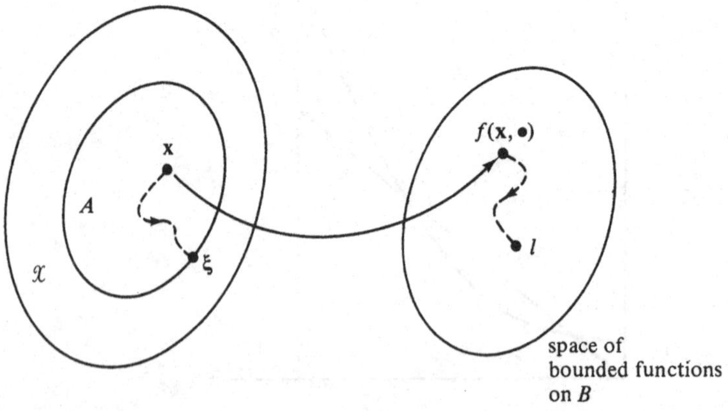

It is important, however, not to forget that the distance between the 'points' $f(\mathbf{x},\bullet)$ and $l$ is defined by

$$u(l, f(\mathbf{x},\bullet)) = \sup_{\mathbf{y}\in B} d(l(\mathbf{y}), f(\mathbf{x}, \mathbf{y})). \tag{6}$$

The above discussion is intended to indicate why uniform convergence is in some respects a more straightforward concept than pointwise convergence although it seems at first sight as though the opposite were the case. However, the discussion also has some practical benefits in that it is often easiest when seeking to establish uniform convergence to begin by calculating or estimating (6).

---

23.14 *Example* Let $A = \mathbb{R}\setminus\{0\}$ and define $g\colon A\times\mathbb{R}\to\mathbb{R}$ by

$$g(x, y) = \frac{x^3 + x^2y + y^3}{x^2 + y^2}.$$

We begin by observing that, for each $y\in\mathbb{R}$,

$$g(x, y)\to y \quad \text{as} \quad x\to 0$$

– i.e. $g(x, y)\to y$ as $x\to 0$ *pointwise* for $y\in\mathbb{R}$. It follows that, if $g(x, y)\to l(y)$ as $x\to 0$ *uniformly* for $y\in\mathbb{R}$, then it must be the case that $l(y) = y$.

Consider

$$u(l, g(x,\bullet)) = \sup_{y\in\mathbb{R}} |l(y) - g(x, y)|.$$

Taking $l(y) = y$, we obtain that, for each $x\in A$,

$$|l(y) - g(x, y)| = \left| y - \frac{x^3 + x^2y + y^3}{x^2 + y^2} \right| = \frac{|x|^3}{x^2 + y^2}.$$

Hence, for each $x\in A$,

$$u(l, g(x,\bullet)) = \sup_{y\in\mathbb{R}} \frac{|x|^3}{x^2 + y^2} = |x|.$$

Since $|x|\to 0$ as $x\to 0$, it follows that $g(x, y)\to y$ as $x\to 0$ *uniformly* for $y\in\mathbb{R}$.

The diagram over illustrates the set of $(x, y)$ for which it is true that $|y - g(x, y)| < \varepsilon$. Observe that the choice $\delta = \varepsilon$ ensures that $0 < |x| < \delta \Rightarrow |y - g(x, y)| < \varepsilon$ for *all* $y\in\mathbb{R}$.

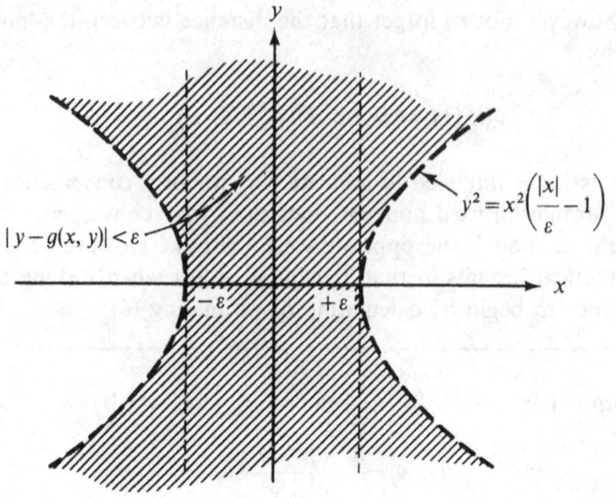


---

23.15 *Example* Let $A=\mathbb{R}\setminus\{0\}$ and define $h\colon A\times\mathbb{R}\to\mathbb{R}$ by

$$h(x,y)=\frac{xy}{x^2+y^2}.$$

We begin by observing that, for each $y\in\mathbb{R}$,

$$h(x,y)\to 0 \quad \text{as} \quad x\to 0$$

– i.e. $h(x,y)\to 0$ as $x\to 0$ *pointwise* for $y\in\mathbb{R}$. It follows that, if $h(x,y)\to l(y)$ as $x\to 0$ *uniformly* for $y\in\mathbb{R}$, then $l(y)=0$.

Observe that, for each $x\in A$,

$$u(0,h(x,\bullet))=\sup_{y\in\mathbb{R}}\left|0-\frac{xy}{x^2+y^2}\right|$$
$$=\tfrac{1}{2}.$$

(The geometric–arithmetic mean inequality gives $|xy|\leqq\frac{1}{2}(x^2+y^2)$ and equality is attained when $x=y$.) Since $u(0,h(x,\bullet))\not\to 0$ as $x\to 0$, it follows that it is not true that $h(x,y)\to 0$ as $x\to 0$ *uniformly* for $y\in\mathbb{R}$.

The diagram below illustrates the set of $(x,y)$ for which it is true that $|h(x,y)|<\varepsilon$ when $0<\varepsilon<\frac{1}{2}$. (See exercise 22.24(3).)

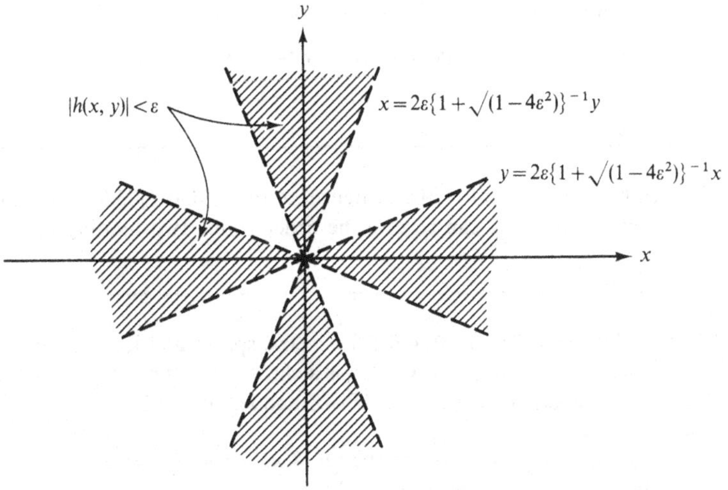


It is clear that, if $0 < \varepsilon < \frac{1}{2}$, then there is *no* value of $\delta > 0$ such that for all $x \in A$ and *all* $y \in \mathbb{R}$

$$0 < |x| < \delta \Rightarrow |h(x, y)| < \varepsilon.$$

If $y \neq 0$, the largest value of $\delta > 0$ for which it is true that $0 < |x| < \delta \Rightarrow h(x, y) < \varepsilon$ is

$$\delta = 2\varepsilon\{1 + \sqrt{(1 - 4\varepsilon^2)}\}^{-1} y.$$

---

23.16† *Theorem* Let $\mathscr{X}$, $\mathscr{Y}$ and $\mathscr{Z}$ be metric spaces and let $\boldsymbol{\xi} \in \mathscr{X}$, $\boldsymbol{\eta} \in \mathscr{Y}$ and $\boldsymbol{\zeta} \in \mathscr{Z}$. Suppose that $A$ is a set in $\mathscr{X}$ and that $B$ is a set in $\mathscr{Y}$. Let $f: S \backslash (\boldsymbol{\xi}, \boldsymbol{\eta}) \to \mathscr{Z}$ where $S = A \times B$ and let $l: B \to \mathscr{Z}$.

Suppose that

$$f(\mathbf{x}, \mathbf{y}) \to l(\mathbf{y}) \quad \text{as} \quad \mathbf{x} \to \boldsymbol{\xi} \tag{7}$$

through the set $A$ *uniformly* for $\mathbf{y} \in B$ and that

$$l(\mathbf{y}) \to \boldsymbol{\zeta} \quad \text{as} \quad \mathbf{y} \to \boldsymbol{\eta} \tag{8}$$

through the set $B$. Then

$$f(\mathbf{x}, \mathbf{y}) \to \boldsymbol{\zeta} \quad \text{as} \quad (\mathbf{x}, \mathbf{y}) \to (\boldsymbol{\xi}, \boldsymbol{\eta})$$

through the set $S$.

*Proof* Let $\varepsilon > 0$ be given. By (7) there exists a $\delta_1 > 0$ such that, for each $\mathbf{x} \in A$ and each $\mathbf{y} \in B$,

$$0 < d_1(\boldsymbol{\xi}, \mathbf{x}) < \delta_1 \Rightarrow d(l(\mathbf{y}), f(\mathbf{x}, \mathbf{y})) < \varepsilon/2.$$

By (8), there exists a $\delta_2 > 0$ such that for each $\mathbf{y} \in B$

$$0 < d_2(\boldsymbol{\eta}, \mathbf{y}) < \delta_2 \Rightarrow d(\boldsymbol{\zeta}, l(\mathbf{y})) < \varepsilon/2.$$

Choose $\delta=\min\{\delta_1, \delta_2\}$. Then, if $0<d_1(\boldsymbol{\xi}, \mathbf{x})<\delta$ and $0<d_2(\boldsymbol{\eta}, \mathbf{y})<\delta$, it follows that

$$\begin{aligned} d(\boldsymbol{\zeta}, f(\mathbf{x}, \mathbf{y})) &\leqq d(\boldsymbol{\zeta}, l(\mathbf{y}))+d(l(\mathbf{y}), f(\mathbf{x}, \mathbf{y})) \\ &<\varepsilon/2+\varepsilon/2=\varepsilon. \end{aligned}$$

---

The next theorem justifies the sentence at the end of §23.5 about the circumstances under which one can reverse the order of the two limiting operations in a repeated limit.

---

23.17† *Theorem* Let $\mathcal{X}$, $\mathcal{Y}$ and $\mathcal{Z}$ be metric spaces and let $\boldsymbol{\xi}\in\mathcal{X}$, $\boldsymbol{\eta}\in\mathcal{Y}$ and $\boldsymbol{\zeta}\in\mathcal{Z}$. Suppose that $A$ is a set in $\mathcal{X}$ for which $\boldsymbol{\xi}$ is a cluster point and that $B$ is a set in $\mathcal{Y}$. Let $f: S\setminus(\boldsymbol{\xi}, \boldsymbol{\eta})\to\mathcal{Z}$ *where* $S=A\times B$. Let $l: B\to\mathcal{Z}$ and let $m: A\to\mathcal{Z}$.

Suppose that

(i) $f(\mathbf{x}, \mathbf{y})\to l(\mathbf{y})$ as $\mathbf{x}\to\boldsymbol{\xi}$ through the set $A$ *uniformly* for $\mathbf{y}\in B$,
(ii) $l(\mathbf{y})\to\boldsymbol{\zeta}$ as $\mathbf{y}\to\boldsymbol{\eta}$ through the set $B$,
(iii) $f(\mathbf{x}, \mathbf{y})\to m(\mathbf{x})$ as $\mathbf{y}\to\boldsymbol{\eta}$ through the set $B$ *pointwise* for $\mathbf{x}\in A$.

Then $m(\mathbf{x})\to\boldsymbol{\zeta}$ as $\mathbf{x}\to\boldsymbol{\xi}$ through the set $A$.

*Proof* The theorem is simply a combination of theorems 23.9 and 23.16.

---

23.18 *Example* Let $g$ be defined as in example 23.14. We have that

(i) $g(x, y)\to y$ as $x\to 0$ *uniformly* for $y\in\mathbb{R}$.
(ii) $y\to 0$ as $y\to 0$.
(iii) $g(x, y)\to x$ as $y\to 0$ for $x\in\mathbb{R}\setminus\{0\}$.

From theorem 23.17 we may conclude that

$$\lim_{y\to 0}\lim_{x\to 0} g(x, y)=\lim_{x\to 0}\lim_{y\to 0} g(x, y).$$

This is easily verified directly.

---

23.19† *Exercise*

(1) Consider the function $f: \mathbb{R}^2\setminus\{(0, 0)\}\to\mathbb{R}$ defined by

$$f(x, y)=\frac{x^2-y^2}{x^2+y^2}.$$

Prove the following:

(i) $f(x, y)\to 1$ as $(x, y)\to(0, 0)$ along the line $y=0$.
(ii) $f(x, y)\to -1$ as $(x, y)\to(0, 0)$ along the line $x=0$.
(iii) $f(x, y)\to 0$ as $(x, y)\to(0, 0)$ along the line $x=y$.

Deduce that $f(x, y)$ does not tend to a limit when $x\to 0$ and $y\to 0$ independently.

(2) Let $f$ be as in question 1. If $0<\varepsilon<1$, prove that $|f(x, y)|<\varepsilon$ if and only if

$$\frac{1-\varepsilon}{1+\varepsilon}<\left(\frac{y}{x}\right)^2<\frac{1+\varepsilon}{1-\varepsilon}.$$

Prove that $f(x, y)\to 0$ provided that $x\to 0$ and $y\to 0$ in such a manner that

$$\frac{1-|x|}{1+|y|}<\left(\frac{y}{x}\right)^2<\frac{1+|y|}{1-|x|}.$$

(3) For each of the functions $f:(\mathbb{R}\setminus\{0\})\times\mathbb{R}\to\mathbb{R}$ given below, decide whether or not it is true that

$$\lim_{x\to 0}\left(\lim_{y\to 0} f(x, y)\right)=\lim_{y\to 0}\left(\lim_{x\to 0} f(x, y)\right).$$

(i) $f(x, y)=\dfrac{x^2-y^2}{1+x^2+y^2}$ (ii) $f(x, y)=\dfrac{x^2y^4}{x^2+y^4}$

(iii) $f(x, y)=\dfrac{x^3+y^4}{x^2+y^4}$ (iv) $f(x, y)=\dfrac{x^3+y^6}{x^2+y^4}$.

(4) In each case considered in question 3, find a function $l\colon \mathbb{R}\to\mathbb{R}$ such that

$$f(x, y)\to l(y) \quad\text{as}\quad x\to 0$$

*pointwise* for $y\in\mathbb{R}$.

(5) In each case considered in question 3, decide whether or not it is true that

$$f(x, y)\to l(y) \quad\text{as}\quad x\to 0$$

*uniformly* for (*a*) $y\in\mathbb{R}$ and (*b*) $y\in[-1, 1]$.

(6) A function $f\colon \mathbb{R}^2\to\mathbb{R}^1$ has the property that

$$(x_1\leqq x_2 \text{ and } y_1\leqq y_2)\Rightarrow f(x_1, y_1)\leqq f(x_2, y_2).$$

Prove that

$$\lim_{x\to\xi+}\left(\lim_{y\to\eta+} f(x, y)\right)=\lim_{y\to\eta+}\left(\lim_{x\to\xi+} f(x, y)\right).$$

---

## 23.20† Uniform continuity

Let $\mathcal{X}$ and $\mathcal{Y}$ be metric spaces and suppose that $f:S\to\mathcal{Y}$ where $S$ is a set in $\mathcal{X}$. To say that $f$ is continuous on $S$ is equivalent to the assertion that, for each $\boldsymbol{\xi}\in S$,

$$f(\mathbf{x})\to f(\boldsymbol{\xi}) \quad\text{as}\quad \mathbf{x}\to\boldsymbol{\xi}$$

through the set $S$. This means, in turn, that

$$\forall\boldsymbol{\xi}\in S\ \forall\varepsilon>0\ \exists\delta>0\ \forall\mathbf{x}\in S,$$
$$d(\boldsymbol{\xi}, \mathbf{x})<\delta \Rightarrow d(f(\boldsymbol{\xi}), f(\mathbf{x}))<\varepsilon.$$

In this statement the value of $\delta>0$ which is asserted to exist may depend both on the value of $\varepsilon>0$ *and* on the value of $\xi\in S$. If, given $\varepsilon>0$, a value of $\delta>0$ can be found which works for *all* $\xi\in S$ simultaneously, then we say that $f$ is *uniformly* continuous on the set $S$. Thus $f$ is *uniformly continuous* on the set $S$ if and only if

$$\forall\varepsilon>0\ \exists\delta>0\ \forall\mathbf{x}\in S\ \forall\xi\in S,$$
$$d(\xi,\mathbf{x})<\delta \Rightarrow d(f(\xi), f(\mathbf{x}))<\varepsilon.$$

---

23.21 *Example* The function $f$: $[0, \infty)\rightarrow[0, \infty)$ defined by

$$f(x)=\sqrt{x}$$

is uniformly continuous on $[0, \infty)$.

*Proof* It is evident from the diagram that $|\sqrt{x}-\sqrt{\xi}|$ is largest for $x\geqq 0$, $\xi\geqq 0$ and $|x-\xi|\leqq\delta$ when $\xi=0$ and $x=\delta$. (This is easily checked analytically by proving that $\sqrt{b}-\sqrt{a}\leqq(b-a)^{1/2}$ when $0\leqq a\leqq b$.)

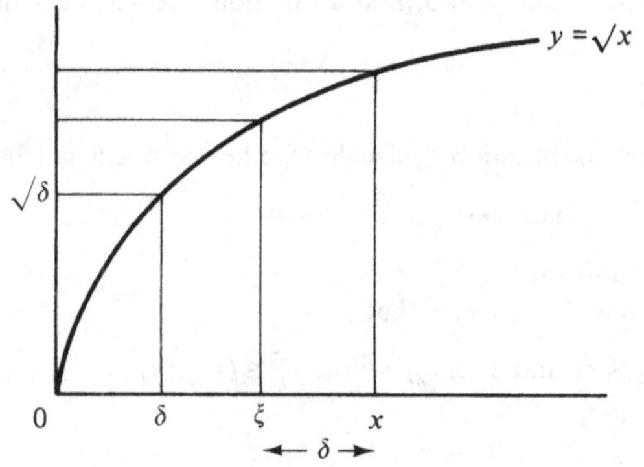


It follows that, given any $\varepsilon>0$, there exists a $\delta>0$ (namely $\delta=\varepsilon^2$) such that, for any $x\geqq 0$ and any $y\geqq 0$,

$$|\xi-x|<\delta\Rightarrow|\sqrt{\xi}-\sqrt{\delta}|=\varepsilon.$$

---

23.22 *Example* The function $f$: $(0, \infty)\rightarrow(0, \infty)$ defined by

$$f(x)=\frac{1}{x}$$

is *not* uniformly continuous on $(0, \infty)$.

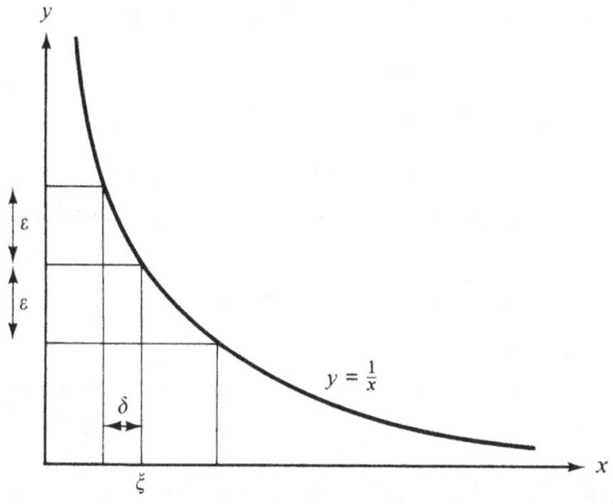


*Proof* Given $\varepsilon > 0$, the largest value of $\delta > 0$ for which

$$|x-\xi| < \delta \Rightarrow \frac{1}{x} - \frac{1}{\xi} < \varepsilon$$

satisfies

$$\frac{1}{\xi - \delta} - \frac{1}{\xi} = \varepsilon.$$

Hence

$$\delta = \frac{\varepsilon \xi^2}{1 + \varepsilon \xi}.$$

Since this expression tends to zero as $\xi \to 0+$ it follows that no $\delta > 0$ can work for *all* $\xi > 0$.

---

23.23† *Theorem* Let $\mathcal{X}$ and $\mathcal{Y}$ be metric spaces and suppose that $f$: $K \to \mathcal{Y}$, where $K$ is a *compact* set in $\mathcal{X}$. If $f$ is continuous on $K$, then $f$ is *uniformly* continuous on $K$.

*Proof* Since $f$ is continuous on $K$, $f(\mathbf{x}) \to f(\mathbf{y})$ as $\mathbf{x} \to \mathbf{y}$ through $K$ for each $\mathbf{y} \in K$. Let $\varepsilon > 0$ be given. Then $\frac{1}{2}\varepsilon > 0$ and so, for each $\mathbf{y} \in K$, there exists a $\delta(\mathbf{y}) > 0$ such that, for each $\mathbf{x} \in K$,

$$d(\mathbf{y}, \mathbf{x}) < \delta(\mathbf{y}) \Rightarrow d(f(\mathbf{y}), f(\mathbf{x})) < \tfrac{1}{2}\varepsilon. \tag{1}$$

Let $\mathcal{U}$ denote the collection of all open balls with centre $\mathbf{y} \in K$ and radius $\frac{1}{2}\delta(\mathbf{y})$. Then $\mathcal{U}$ covers $K$. From the definition of a compact set (§20.4) it follows that a finite subcollection $\mathcal{F}$ of $\mathcal{U}$ covers $K$. Let $\delta > 0$ be the radius of the open ball of minimum radius in the subcollection $\mathcal{F}$.

Now suppose that $\mathbf{x}$ and $\boldsymbol{\xi}$ are any points of $K$ which satisfy $d(\boldsymbol{\xi}, \mathbf{x}) < \delta$. Since $\mathcal{F}$ covers $K$, $\boldsymbol{\xi}$ belongs to an open ball in $\mathcal{F}$. Let the centre of this ball be $\mathbf{y}$. Its radius is then $\frac{1}{2}\delta(\mathbf{y})$.

We then have that $d(\mathbf{y}, \boldsymbol{\xi}) < \frac{1}{2}\delta(\mathbf{y}) < \delta(\mathbf{y})$ and so, by (1),

$$d(f(\mathbf{y}), f(\boldsymbol{\xi})) < \tfrac{1}{2}\varepsilon. \tag{2}$$

Also, $d(\mathbf{y}, \mathbf{x}) \leqq d(\mathbf{y}, \boldsymbol{\xi}) + d(\boldsymbol{\xi}, \mathbf{x}) < \frac{1}{2}\delta(\mathbf{y}) + \delta \leqq \delta(\mathbf{y})$. Thus, by (1) again,

$$d(f(\mathbf{y}), f(\mathbf{x})) < \tfrac{1}{2}\varepsilon. \tag{3}$$

Combining (2) and (3), we obtain that

$$d(f(\boldsymbol{\xi}), f(\mathbf{x})) < \varepsilon.$$

We have therefore shown that, for any $\varepsilon > 0$, there exists a $\delta > 0$ such that, for each $\mathbf{x} \in K$ and each $\boldsymbol{\xi} \in K$,

$$d(\boldsymbol{\xi}, \mathbf{x}) < \delta \Rightarrow d(f(\boldsymbol{\xi}), f(\mathbf{x})) < \varepsilon$$

and so the proof of the theorem is complete.

---

23.24† *Exercise*

(1) Determine which of the following functions $f$: $(0, 1) \to \mathbb{R}$ are uniformly continuous on the interval (0, 1).
(i) $f(x) = x^2$ (ii) $f(x) = (1-x)^{-2}$
(iii) $f(x) = |x - \frac{1}{2}|$ (iv) $f(x) = x^{-1/3}$.

(2) Let $\mathscr{X}$ and $\mathscr{Y}$ be metric spaces and suppose that $f: S \to \mathscr{Y}$ where $S$ is a set in $\mathscr{X}$. Let $g_1$: $[0, 1] \to S$ and $g_2$: $[0, 1] \to S$ and suppose that $d(g_1(t), g_2(t)) \to 0$ as $t \to 0+$. If $f$ is uniformly continuous on $S$, prove that

$$d(f(g_1(t)), f(g_2(t))) \to 0 \quad \text{as} \quad t \to 0+.$$

Interpret this result geometrically. Show that the result need *not* be true if $f$ is only continuous on $S$.

(3) Let $\{a_0, a_1, \ldots, a_n\}$ and $\{b_1, b_2, \ldots, b_n\}$ be two sets of real numbers and suppose that $0 = a_0 < a_1 < a_2 < \ldots < a_n = 1$. A function $g$: $[0, 1] \to \mathbb{R}$ which satisfies

$$a_{k-1} \leqq x < a_k \Rightarrow g(x) = b_k \quad (k = 1, 2, \ldots, n)$$

is said to be a *step function.*

Prove that any continuous function $f$:$[0, 1] \to \mathbb{R}$ can be uniformly approximated on [0, 1] by a step function – i.e. given any $\varepsilon > 0$, there exists a step function $g$: $[0, 1] \to \mathbb{R}$ such that

$$u(f, g) = \sup_{0 \leqq x \leqq 1} |f(x) - g(x)| < \varepsilon.$$

---

# 24 POINTS AT INFINITY

## 24.1 Introduction

The notion of 'infinity' is one which has received much attention from philosophers, theologians and assorted mystics of one persuasion or another throughout the ages. By and large, however, their writings have little useful to offer except the observation that 'infinity' can be used to 'prove' almost anything provided that one is willing to argue sufficiently badly.

We have already taken note of mathematical howlers of the type

$$\begin{aligned} \infty + \infty &= \infty \\ \therefore\ 2 \times \infty &= \infty \\ \therefore\ 2 &= 1. \end{aligned}$$

Here the fallacy is readily traceable to the fact that the symbol $\infty$ is treated as though it were a real number and, whatever $\infty$ may represent in a particular context, it certainly *never* represents a real number. Philosophical howlers tend to be slightly more subtle and to come attractively gift-wrapped in verbal packaging. But they share the same fatal flaw as the elementary mathematical howlers in that they assume that 'infinity' can be treated as though it were some more familiar object without offering any justification for this assumption.

As mathematicians, we know that there is no point in seeking to uncover the secrets of the universe by deductive reasoning alone. The best we can hope to do is to set up a consistent axiom system from which one can deduce the existence and the properties of objects which it is sensible to interpret as 'infinite'. These objects may or may not have a referent in the 'real world' but this is none of our concern. Such questions can be safely left to the philosophers.

As soon as one tackles the problem from this point of view, it becomes apparent that the word 'infinity' does not apply to a single concept. It applies to a whole bundle of rather diverse concepts and before one can say anything useful on the subject one first has to specify what *sort* of 'infinity' one has in mind.

Chapter 12 contains an account of one approach to the problem of infinity. This theory is the creation of the great German mathematician Cantor. As we have seen, he offered a precise definition of an infinite set and then proceeded to the observation that some infinite sets are 'larger' than others. This led to the introduction of the transfinite cardinals $\aleph_0, \aleph_1, \aleph_2, \ldots$ for the classification of infinite sets. The set $\{a, b, c\}$ has cardinality 3 (because it has three elements). The smallest type of infinite set (e.g. $\mathbb{N}$ or $\mathbb{Q}$) has cardinality $\aleph_0$ (aleph-zero). The next largest type of infinite set has cardinality $\aleph_1$ and so on. Note, incidentally, the richness of the structures discovered by Cantor. Contrary to popular opinion, the removal of romantic misconceptions from intuitive ideas seldom leads to something less exciting (unless the original idea is essentially empty).

One may think of Cantor's infinities as 'set-theoretic infinities'. They are useful for describing the number of elements in a set. But there are other essentially distinct types of infinity. As an example we mention the 'geometric infinities' used in translating statements from Euclidean geometry into statements of projective geometry. Here the Euclidean plane $\mathbb{R}^2$ is augmented by inventing an extra 'line' called the 'line at infinity' on which all parallel lines meet. A pair of parallel lines can then be treated just like any other pair of lines with a consequent gain in simplicity.

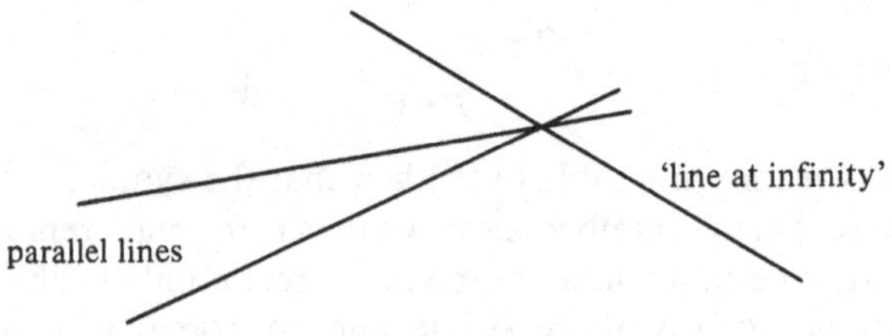


In this chapter we shall be concerned with what might be called 'topological infinities'. These are the infinities which appear in statements like

$$f(x) \to +\infty \quad \text{as} \quad x + -\infty.$$

The theory of such infinities is rather humdrum compared with the theories mentioned above. However, it is a theory which requires careful attention since the opportunities for confusion are considerable. Our treatment will be an elementary one but some readers may find it helpful to skip through or to review the topological ideas introduced in chapter 21.

## 24.2 One-point compactification of the reals

The real number system $\mathbb{R}$ has many satisfactory features but there is one respect in which it is not so satisfactory. As we have seen, compact sets have some very useful properties but $\mathbb{R}$ is *not* compact because it is not bounded (theorem 20.12). What can be done about this?

The natural mathematical response to such a problem is to ask whether it is possible to fit $\mathbb{R}$ inside some larger system $\mathcal{X}$ which *is* compact. It turns out that it is quite easy to find an appropriate $\mathcal{X}$. Indeed, one can find an $\mathcal{X}$ which has only one more point than $\mathbb{R}$. This is called the *one-point compactification* of $\mathbb{R}$. For most purposes, however, it is very much more useful to employ a *two-point compactification* of $\mathbb{R}$. (See §24.4.)

We begin by considering the one-point compactification. This consists of a compact space $\mathbb{R}^{\#}$ which contains $\mathbb{R}$ as a subspace but has only one point more than $\mathbb{R}$. For reasons which will soon be apparent we denote this extra point by $\infty$.

Recall from §21.4 that $\mathbb{R}$ is topologically equivalent to a circle $C$ in $\mathbb{R}^2$ with one point $N$ deleted. The stereographic projection $f$: $\mathbb{R} \to C \backslash N$ indicated below provides an appropriate homeomorphism.

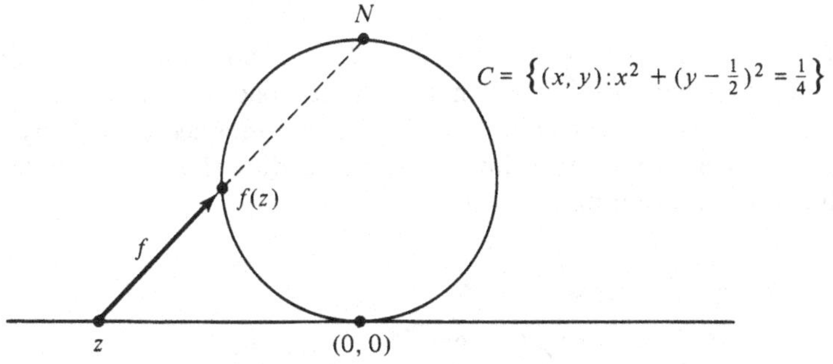


It is now perfectly natural to invent a new object to be mapped onto $N$. What we use for this new object is entirely irrelevant provided only that we do *not* use a real number. Whatever our choice for this new object, the symbol we use to denote it is $\infty$ and we write $\mathbb{R}^{\#} = \mathbb{R} \cup \{\infty\}$. The function $f$: $\mathbb{R} \to C \backslash \{N\}$ is extended to be a function $g: \mathbb{R}^{\#} \to C$ by defining

$$g(x) = \begin{cases} f(x) & (x \in \mathbb{R}) \\ N & (x = \infty) \end{cases}$$

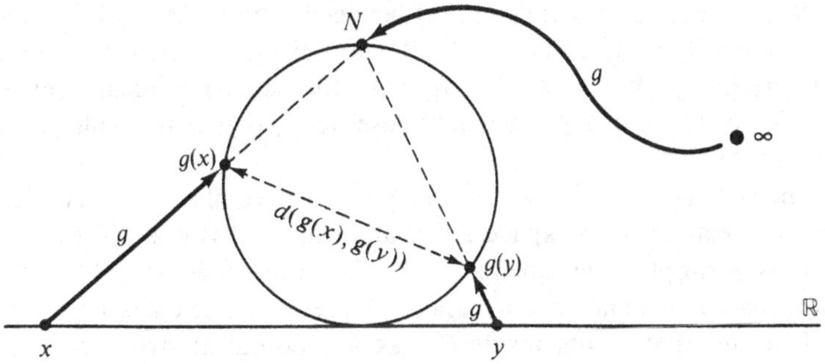

We want $\mathbb{R}^\#$ to be compact. But unless $\mathbb{R}^\#$ is a metric (or a topological) space this assertion will be meaningless. We make the obvious choice of a metric $c: \mathbb{R}^\# \to \mathbb{R}$ for the space $\mathbb{R}^\#$ by defining

$$c(x, y) = d(g(x), g(y)). \tag{1}$$

Thus the distance between $x$ and $y$ regarded as elements of $\mathbb{R}^\#$ is the Euclidean distance between the points $g(x)$ and $g(y)$ in $\mathbb{R}^2$. If $x$ and $y$ are real numbers,

$$c(x, y) = \frac{|x-y|}{(1+x^2)^{1/2}(1+y^2)^{1/2}}.$$

If $z$ is a real number,

$$c(z, \infty) = c(\infty, z) = \frac{1}{(1+z^2)^{1/2}}.$$

In view of (1), the function $g: \mathbb{R}^\# \to C$ is a homeomorphism. Since $C$ is compact it therefore follows that $\mathbb{R}^\#$ is compact (theorem 19.13).

It may be helpful to think of $\mathbb{R}^\#$ as being obtained from $\mathbb{R}$ by squeezing the real line into an open line segment, bending this into a circle and then plugging the resulting gap with $\infty$.

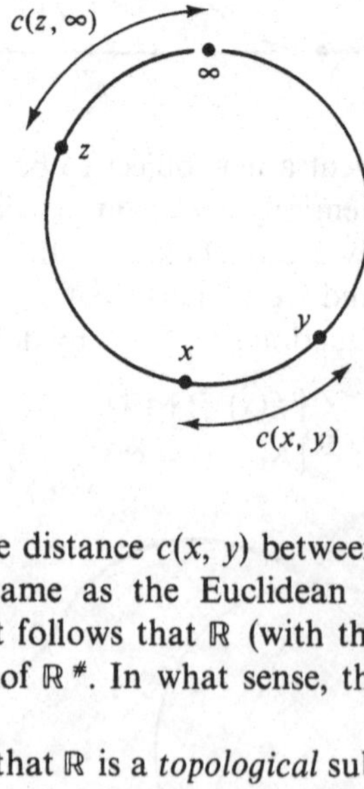

schematic diagram of $\mathbb{R}^\#$

Notice that the distance $c(x, y)$ between $x$ and $y$ regarded as elements of $\mathbb{R}^\#$ is *not* the same as the Euclidean distance between $x$ and $y$ – i.e. $c(x, y) \neq |x-y|$. It follows that $\mathbb{R}$ (with the usual Euclidean metric) is *not* a metric subspace of $\mathbb{R}^\#$. In what sense, then, is it reasonable to say that $\mathbb{R}$ 'fits inside' $\mathbb{R}^\#$?

The answer is that $\mathbb{R}$ is a *topological* subspace of $\mathbb{R}^\#$. In fact, a subset of $\mathbb{R}$ is open relative to the space $\mathbb{R}^\#$ if and only if it is open in the usual sense. This is a simple consequence of the fact that $f: \mathbb{R} \to C \setminus \{N\}$ is a homeomorphism and hence both $f$ and $f^{-1}$ preserve open sets (theorem 21.7).

The fact that $\mathbb{R}$ fits inside $\mathbb{R}^\#$ as a topological structure but not as a metric structure means that it is a mistake to attach much significance to the metric $c$: $\mathbb{R}^{\#} \to \mathbb{R}$. This is useful only as an auxiliary device. What matters is the topology of $\mathbb{R}^{\#}$ – i.e. the nature of its open sets. Since our convergence definitions have been framed in terms of open sets, this is, in any case, precisely what we need to know for applications.

The easiest way to describe the open sets of $\mathbb{R}^{\#}$ involves an abuse of the term 'open ball'. A set $G$ in a metric space $\mathcal{X}$ is open if and only if, for each $\xi \in G$, there exists an open ball $B$ with centre $\xi$ such that $B \subset G$ (theorem 14.12). This criterion remains valid for an open set $G$ in $\mathbb{R}^{\#}$ if we think of an 'open ball' in $\mathbb{R}^{\#}$ as an interval of the form $(\xi - \delta, \xi + \delta)$ when its centre is a real number $\xi$ and as a set of the form

$$\{\infty\} \cup \{x : x \in \mathbb{R} \text{ and } |x| > X\}$$

when its centre is $\infty$. (The abuse lies in using the Euclidean metric when the centre is a real number and the $\mathbb{R}^{\#}$ metric when the centre is $\infty$. When indulging in this abuse we shall write 'open ball' in inverted commas.)

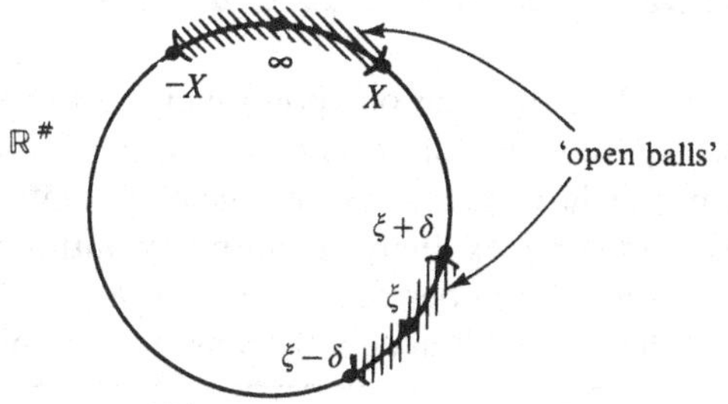


## 24.3 The Riemann sphere and the Gaussian plane

The discussion given above of the one-point compactification of $\mathbb{R}$ works equally well when employed with $\mathbb{R}^n$. The open sets $G$ in the space

$$\mathbb{R}^{n\#} = \mathbb{R}^n \cup \{\infty\}$$

are those with the property that, for each $\xi \in G$, there exists an 'open ball' $B$ with centre $\xi$ such that $B \subset G$. Here an 'open ball' is to be understood as a Euclidean ball when its centre $\xi \in \mathbb{R}^n$ – i.e.

$$B = \{\mathbf{x} : \|\mathbf{x} - \boldsymbol{\xi}\| < \delta\}$$

and as a set of the form

$$B = \{\mathbf{x} : \|\mathbf{x}\| > X\} \cup \{\infty\}$$

when its centre is $\infty$.

The case $n = 2$ is particularly important since $\mathbb{R}^2$ can be identified with the system $\mathbb{C}$ of complex numbers (§10.20). The one-point compactification of $\mathbb{R}^2$ is called the *Riemann sphere* or the *Gaussian plane* depending on whether one prefers to think of $\mathbb{R}^2$ as squeezed and wrapped into a sphere with $\infty$ plugging the gap or as a plane with $\infty$ floating ethereally in some nameless limbo. The link between these viewpoints is, as before, the stereographic projection. (See §21.4.)

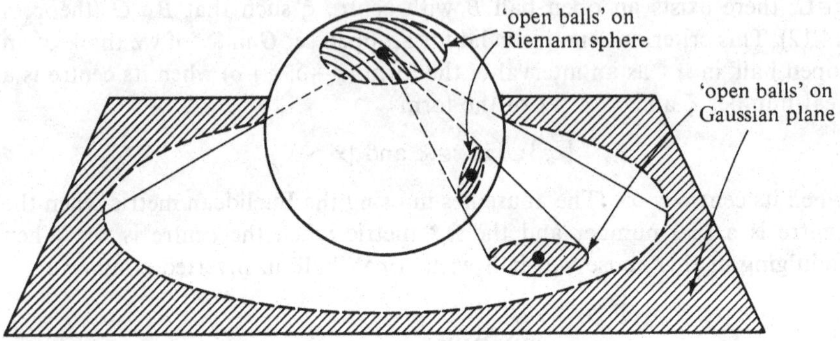


We mentioned in §24.2 that the one-point compactification of $\mathbb{R}$ is not the most useful possible compactification of $\mathbb{R}$. But this ceases to be true for $\mathbb{R}^n$ when $n \geqq 2$. In particular, one *always* uses the one-point compactification of $\mathbb{R}^2$ when working in complex analysis and many authors on this subject write as though $\infty$ were a member of $\mathbb{C}$ – i.e. they use $\mathbb{C}$ where we would write $\mathbb{C}^\#$. However, they are seldom entirely consistent in this practice.

## 24.4 Two-point compactification of the reals

Our discussion of the one-point compactification was based on the fact that the stereographic projection provides a topological equivalence between $\mathbb{R}$ and $C\setminus\{N\}$. A somewhat more natural topological equivalence, however, is that which holds between $\mathbb{R}$ and the open interval $(-1, 1)$. (See §21.3.) The diagram below illustrates a homeomorphism which can be used to demonstrate this topological equivalence.

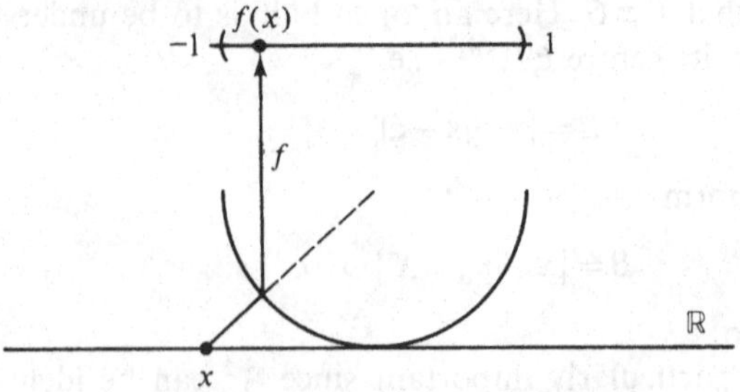

It is natural here to augment ℝ by inventing *two* further objects which we shall denote by $-\infty$ and $+\infty$ (rather than one extra object as in §24.2). We shall write

$$[-\infty, +\infty]=\mathbb{R}\cup\{-\infty, +\infty\}.$$

We then define an open set in $[-\infty, +\infty]$ to be a set $G$ with the property that, for each $\xi\in G$, there exists an 'open ball' $B$ with centre $\xi$ such that $B\subset G$. Here an 'open ball' is an interval of the form $(\xi-\delta, \xi+\delta)$ when its centre is a real number $\xi$ but is a set of the form

$$(X, +\infty)\cup\{+\infty\}$$

when its centre is $+\infty$ and a set of the form

$$(-\infty, Y)\cup\{-\infty\}$$

when its centre is $-\infty$.

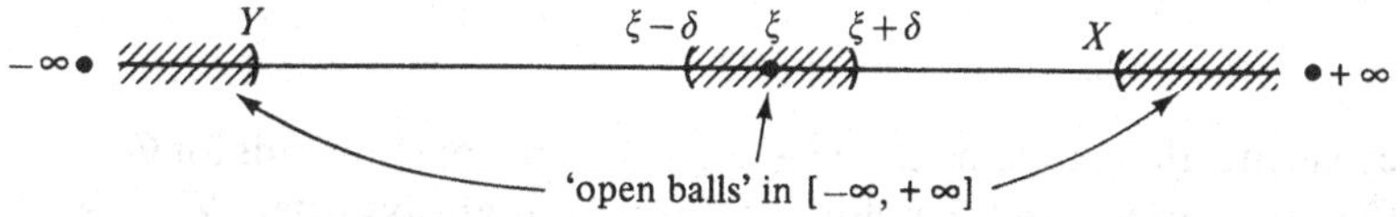


This definition for an open set in $[-\infty, +\infty]$ makes the function $g:[-\infty, +\infty]\to[-1, 1]$ illustrated below a homeomorphism. Thus $[-\infty, +\infty]$ is topologically equivalent to $[-1, 1]$. In particular, $[-\infty, +\infty]$ is compact.

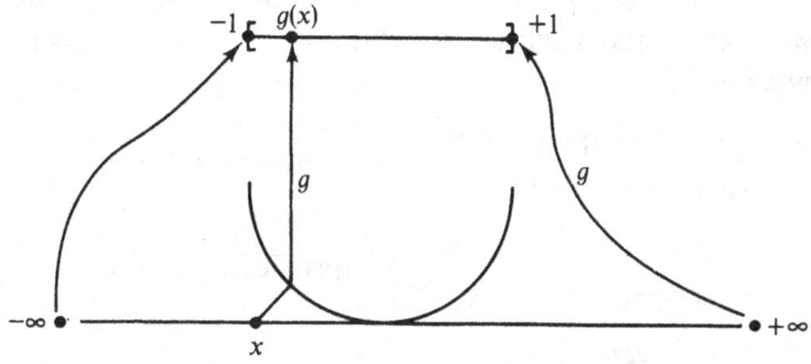


Moreover, ℝ is a topological subspace of $[-\infty, +\infty]$. In fact, a subset of ℝ is open relative to $[-\infty, +\infty]$ if and only if it is open in the usual sense.

When discussing the one-point compactification of ℝ, we made no attempt to extend anything but the topological structure of ℝ to $\mathbb{R}^{\#}$. In particular, there is no way in which the order structure of ℝ can sensibly be extended to $\mathbb{R}^{\#}$. But this is not true of the two-point compactification of ℝ and the order structure of $\mathbb{R}$ extends in a very natural way to $[-\infty, +\infty]$.

An ordering $\leqq$ on $[-\infty, +\infty]$ is obtained by requiring that, for all $x \in [-\infty, +\infty]$,

$$-\infty \leqq x \leqq +\infty$$

and that, for all real numbers $x$ and $y$, $x \leqq y$ if and only if this is true in the usual sense.

This usage has considerable notational advantages. For example, in $[-\infty, +\infty]$ *all* non-empty sets have suprema and infima. If $S$ is a non-empty set of real numbers which is bounded above in $\mathbb{R}$ (i.e. which has a real number as an upper bound), then sup $S$ has its usual meaning. If $S$ is unbounded above in $\mathbb{R}$, then its smallest upper bound in $[-\infty, +\infty]$ is $+\infty$ and hence

$$\sup S = +\infty.$$

Thus $\sup S = +\infty$ is a shorthand way of saying that $S$ is unbounded above in $\mathbb{R}$. Note also that

$$\sup \emptyset = -\infty.$$

The reason is that *all* elements of $[-\infty, +\infty]$ are upper bounds for $\emptyset$ (because $x \in \emptyset$ is always false and therefore $x \in \emptyset \Rightarrow x \leqq y$ is always true). With $+\infty$ and $-\infty$ available, one can therefore use the notation sup $S$ without first needing to check that $S$ is non-empty and bounded above. Similar remarks, of course, apply in the case of inf $S$.

There is also some advantage in extending the algebra of $\mathbb{R}$ to $[-\infty, +\infty]$. One cannot, of course, extend *all* of the algebraic structure of $\mathbb{R}$ to $[-\infty, +\infty]$ since then $[-\infty, +\infty]$ would be a continuum ordered field and we know from §9.21 that $\mathbb{R}$ is the only such object. However, one can extend *some* of the algebraic structure of $\mathbb{R}$ to $[-\infty, +\infty]$. We make the following definitions.

(i) $\left.\begin{array}{l} x+(+\infty)=(+\infty)+x=+\infty \\ x-(-\infty)=-(-\infty)+x=+\infty \end{array}\right\}$ (provided $x \neq -\infty$)

(ii) $\left.\begin{array}{l} x+(-\infty)=(-\infty)+x=-\infty \\ x-(+\infty)=-(+\infty)+x=-\infty \end{array}\right\}$ (provided $x \neq +\infty$)

(iii) $\left.\begin{array}{l} x(+\infty)=(+\infty)x=+\infty \\ x(-\infty)=(-\infty)x=-\infty \end{array}\right\}$ (provided $x > 0$)

(iv) $\left.\begin{array}{l} x/+\infty=0 \\ x/-\infty=0 \end{array}\right\}$ (provided $x \neq \pm\infty$).

Perhaps the most significant feature of these definitions are those items which are omitted. Observe that $(+\infty)+(-\infty)$, $(+\infty)-(+\infty)$, $0(+\infty)$,

$0(-\infty)$, $+\infty/+\infty$ and other expressions are *not* defined. There is no sensible way to attach a meaning to these collections of symbols and we do not attempt to do so. This fact means that there is little point in attempting algebraic manipulations involving $+\infty$ and $-\infty$. Definitions (i)–(iv) should therefore be regarded as handy conventions rather than as a basis for a serious mathematical theory. Note also that division by zero remains unacceptable.

As an example of the use to which these conventions may be put, consider two non-empty sets $S$ and $T$ of real numbers and two non-empty sets $P$ and $Q$ of positive real numbers. We then have that

(i) $\sup_{x\in S}(-x) = -\left(\inf_{x\in S} x\right)$

(ii) $\sup_{(x,y)\in S\times T}(x+y) = \left(\sup_{x\in S} x\right) + \left(\sup_{y\in T} y\right)$

(iii) $\sup_{(x,y)\in P\times Q}(xy) = \left(\sup_{x\in P} x\right)\left(\sup_{y\in Q} y\right)$

without any need for assumptions about the boundedness of the sets concerned provided that the suprema and infima are allowed the values $+\infty$ or $-\infty$ (See exercise 9.10(9).) Note, however, that some vigilance is necessary if one or more of the sets is empty. For example, if $S=\emptyset$ and $T$ is unbounded above, the right-hand side of (ii) becomes meaningless.

The system $[-\infty, +\infty]$ with the structure described in this section is called the *extended real number system*. It should be pointed out that some authors use $\mathbb{R}^{\#}$ (which we have used for the one-point compactification) to stand for $[-\infty, +\infty]$. Since other authors use $\mathbb{R}^{\#}$ for yet other purposes, some element of confusion is inevitable in any case.

## 24.5 Convergence and divergence

Let $S$ be a set of real numbers. Usually it would be convenient to regard $S$ as a set in the space $\mathbb{R}$ but, in what follows, we shall wish to regard $S$ as a set in the space $[-\infty, +\infty]$. Next consider a function $f: S\to[-\infty, +\infty]$ for which $f(S)\subset\mathbb{R}$. Again, one would normally regard $f$ as taking values in $\mathbb{R}$ but here it is convenient to regard $f$ as taking values in $[-\infty, +\infty]$.

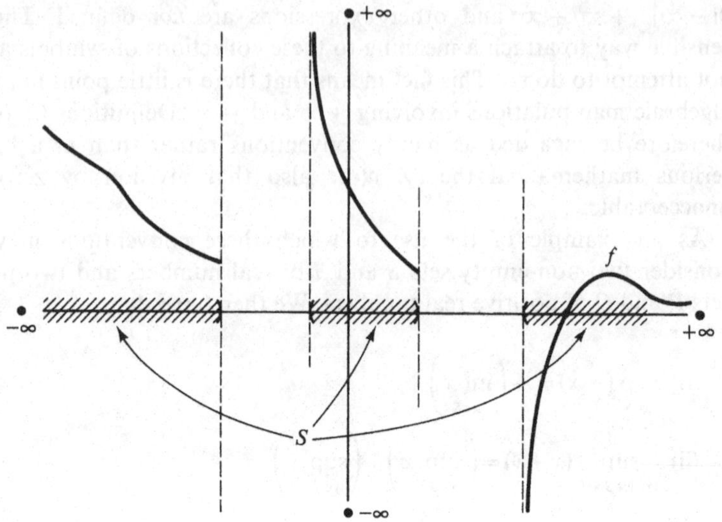


If $\xi \in [-\infty, +\infty]$ and $\eta \in [-\infty, +\infty]$, it makes sense to ask whether or not $f(x) \to \eta$ as $x \to \xi$ through the set $S$. The definition is the same as always. (See §22.3.) For any open set $G$ in $[-\infty, +\infty]$ containing $\eta$, there must exist an open set $H$ in $[-\infty, +\infty]$ containing $\xi$ such that

$$x \in H \cap S \Rightarrow f(x) \in G$$

provided that $x \neq \xi$.

As we found in §22.4, it is more helpful for technical purposes to rephrase this definition in terms of 'open balls' in $[-\infty, +\infty]$. We obtain that $f(x) \to \eta$ as $x \to \xi$ through $S$ if and only if:

For each 'open ball' $E$ in $[-\infty, +\infty]$ with centre $\eta$, there exists an 'open ball' $\Delta$ in $[-\infty, +\infty]$ with centre $\xi$ such that

$$x \in \Delta \cap S \Rightarrow f(x) \in E$$

provided $x \neq \xi$.

If $\xi$ and $\eta$ are real numbers, this definition is exactly the same as the familiar definition of §22.4. But if $\xi$ or $\eta$ are $+\infty$ or $-\infty$ then something new is obtained. If we insert the relevant definition for an 'open ball' given in §24.4 and bear in mind that $f$ takes only real values, we obtain the criteria listed in the table below. In this table $l$ and $a$ are to be understood as real numbers.

There is little point in committing these criteria to memory since it is easy to work out in any particular case what the form of the definition must be. There is some point, however, in acquiring a feeling for the intuitive meaning of the different statements.

Consider, for example, the statement '$f(x) \to l$ as $x \to +\infty$'. One can think of $f(x)$ as an approximation to $l$. Then $|f(x) - l|$ is the error involved in using

$f(x)$ as an approximation for $l$. The statement '$f(x) \to l$ as $x \to +\infty$' then tells us that this error can be made as small as we choose (i.e. less than $\varepsilon$) by making $x$ sufficiently large (i.e. greater than $X$). Similarly, '$f(x) \to -\infty$ as $x \to -\infty$' means that we can make $f(x)$ as negative as we choose (i.e. less than $Y$) by making $x$ sufficiently negative (i.e. less than $X$).

| | |
|---|---|
| $f(x) \to l$ as $x \to a$ through $S$ | $\forall \varepsilon > 0\ \exists \delta > 0\ \forall x \in S$, $0 < \lvert x-a \rvert < \delta \Rightarrow \lvert f(x) - l \rvert < \varepsilon$ |
| $f(x) \to l$ as $x \to +\infty$ through $S$ | $\forall \varepsilon > 0\ \exists X\ \forall x \in S$, $x > X \Rightarrow \lvert f(x) - l \rvert < \varepsilon$ |
| $f(x) \to l$ as $x \to -\infty$ through $S$ | $\forall \varepsilon > 0\ \exists Y\ \forall x \in S$, $x < Y \Rightarrow \lvert f(x) - l \rvert < \varepsilon$ |
| $f(x) \to +\infty$ as $x \to a$ through $S$ | $\forall X\ \exists \delta > 0\ \forall x \in S$, $0 < \lvert x-a \rvert < \delta \Rightarrow f(x) > X$ |
| $f(x) \to -\infty$ as $x \to a$ through $S$ | $\forall Y\ \exists \delta > 0\ \forall x \in S$, $0 < \lvert x-a \rvert < \delta \Rightarrow f(x) < Y$ |
| $f(x) \to +\infty$ as $x \to +\infty$ through $S$ | $\forall X\ \exists Y\ \forall x \in S$, $x > Y \Rightarrow f(x) > X$ |
| $f(x) \to +\infty$ as $x \to -\infty$ through $S$ | $\forall X\ \exists Y\ \forall x \in S$, $x < Y \Rightarrow f(x) > X$ |
| $f(x) \to -\infty$ as $x \to +\infty$ through $S$ | $\forall Y\ \exists X\ \forall x \in S$, $x > X \Rightarrow f(x) < Y$ |
| $f(x) \to -\infty$ as $x \to -\infty$ through $S$ | $\forall Y\ \exists X\ \forall x \in S$, $x < X \Rightarrow f(x) < Y$ |

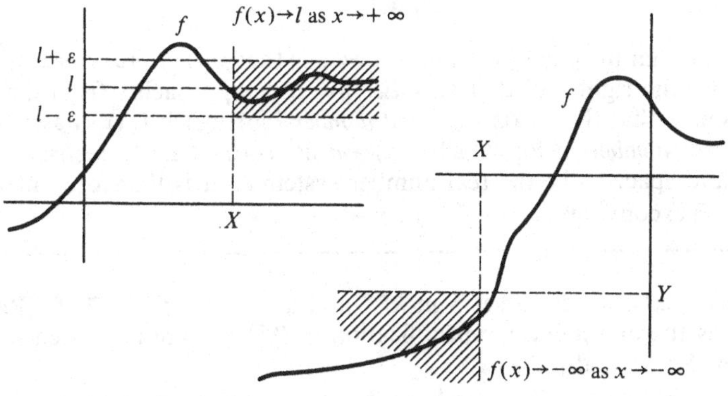


24.6 *Example* Consider the function $f: \mathbb{R} \setminus \{0\} \to \mathbb{R}$ defined by $f(x) = 1/x$. We have that

(i) $1/x \to 0$ as $x \to +\infty$

(ii) $1/x \to +\infty$ as $x \to 0+$

(iii) $1/x \to -\infty$ as $x \to 0-$
(iv) $1/x \to 0$ as $x \to -\infty$.

To prove (i) we have to show that, for each $\varepsilon > 0$, there exists an $X$ such that

$$x > X \Rightarrow \left|\frac{1}{x}\right| < \varepsilon.$$

Clearly the choice $X = 1/\varepsilon$ suffices. To prove (iii), we have to show that, for each $Y$, there exists a $\delta > 0$ such that

$$-\delta < x < 0 \Rightarrow \frac{1}{x} < Y.$$

Any value of $\delta > 0$ is adequate for this purpose when $Y \geqq 0$. When $Y < 0$, the choice $\delta = -1/Y$ suffices.

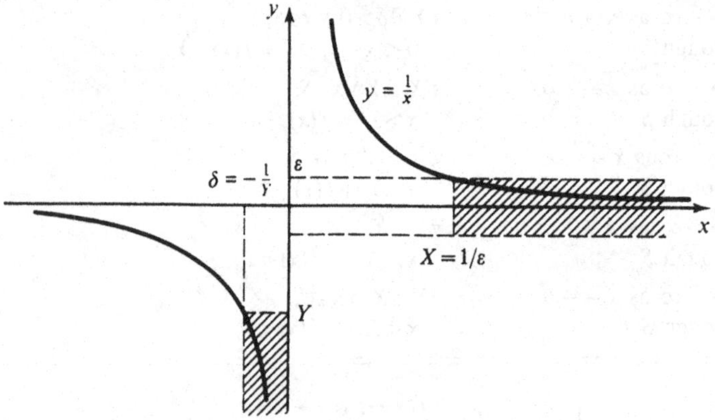


Note that in spite of (ii) and (iii) we say that $f(x)$ *diverges* as $x$ approaches 0 from the right and that $f(x)$ *diverges* as $x$ approaches from the left. The reason is that there exists no *real number* $l$ for which $f(x) \to l$ as $x \to 0+$ and no *real number* $m$ for which $f(x) \to m$ as $x \to 0-$. If we restrict our background space to be the real number system $\mathbb{R}$, it is therefore untrue to say that $f(x)$ converges as $x \to 0+$ or $x \to 0-$.

---

We next consider a set $S$ in $\mathbb{R}^{n\#}$ and a function $f: S \to \mathbb{R}^{m\#}$. (Recall that $\mathbb{R}^{n\#}$ is the one-point compactification of $\mathbb{R}^n$.) We are interested in the case when $S \subset \mathbb{R}^n$ and $f(S) \subset \mathbb{R}^m$.

If $\boldsymbol{\xi} \in \mathbb{R}^{n\#}$ and $\boldsymbol{\eta} \in \mathbb{R}^{m\#}$, we have that $f(\mathbf{x}) \to \boldsymbol{\eta}$ as $\mathbf{x} \to \boldsymbol{\xi}$ through $S$ if and only if:

For each 'open ball' $E$ in $\mathbb{R}^{m\#}$ with centre $\boldsymbol{\eta}$, there exists an 'open ball' $\Delta$ in $\mathbb{R}^{n\#}$ with centre $\boldsymbol{\xi}$ such that

$$\mathbf{x} \in \Delta \cap \mathbf{S} \Rightarrow f(\mathbf{x}) \in E$$

provided $\mathbf{x} \neq \boldsymbol{\xi}$.

If we insert the relevant definition for an 'open ball' given in §24.2 and bear in mind that $f$ takes values only in $\mathbb{R}^m$, we obtain the criteria listed in the table below. In this table, $\mathbf{l}$ and $\mathbf{a}$ are to be understood as vectors in $\mathbb{R}^m$ and $\mathbb{R}^n$ respectively.

| | |
|---|---|
| $f(\mathbf{x})\to\mathbf{l}$ as $\mathbf{x}\to\mathbf{a}$ through $S$ | $\forall\varepsilon>0\ \exists\delta>0\ \forall\mathbf{x}\in S$ $0<\\|\mathbf{x}-\mathbf{a}\\|<\delta \Rightarrow \\|f(\mathbf{x})-\mathbf{l}\\|<\varepsilon$ |
| $f(\mathbf{x})\to\mathbf{l}$ as $\mathbf{x}\to\infty$ through $S$ | $\forall\varepsilon>0\ \exists X\ \forall\mathbf{x}\in S$ $\\|\mathbf{x}\\|>X \Rightarrow \\|f(\mathbf{x})-\mathbf{l}\\|<\varepsilon$ |
| $f(\mathbf{x})\to\infty$ as $\mathbf{x}\to\mathbf{a}$ through $S$ | $\forall X\ \exists\delta>0\ \forall\mathbf{x}\in S$ $0<\\|\mathbf{x}-\mathbf{a}\\|<\delta \Rightarrow \\|f(\mathbf{x})\\|>X$ |
| $f(\mathbf{x})\to\infty$ as $\mathbf{x}\to\infty$ through $S$ | $\forall X\ \exists Y\ \forall\mathbf{x}\in S$ $\\|\mathbf{x}\\|>Y \Rightarrow \\|f(\mathbf{x})\\|>X$ |

24.7 *Example* Consider the function $f\colon \mathbb{R}^2\setminus\{(0, 0)\}\to\mathbb{R}^2$ defined by

$$f(x, y)=\left(\frac{x}{x^2+y^2}, \frac{-y}{x^2+y^2}\right).$$

We have that

(i) $$\left(\frac{x}{x^2+y^2}\ \frac{-y}{x^2+y^2}\right)\to(0, 0) \quad \text{as} \quad (x, y)\to\infty$$

(ii) $$\left(\frac{x}{x^2+y^2}\ \frac{-y}{x^2+y^2}\right)\to\infty \quad \text{as} \quad (x, y)\to(0, 0).$$

*Proof* To prove (i), we have to show that, for any $\varepsilon>0$, there exists an $X$ such that

$$\|(x, y)\|>X \Rightarrow \|f(x, y)-0\|<\varepsilon$$

– i.e. $$(x^2+y^2)>X^2 \Rightarrow \left(\frac{x}{x^2+y^2}\right)^2+\left(\frac{-y}{x^2+y^2}\right)^2<\varepsilon^2$$

– i.e. $$(x^2+y^2)>X^2 \Rightarrow \frac{1}{x^2+y^2}<\varepsilon^2.$$

The choice $X=1/\varepsilon$ clearly suffices. To prove (ii) we have to show that, for any $X$, there exists a $\delta>0$ such that

$$0<\|(x, y)-(0, 0)\|<\delta \Rightarrow \|f(x, y)\|>X.$$

If $X\leqq 0$ any value of $\delta>0$ is adequate. Otherwise we require a $\delta>0$ such that

$$0<x^2+y^2<\delta^2 \Rightarrow \frac{1}{x^2+y^2}>X^2$$

and the choice $\delta = 1/X$ clearly suffices.

The diagram illustrates (i). The function value $f(x, y)$ lies in the disc with centre (0, 0) and radius $\varepsilon > 0$ provided $(x, y)$ lies *outside* the disc with centre (0, 0) and radius $X$.

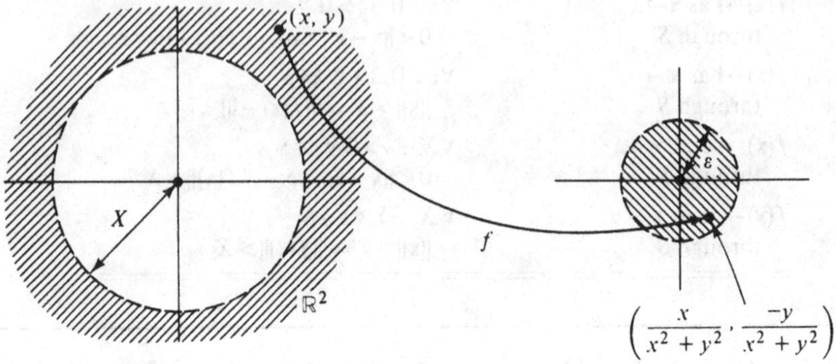


This example, incidentally, is easier if complex notation is used. Since

$$\frac{1}{x+iy} = \frac{1}{(x+iy)}\frac{(x-iy)}{(x-iy)} = \frac{x-iy}{x^2+y^2},$$

we can identify the given function with the function $f\colon \mathbb{C}\setminus\{0\}\to\mathbb{C}$ defined by $f(z) = 1/z$. To prove (i) we then have to show that, for any $\varepsilon > 0$, there exists an $X$ such that

$$|z| > X \Rightarrow \frac{1}{|z|} < \varepsilon.$$

---

When working with $\mathbb{R}^1$ there is sometimes room for confusion over whether the one-point or two-point compactification is in use especially since it is not uncommon for authors to write $\infty$ where we have been writing $+\infty$. Indeed, in earlier chapters, we have been using the notation $(a, \infty) = \{x : x > 0\}$ rather than $(a, +\infty)$ since the latter notation seems painfully pedantic.

In those exceptional cases when one wishes to make use of the one-point compactification of $\mathbb{R}$ it is therefore as well to employ notation of the type

$$f(x) \to l \quad \text{as} \quad |x| \to \infty$$

or

$$|f(x)| \to \infty \quad \text{as} \quad x \to \xi$$

in order to avoid any possible confusion. Similar notation is also sometimes useful when working in $\mathbb{R}^n$ in order to emphasise the use of the one-point compactification.

---

## 24.8 Combination theorems

In the previous section we have explained how some problems concerning the divergence of functions $f: \mathbb{R}^n \to \mathbb{R}^m$ can be replaced by convergence problems by replacing $\mathbb{R}$ by $[-\infty, +\infty]$ and $\mathbb{R}^n$ by $\mathbb{R}^{n\#}$. A natural next question is to ask to what extent the combination theorems of §22.25 carry over to the new situation.

There is no problem at all when $[-\infty, +\infty]$ or $\mathbb{R}^{n\#}$ replaces the metric space $\mathcal{X}$ in these theorems. Thus, for example, theorem 22.30 shows that, if $f_1(x) \to l_1$ as $x \to -\infty$ and $f_2(x) \to l_2$ as $x \to -\infty$, then

$$af_1(x) + bf_2(x) \to al_1 + bl_2 \quad \text{as} \quad x \to -\infty.$$

However, there are a number of cases not covered by the theorems of §22.25 The most important of these cases is the subject of the next result.

---

24.9 *Proposition* Let $\mathcal{X}$ be a metric (or a topological) space and let $f_1: S \to [-\infty, +\infty]$ and $f_2: S \to [-\infty, +\infty]$, where $S$ is a set in $\mathcal{X}$. Let $\xi \in \mathcal{X}$ and suppose that

$$f_1(\mathbf{x}) \to \eta_1 \quad \text{as} \quad \mathbf{x} \to \xi$$

and
$$f_2(\mathbf{x}) \to \eta_2 \quad \text{as} \quad \mathbf{x} \to \xi$$

through the set $S$. Then, if $a$ and $b$ are real numbers

(i) $af_1(\mathbf{x}) + bf_2(\mathbf{x}) \to a\eta_1 + b\eta_2$
(ii) $f_1(\mathbf{x})f_2(\mathbf{x}) \to \eta_1\eta_2$
(iii) $f_1(\mathbf{x})/f_2(\mathbf{x}) \to \eta_1/\eta_2$

as $\mathbf{x} \to \xi$ through the set $S$, in each case for which the right-hand side is a meaningful expression in the sense of §24.4.

---

It is sometimes useful in addition to note that, if $0 < \eta_1 \leqq +\infty, f_2(\mathbf{x}) > 0$ for $\mathbf{x} \in S$ and $\eta_2 = 0$, then

$$f_1(\mathbf{x})/f_2(\mathbf{x}) \to +\infty \quad \text{as} \quad \mathbf{x} \to \xi$$

through $S$. Similarly, if $-\infty \leqq \eta_1 < 0, f_2(\mathbf{x}) > 0$ for $\mathbf{x} \in S$ and $\eta_2 = 0$, then

$$f_2(\mathbf{x})/f_2(\mathbf{x}) \to -\infty \quad \text{as} \quad \mathbf{x} \to \xi$$

through $S$.

It is dangerous to assume the truth of results other than those explicitly mentioned above. For example, it is tempting to assume that, if $f_1(x) \to +\infty$ as $x \to 0+$ and $f_2(x) \to +\infty$ as $x \to 0+$, then $f_1(x)/f_2(x) \to 1$ as $x \to 0+$. However, as the examples below indicate, this result is false in general.

(i) $f_1(x) = x^2, \quad f_2(x) = x$

(ii) $f_1(x)=2x, \quad f_2(x)=x$
(iii) $f_1(x)=x, \quad f_2(x)=x^2$.

---

24.10 *Exercise*

(1) Prove the following results:

(i) $\sup_{x>0}(2x+3)=+\infty$ (ii) $\sup_{x^2<0}(2x+3)=-\infty$

(iii) $\inf_{x>0,\,y<0}(xy)=-\infty$ (iv) $\sup_{x>0,\,y>0}(y-x)=+\infty$.

(2) Let $f:\mathbb{R}\setminus\{1\}\to\mathbb{R}$ be defined by

$$f(x)=\frac{x+1}{x-1}.$$

Prove the following results:

(i) $f(x)\to 1$ as $x\to+\infty$
(ii) $f(x)\to 1$ as $x\to-\infty$
(iii) $f(x)\to 1$ as $|x|\to\infty$
(iv) $f(x)\to+\infty$ as $x\to 1+$
(v) $f(x)\to-\infty$ as $x\to 1-$
(vi) $|f(x)|\to\infty$ as $x\to 1$
(vii) $xf(x)\to+\infty$ as $x\to+\infty$
(viii) $xf(x)\to-\infty$ as $x\to-\infty$
(ix) $|xf(x)|\to\infty$ as $|x|\to\infty$.

(3) (i) Let $f:\mathbb{R}^2\to\mathbb{R}$ be defined by $f(x,y)=(1+x^2+y^2)^{-1}$. Prove that $f(x,y)\to 0$ as $(x,y)\to\infty$. Consider also the function $g:\mathbb{R}^2\to\mathbb{R}$ defined by $g(x,y)=(1+(x-y)^2)^{-1}$. Prove that $g(x,y)\to 1$ as $(x,y)\to\infty$ along the line $x=y$. Does $g(x,y)$ approach a limit as $(x,y)\to\infty$?
(ii) The function $h:(0,\infty)\to\mathbb{R}^2$ given by $h(t)=(1/t,t)$ defines a curve in $\mathbb{R}^2$. Sketch this curve. Prove that

(*a*) $h(t)\to\infty$ as $t\to 0+$
(*b*) $h(t)\to\infty$ as $t\to+\infty$.

(4) A function $f:\mathbb{R}\to\mathbb{R}$ has the property that $f(x)\to l$ as $x\to+\infty$ and $f(x)\to m$ as $x\to-\infty$ where $l$ and $m$ are real numbers. Define $g:[-\infty,+\infty]\to\mathbb{R}$ by

$$g(x)=\begin{cases} f(x) & (x\in\mathbb{R})\\ l & (x=+\infty)\\ m & (x=-\infty).\end{cases}$$

If $f$ is continuous on $\mathbb{R}$, explain why $g$ is continuous on $[-\infty,\infty]$. Deduce the following results:

(i) The set $f(\mathbb{R})\cup\{l,m\}$ is compact.
(ii) The function $f$ is bounded on $\mathbb{R}$.
(iii) If there exists an $x\in\mathbb{R}$ such that $f(x)\geqq l$ and $f(x)\geqq m$,

then $f$ achieves a maximum on $\mathbb{R}$.

†(iv) The function $f$ is *uniformly* continuous on $\mathbb{R}$ (§23.20).

(5) The metric space $\mathbb{R}^\#$ with metric $c: \mathbb{R}^\# \to \mathbb{R}$ was introduced in §24.2. Explain why a set $G$ in $\mathbb{R}^\#$ is open if and only if, for each $\xi \in G$, there exists an 'open ball' $B$ with centre $\xi$ such that $B \subset G$. Here 'open ball' has the meaning assigned in §24.2.

(6) Prove proposition 24.9 Suppose that $f_1(x) \to +\infty$ as $x \to -\infty$ and $f_2(x) \to +\infty$ as $x \to -\infty$. Give examples to show that neither of the following statements need be true.

(i) $f_1(x) - f_2(x) \to 0$ as $x \to -\infty$
(ii) $f_1(x)/f_2(x) \to 1$ as $x \to -\infty$.

---

## 24.11† Complex functions

We have studied the one-point compactification of $\mathbb{R}^2$ and observed that it should always be assumed that this compactification is in use when $\mathbb{R}^2$ is identified with the system $\mathbb{C}$ of complex numbers. Since $\mathbb{C}$ has a rich algebraic structure (i.e. $\mathbb{C}$ is a field), it is natural to seek to extend some of this structure to $\mathbb{C}^\#$ just as we extended some of the algebraic structure of $\mathbb{R}$ to $[-\infty, +\infty]$. There is no point, of course, in seeking to extend the order structure of $\mathbb{C}$ since $\mathbb{C}$ has no order structure. (See §10.20.)

In this context, we make the following definitions:

(i) $x + \infty = \infty + x = \infty \quad (x \neq \infty)$
(ii) $x \cdot \infty = \infty \cdot x = \infty \quad (x \neq 0)$
(iii) $x/\infty = 0 \quad (x \neq \infty)$
(iv) $x/0 = \infty \quad (x \neq 0)$.

With these definitions proposition 24.9 remains valid when $\mathbb{C}^\#$ replaces $[-\infty, +\infty]$. Note that $\infty + \infty$, $0 \cdot \infty$, $\infty/\infty$ and $0/0$ are *not* defined. In particular, we do *not* define $\infty + \infty = \infty$. For example, $z \to \infty$ as $z \to \infty$ and $-z \to \infty$ as $z \to \infty$ but $z + (-z) \to 0$ as $z \to \infty$.

Even more than in §24.4, it is necessary to warn against attempting algebraic manipulations on the basis of definitions (i)–(iv). Item (iv), for example, does not represent a *genuine* 'division by zero' but merely indicates a handy convention.

---

## 24.12† Product spaces

The one-point compactification of $\mathbb{R}^n$ is not the only possible compactification of $\mathbb{R}^n$. An obvious and important alternative is to regard $\mathbb{R}^n$ as sitting inside the space $[-\infty, +\infty]^n$. In the case of $\mathbb{R}^2$ this amounts to thinking of the plane as squeezed into the interior of a square and regarding the edges of the square as the 'points at infinity'.

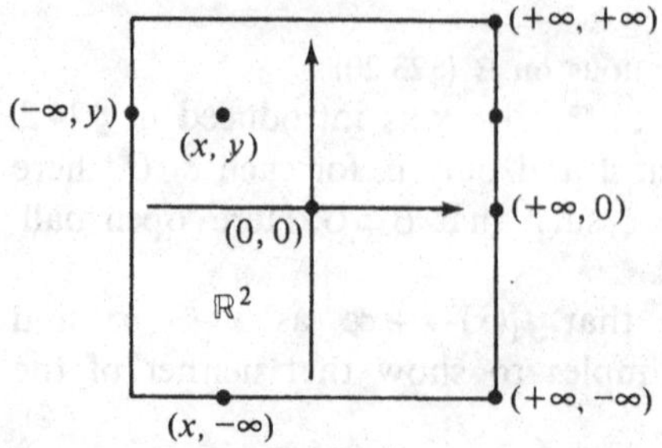


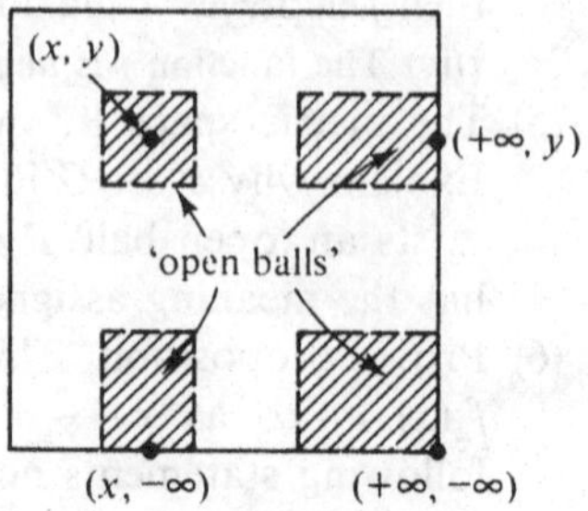


*Schematic diagram of* $[-\infty, +\infty]^2$

In $[-\infty, +\infty]^n$, of course, one uses the product topology (see §21.18). One can describe this by defining an 'open ball' $B$ with centre $\boldsymbol{\xi}=(\xi_1, \xi_2, \ldots, \xi_n)$ in $[-\infty, +\infty]^n$ to be a set of the form

$$B=B_1 \times B_2 \times \ldots \times B_n,$$

where each set $B_i$ $(i=1, 2, \ldots, n)$ is an 'open ball' in $[-\infty, +\infty]$ with centre $\xi_i$. A set $G$ is then open in $[-\infty, +\infty]^n$ if and only if each $\boldsymbol{\xi}\in G$ is the centre of an 'open ball' $B$ such that $B\subset G$. Note that $\mathbb{R}^n$ is a topological subspace of $[-\infty, +\infty]^n$. In fact a subset of $\mathbb{R}^n$ is open relative to $[-\infty, +\infty]^n$ if and only if it is open in the usual sense.

Next consider a set $S$ in $[-\infty, +\infty]^n$ and a function $f\colon S\to[-\infty, +\infty]^m$. We are interested in the case when $S\subset\mathbb{R}^n$ and $f(S)\subset\mathbb{R}^m$.

If $\boldsymbol{\xi}\in[-\infty, +\infty]^n$ and $\boldsymbol{\eta}\in[-\infty, +\infty]^m$, we have that $f(\mathbf{x})\to\boldsymbol{\eta}$ as $\mathbf{x}\to\boldsymbol{\xi}$ through $S$ if and only if:

For each 'open ball' $E$ in $[-\infty, +\infty]^m$, with centre $\boldsymbol{\eta}$ there exists an 'open ball' $\Delta$ in $[-\infty, +\infty]^n$ with centre $\boldsymbol{\xi}$ such that

$$\mathbf{x}\in\Delta\cap S\Rightarrow f(\mathbf{x})\in E$$

provided $\mathbf{x}\neq\boldsymbol{\xi}$.

The table below gives some sample criteria in the case $m=1$ and $n=2$. In this table $l$, $a_1$ and $a_2$ are to be understood as real numbers.

| | |
|---|---|
| $f(\mathbf{x})\to l$ as $\mathbf{x}\to\mathbf{a}$ through $S$ | $\forall\varepsilon>0\ \exists\delta>0\ \forall\mathbf{x}\in S$, $\lvert x_1-a_1\rvert<\delta$ and $\lvert x_2-a_2\rvert<\delta\Rightarrow\lvert f(\mathbf{x})-l\rvert<\varepsilon$ provided $\mathbf{x}\neq\mathbf{a}$. |
| $f(\mathbf{x})\to l$ as $\mathbf{x}\to(+\infty, a_2)$ through $S$ | $\forall\varepsilon>0\ \exists X\ \exists\delta>0\ \forall\mathbf{x}\in S$, $x_1>X$ and $\lvert x_2-a_2\rvert<\delta\Rightarrow\lvert f(\mathbf{x})-l\rvert<\varepsilon$ |
| $f(\mathbf{x})\to+\infty$ as $\mathbf{x}\to(a_1, -\infty)$ through $S$ | $\forall X\ \exists\delta>0\ \exists Y\ \forall\mathbf{x}\in S$, $\lvert x_1-a_1\rvert<\delta$ and $x_2<Y\Rightarrow f(\mathbf{x})>X$ |
| $f(\mathbf{x})\to-\infty$ as $\mathbf{x}\to(+\infty, +\infty)$ through $S$ | $\forall Y\ \exists X\ \forall\mathbf{x}\in S$ $x_1>X$ and $x_2>X\Rightarrow f(\mathbf{x})<Y$ |

As we know from §23.1, the first criterion in this table is equivalent to the usual definition of the statement '$f(\mathbf{x})\to\mathbf{l}$ as $\mathbf{x}\to\mathbf{a}$ through $S$'.

24.13 *Example* Consider the function $f: \mathbb{R}^2 \to \mathbb{R}$ defined by

$$f(x, y) = \frac{(xy)^2}{1+(xy)^2}.$$

This does *not* tend to a limit as $(x, y) \to \infty$. We have that $f(x, y) \to \frac{1}{2}$ as $(x, y) \to \infty$ along the curve $xy = 1$ but $f(x, y) \to 0$ as $(x, y) \to \infty$ along the line $x = 0$. However,

$$\lim_{(x, y) \to (+\infty, +\infty)} \frac{(xy)^2}{1+(xy)^2} = 1. \qquad (1)$$

To prove this we have to show that, given any $\varepsilon > 0$, there exists an $X$ such that, for any $x > X$ and any $y > X$,

$$\left| \frac{(xy)^2}{1+(xy)^2} - 1 \right| < \varepsilon.$$

But

$$\left| \frac{(xy)^2}{1+(xy)^2} - 1 \right| = \frac{1}{1+(xy)^2} < \frac{1}{(xy)^2}$$

and hence the choice $X = \varepsilon^{-1/4}$ clearly suffices.

It is worth noting that, although (1) exists,

$$\lim_{(x, y) \to \infty} \frac{(xy)^2}{1+(xy)^2} \qquad (2)$$

does *not* exist. The reason for the difference is that in considering (1) we look at regions like that indicated on the left below while, when considering (2), we look at regions like that on the right.

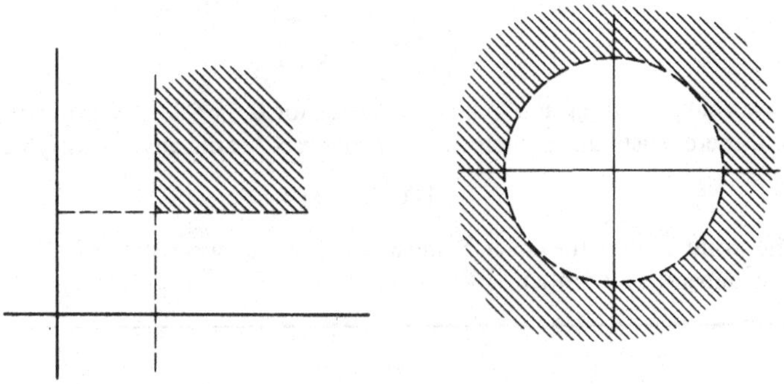

Finally, note that many authors would write (1) in the form

$$\lim_{\substack{x \to \infty \\ y \to \infty}} \frac{(xy)^2}{1+(xy)^2} = 1.$$

One has to be careful not to confuse this notation with (2) (nor to confuse it with the notation for repeated limits). (See §23.5.)

**24.14† *Exercise***

(1) Prove that

(i) $$\lim_{x\to+\infty}\left(\lim_{y\to+\infty}\frac{x-y}{x+y}\right)=-1$$

(ii) $$\lim_{y\to+\infty}\left(\lim_{x\to+\infty}\frac{x-y}{x+y}\right)=1.$$

Discuss the existence of

(iii) $$\lim_{(x,\,y)\to(+\infty,\,+\infty)}\left(\frac{x-y}{x+y}\right).$$

(2) Let $f\colon \mathbb{R}^2\to\mathbb{R}^2$ and consider the following statements

(i) $$\lim_{x\to+\infty}\left(\lim_{y\to+\infty} f(x, y)\right)=\mathbf{l}$$

(ii) $$\lim_{y\to+\infty}\left(\lim_{x\to+\infty} f(x, y)\right)=\mathbf{l}$$

(iii) $$\lim_{(x,\,y)\to(+\infty,\,+\infty)} f(x, y)=\mathbf{l}$$

(iv) $$\lim_{(x,\,y)\to\infty} f(x, y)=\mathbf{l}.$$

Show that (iv)⇒(iii) and (iv)⇒(ii). Under what circumstances is it true that (iii)⇒(ii) and (iii)⇒(i)? Under what circumstances is it true that (ii)⇒(i)? [*Hint:* See theorems 23.9 and 23.17.]

(3) Consider the function $f\colon \mathbb{R}^2\to\mathbb{R}$ defined by

$$f(x, y)=\frac{yx+1}{x+1}.$$

Prove that $f(x, y)\to y$ as $x\to+\infty$ *uniformly* for $0\leqq y\leqq 1$ – i.e. prove that, for any $\varepsilon>0$, there exists an $X$ such that, for any $x>X$ and any $y$ satisfying $0\leqq y\leqq 1$,

$$|f(x, y)-y|<\varepsilon.$$

(The point is that the *same* $X$ must work for *all* $y$ satisfying $0\leqq y\leqq 1$. Thus $X$ must depend only on $\varepsilon$.)

# 25 SEQUENCES

## 25.1 Introduction

In this chapter we shall study the general properties of sequences of points in a metric space. As we shall see, this topic allows us to tie together many of the ideas introduced in earlier chapters and, to some extent, this chapter will therefore provide a useful opportunity for some revision of these ideas. Sequences of real numbers are particularly important but we have not felt it necessary to stress the importance of their special properties since most readers are likely to have studied these special properties at length elsewhere.

Formally, a *sequence* of points in a metric space $\mathcal{Y}$ is a function $f: \mathbb{N} \to \mathcal{Y}$. We call $f(k)$ the $k$th *term* of the sequence. The notation

$$\langle y_k \rangle$$

will be used to denote the sequence whose $k$th term is $y_k$. Thus $\langle y_k \rangle$ denotes the function $f: \mathbb{N} \to \mathcal{Y}$ defined by $f(k) = y_k$ $(k \in \mathbb{N})$.

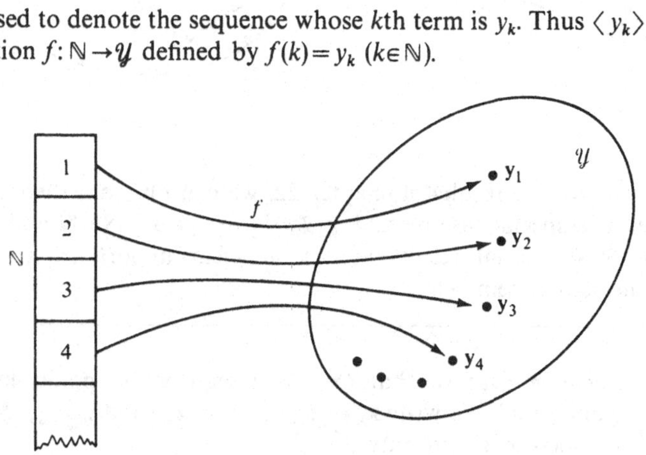


Intuitively, it is useful to think of a sequence $\langle y_k \rangle$ as a list. We therefore sometimes refer to a sequence by listing its first few terms. The first few terms of the sequence $\langle k^2+1 \rangle$ for example are

$$2, 5, 10, 17, 26, 37, \ldots$$

## 25.2 Convergence of sequences

Suppose that $\langle y_k \rangle$ is a sequence of points in a metric space $\mathcal{Y}$ and that $\boldsymbol{\eta} \in \mathcal{Y}$. We say that $\langle y_k \rangle$ converges to $\boldsymbol{\eta}$ and write

$$\mathbf{y}_k \to \boldsymbol{\eta} \quad \text{as} \quad k \to \infty$$

if and only if

$$\boxed{\begin{aligned} &\forall \varepsilon > 0 \ \exists K, \\ &\quad k > K \Rightarrow d(\boldsymbol{\eta}, \mathbf{y}_k) < \varepsilon. \end{aligned}}$$

This definition is nothing new. If $f: \mathbb{N} \to \mathcal{Y}$ satisfies $f(k) = \mathbf{y}_k$, the definition simply asserts that $f(k) \to \boldsymbol{\eta}$ as $k \to +\infty$ through the set $\mathbb{N}$ (see §24.5).

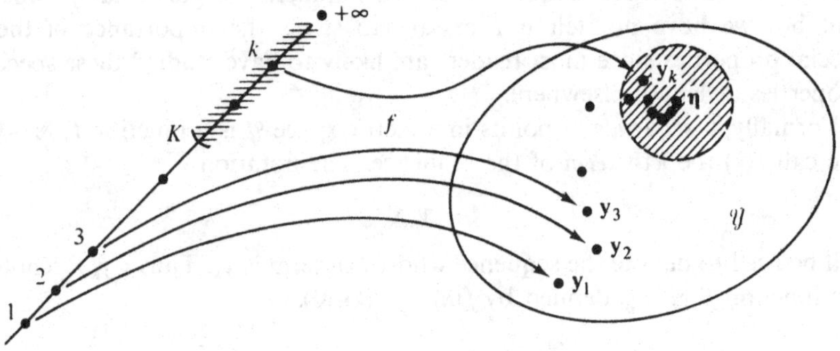


From each of the results of chapter 22, we can obtain a theorem about sequences as a special case by taking $\mathcal{X} = [-\infty, +\infty]$, $S = \mathbb{N}$ and $\boldsymbol{\xi} = +\infty$. Some of these special results are of sufficient importance to make it worthwhile listing them below.

---

25.3 *Theorem* Suppose that $\langle \mathbf{x}_k \rangle$ is a sequence of points in $\mathbb{R}^n$ and that $\boldsymbol{\xi}$ is a point in $\mathbb{R}^n$. We write $\mathbf{x}_k = (x_{1k}, x_{2k}, \ldots, x_{nk})$ and $\boldsymbol{\xi} = (\xi_1, \xi_2, \ldots, \xi_n)$. Then $\mathbf{x}_k \to \boldsymbol{\xi}$ as $k \to \infty$ if and only if

$$x_{jk} \to \xi_j \quad \text{as} \quad k \to \infty$$

for each $j = 1, 2, \ldots, n$.

*Proof* Theorem 22.28.

---

25.4 *Example* Consider the sequence

$$\langle \mathbf{x}_k \rangle = \left\langle \left( \frac{\cos k}{k}, \frac{\sin k}{k} \right) \right\rangle$$

of points in $\mathbb{R}^2$. We have that

$$\left| \frac{\cos k}{k} \right| \leq \frac{1}{k} \to 0 \quad \text{as} \quad k \to \infty$$

and

$$\left| \frac{\sin k}{k} \right| \leq \frac{1}{k} \to 0 \quad \text{as} \quad k \to \infty.$$

Hence, by theorem 25.3, $\mathbf{x}_k \to (0, 0)$ as $k \to \infty$.

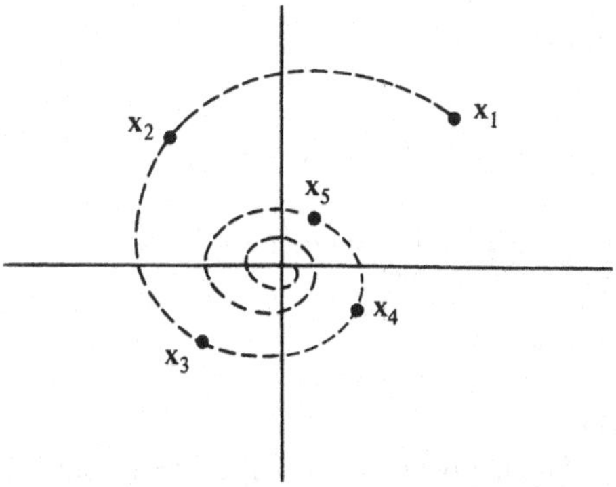


---

25.5 *Theorem* Suppose that $\langle \mathbf{x}_k \rangle$ and $\langle \mathbf{y}_k \rangle$ are sequences of points in $\mathbb{R}^n$ and that

$$\mathbf{x}_k \to \boldsymbol{\xi} \quad \text{as} \quad k \to \infty$$

and

$$\mathbf{y}_k \to \boldsymbol{\eta} \quad \text{as} \quad k \to \infty.$$

Then, for any real numbers $a$ and $b$,

$$a\mathbf{x}_k + b\mathbf{y}_k \to a\boldsymbol{\xi} + b\boldsymbol{\eta} \quad \text{as} \quad k \to \infty.$$

*Proof* Theorem 22.30.

---

25.6 *Theorem* Suppose that $\langle x_k \rangle$ and $\langle y_k \rangle$ are sequences of real numbers and that

$$x_k \to \xi \quad \text{as} \quad k \to \infty$$

and

$$y_k \to \eta \quad \text{as} \quad k \to \infty.$$

Then;

(i) $x_k y_k \to \xi\eta$ as $k \to \infty$ and, provided that $\eta \neq 0$,

(ii) $\dfrac{x_k}{y_k} \to \dfrac{\xi}{\eta}$ as $k \to \infty$.

*Proof* Theorems 22.31 and 22.32.

---

This theorem remains true if $\mathbb{R}$ is replaced throughout by $\mathbb{C}$.

---

**25.7 Convergence of functions and sequences**

25.8 *Theorem* Let $\mathcal{X}$ and $\mathcal{Y}$ be metric spaces and suppose that $f: S \to \mathcal{Y}$ where $S$ is a set in $\mathcal{X}$. Then the statements

(1) $f(\mathbf{x}) \to \boldsymbol{\eta}$ as $\mathbf{x} \to \boldsymbol{\xi}$ through $S$;

and (2) For each sequence $\langle \mathbf{x}_k \rangle$ of points of $S \backslash \{\boldsymbol{\xi}\}$,

$$\left(\lim_{k\to\infty} \mathbf{x}_k = \boldsymbol{\xi}\right) \Rightarrow \left(\lim_{k\to\infty} f(\mathbf{x}_k) = \boldsymbol{\eta}\right);$$

are equivalent.

*Proof* (i) (1)$\Rightarrow$(2).

Suppose that (1) holds. Then, given any $\varepsilon > 0$, there exists a $\delta > 0$ such that for any $\mathbf{x} \in S \backslash \{\boldsymbol{\xi}\}$

$$d(\boldsymbol{\xi}, \mathbf{x}) < \delta \Rightarrow d(\boldsymbol{\eta}, f(\mathbf{x})) < \varepsilon. \tag{3}$$

Now suppose that $\langle \mathbf{x}_k \rangle$ is a sequence of points of $S \backslash \{\boldsymbol{\xi}\}$ such that $\mathbf{x}_k \to \boldsymbol{\xi}$ as

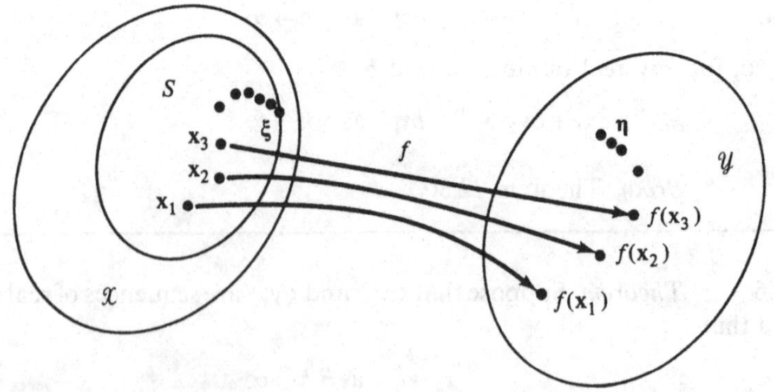


$k \to \infty$. Then there exists a $K$ such that

$$k > K \Rightarrow d(\boldsymbol{\xi}, \mathbf{x}_k) < \delta. \tag{4}$$

Combining (3) and (4), we obtain that

$$k>K \Rightarrow d(\boldsymbol{\eta}, f(\mathbf{x}_k))<\varepsilon$$

and hence $f(\mathbf{x}_k)\to\boldsymbol{\eta}$ as $k\to\infty$.

(ii) not (1) $\Rightarrow$ not (2) (see §2.10).

Suppose that it is false that $f(\mathbf{x})\to\boldsymbol{\eta}$ as $\mathbf{x}\to\boldsymbol{\xi}$ through $S$. If it is false that

$$\forall\varepsilon>0\ \exists\delta>0\ \forall\mathbf{x}\in S\backslash\{\boldsymbol{\xi}\},\ (d(\boldsymbol{\xi},\mathbf{x})<\delta \Rightarrow d(\boldsymbol{\eta}, f(\mathbf{x}))<\varepsilon),$$

then it is true that

$$\exists\varepsilon>0\ \forall\delta>0\ \exists\mathbf{x}\in S\backslash\{\boldsymbol{\xi}\},\ (d(\boldsymbol{\xi},\mathbf{x})<\delta \text{ and } d(\boldsymbol{\eta}, f(\mathbf{x}))\geqq\varepsilon).$$

This latter expression is true for some fixed $\varepsilon>0$ and *all* $\delta>0$. Given $\delta=1/k$ ($k=1, 2, 3,\ldots$), we can therefore find $\mathbf{x}_k\in S\backslash\{\boldsymbol{\xi}\}$ such that

$$d(\boldsymbol{\xi}, \mathbf{x}_k)<1/k \quad\text{and}\quad d(\boldsymbol{\eta}, f(\mathbf{x}_k))\geqq\varepsilon.$$

From the first of these inequalities it follows that $\mathbf{x}_k\to\boldsymbol{\xi}$ as $k\to\infty$ and, from the second, it follows that $f(\mathbf{x}_k)\not\to\boldsymbol{\eta}$ as $k\to\infty$.

---

25.9 *Corollary* Let $\mathcal{X}$ and $\mathcal{Y}$ be metric spaces and suppose that $f: S\to\mathcal{Y}$ where $S$ is a set in $\mathcal{X}$. Then $f$ is continuous on $S$ if and only if, for each sequence $\langle\mathbf{x}_k\rangle$ of points of $S$ which converges to a point of $S$,

$$\lim_{k\to\infty} f(\mathbf{x}_k)= f\left(\lim_{k\to\infty} \mathbf{x}_k\right).$$

---

25.10 *Example* In §9.2 we explained how Archimedes trapped the area of a quarter-circle of radius 1 between two sequences $\langle A_n\rangle$ and $\langle B_n\rangle$ of rational numbers. The terms of these sequences satisfy $A_1=\frac{1}{2}$, $B_1=1$ and

(i) $A_{n+1}=\{A_nB_n\}^{1/2}$

(ii) $B_{n+1}^{-1}=2\{A_{n+1}^{-1}+B_n^{-1}\}$.

Using the fact that the geometric and harmonic mean of two real numbers lies between these numbers, it is trivial to prove that $\frac{1}{2}\leqq A_n\leqq A_{n+1}\leqq B_{n+1}\leqq B_n\leqq 1$. From this it follows that $\langle A_n\rangle$ and $\langle B_n\rangle$ converge (see exercise 25.11(4)) and that their respective limits $A$ and $B$ satisfy $\frac{1}{2}\leqq A\leqq B\leqq 1$.

Applying theorem 25.5 to the equation (ii), we obtain that

$$B^{-1}=2(A^{-1}+B^{-1})$$

and hence $A=B$. It therefore makes sense to interpret the common limit of the two sequences as the area of the quarter-circle. To *prove* that the common limit is $\pi/4$, however, is considerably more difficult.

The function $f: [0, \infty)\to\mathbb{R}$ defined by $f(x)=\sqrt{x}$ is continuous on $[0, \infty)$. We have that $A_nB_n \to AB$ as $n \to \infty$ and hence, by corollary 25.9,

$$\sqrt{(A_nB_n)} \to \sqrt{(AB)} \quad \text{as} \quad n \to \infty.$$

Since $A_{n+1} = \sqrt{(A_nB_n)}$, it follows that $A = \sqrt{(AB)}$ and hence the result $A = B$ is obtained again.

---

### 25.11 *Exercise*

(1) Determine whether or not the sequence $\langle \mathbf{x}_k \rangle$ converges in $\mathbb{R}^2$ in each of the following cases.

(i) $\mathbf{x}_k = \left(\frac{k+1}{k+2}, \frac{k+2}{k+1}\right)$ (ii) $\mathbf{x}_k = \left(1, \frac{1}{k}\right)$

(iii) $\mathbf{x}_k = \left(\frac{1}{k}, k\right)$ (iv) $\mathbf{x}_k = (\cos k, \sin k)$.

(2) Write down definitions for the statements '$x_k \to +\infty$ as $k \to \infty$' and '$x_k \to -\infty$ as $k \to \infty$' in the case when $\langle x_k \rangle$ is a sequence of real numbers. Show that $k^2 \to +\infty$ as $k \to \infty$ and that $-\sqrt{k} \to -\infty$ as $k \to \infty$. Discuss the sequences $\langle (-1)^k \rangle$ and $\langle k(-1)^k \rangle$.

(3) Let $\langle \mathbf{x}_k \rangle$ be a sequence of points in $\mathbb{R}^n$. Write down a definition for the statement '$\mathbf{x}_k \to \infty$ as $k \to \infty$' based on the one-point compactification of $\mathbb{R}^n$. For which of the sequences of question 1 is it true that $\mathbf{x}_k \to \infty$ as $k \to \infty$?

(4) A sequence $\langle x_k \rangle$ of real numbers is said to *increase* if and only if $x_k \leqq x_{k+1}$ for each $k \in \mathbb{N}$. Prove that an increasing sequence $\langle x_k \rangle$ of real numbers is unbounded above if and only if $x_k \to +\infty$ as $k \to \infty$. Prove that an increasing sequence of $\langle x_k \rangle$ of real numbers is bounded above if and only if $x_k \to x$ as $k \to \infty$ and that the limit $x$ is the supremum of the sequence. What are the corresponding results for decreasing sequences?

(5) A sequence $\langle x_k \rangle$ is defined by $x_1 = 1$ and $x_{k+1} = f(x_k)$ where $f:[0, \infty) \to \mathbb{R}$ is defined by

$$f(x) = \frac{x+1}{x+2}.$$

Prove that $0 < x_{k+1} \leqq x_k$ $(k = 1, 2, \ldots)$ and explain why it follows that $\langle x_k \rangle$ converges. If $x_k \to x$ as $k \to \infty$, justify the conclusion that $x = f(x)$ and hence show that

$$x = \frac{\sqrt{5}-1}{2}.$$

(6) Prove that the sequence $\langle x_k \rangle$ of rational numbers defined by $x_1 = 2$ and

$$x_{k+1} = \frac{1}{2}\left(x_k + \frac{2}{x_k}\right)$$

decreases and is bounded below. Deduce that $\langle x_k \rangle$ converges and prove that the limit is $\sqrt{2}$.

---

## 25.12 Sequences and closure

We begin with the following lemma.

---

25.13 *Lemma* Let $S$ be a non-empty set in a metric space $\mathcal{X}$ and let $\xi \in \mathcal{X}$. Then $d(\xi, S) = 0$ if and only if there exists a sequence $\langle \mathbf{x}_k \rangle$ of points of $S$ such that $\mathbf{x}_k \to \xi$ as $k \to \infty$.

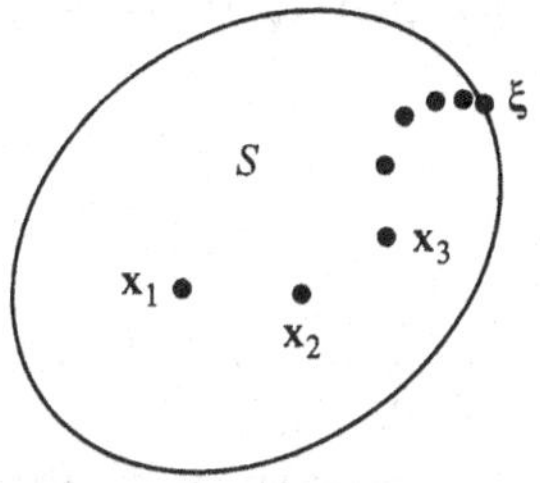


*Proof* (i) Suppose that $d(\xi, S) = 0$. By theorem 13.22, given any $\varepsilon > 0$, there exists an $\mathbf{x} \in S$ such that $d(\xi, \mathbf{x}) < \varepsilon$. Since $1/k > 0$, it follows that there exists an $\mathbf{x}_k \in S$ such that

$$d(\xi, \mathbf{x}_k) < 1/k \to 0 \quad \text{as} \quad k \to \infty.$$

Hence $\mathbf{x}_k \to \xi$ as $k \to \infty$.

(ii) Suppose that there exists a sequence $\langle \mathbf{x}_k \rangle$ of points of $S$ such that $\mathbf{x}_k \to \xi$ as $k \to \infty$. Then, given any $\varepsilon > 0$, there exists a $K$ such that $k > K \Rightarrow d(\xi, \mathbf{x}_k) < \varepsilon$. In particular, $d(\xi, \mathbf{x}_{K+1}) < \varepsilon$. It follows from theorem 13.22 that $d(\xi, S) = 0$.

---

25.14 *Theorem* Let $S$ be a set in a metric space $\mathcal{X}$ and let $\xi \in \mathcal{X}$. Then $\xi \in \bar{S}$ if and only if there exists a sequence $\langle \mathbf{x}_k \rangle$ of points of $S$ such that $\mathbf{x}_k \to \xi$ as $k \to \infty$.

*Proof* By theorem 15.5, $\xi \in \bar{S} \Leftrightarrow d(\xi, S)=0$ and so the theorem follows immediately from lemma 25.13.

---

25.15 *Corollary* A set $S$ in a metric space $\mathscr{X}$ is closed if and only if each convergent sequence of points of $S$ converges to a point of $S$.

*Proof* By exercise 15.3(1), $S$ is closed if and only if $S=\bar{S}$.

---

25.16† *Note* It is of some importance to take note of the fact that theorem 25.14 and corollary 25.15 are *false* for a general *topological* space $\mathscr{X}$. Analogues of these results are valid but the notion of a sequence must be replaced by the more general notion of a *net*.

---

25.17 *Exercise*

(1) If $\langle \mathbf{x}_k \rangle$ is a sequence of points in $\mathbb{R}^n$ such that $\mathbf{x}_k \to \xi$ as $k \to \infty$, explain why

(i) $\langle \mathbf{x}_k, \mathbf{u} \rangle \to \langle \xi, \mathbf{u} \rangle$ as $k \to \infty$

(ii) $\|\mathbf{x}_k\| \to \|\xi\|$ as $k \to \infty$.

[*Hint*: Use corollary 25.9.] Deduce from corollary 25.15 that the sets $S=\{\mathbf{x} : \langle \mathbf{x}, \mathbf{u} \rangle \leqq c\}$ and $T=\{\mathbf{x} : \|\mathbf{x}\| \leqq r\}$ are closed.

(2) Let $A$ and $B$ be two non-empty sets in a metric space $\mathscr{X}$. Prove that $A$ and $B$ are contiguous if and only if a sequence of points in one of the sets converges to a point of the other.

Use this result and corollary 25.9 to prove that, if $S$ is a connected set in $\mathscr{X}$ and $f: S \to \mathscr{Y}$ is continuous on $S$, then $f(S)$ is connected. (See theorem 17.9.)

(3) Let $S$ be a set in a metric space $\mathscr{X}$ and let $\xi \in \mathscr{X}$. Prove that $\xi$ is a cluster point of $S$ if and only if there exists a sequence $\langle \mathbf{x}_k \rangle$ of *distinct* points of $S$ such that $\mathbf{x}_k \to \xi$ as $k \to \infty$. (Distinct means that $x_j = x_k \Leftrightarrow j=k$.)

---

**25.18 Subsequences**

Suppose that $g: \mathbb{N} \to \mathbb{N}$ is strictly increasing. Then the sequence $f \circ g: \mathbb{N} \to \mathscr{Y}$ is said to be a *subsequence* of the sequence $f: \mathbb{N} \to \mathscr{Y}$.
If we think of the subsequence illustrated in the diagram as a list, then its first few terms are

$$y_2, y_3, y_5, y_7, \ldots$$

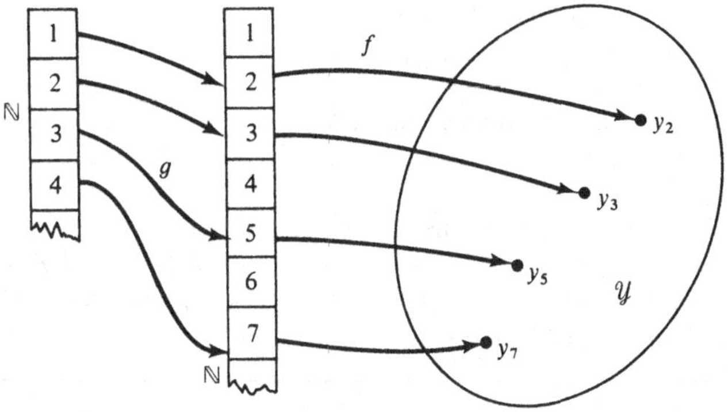


Note that this is obtained from the original sequence

$$y_1, y_2, y_3, y_4, y_5, y_6, y_7, \ldots$$

by omitting some of the terms.

If $\langle y_k \rangle$ is a sequence and $\langle k_l \rangle$ is a strictly increasing sequence of natural numbers, then

$$\langle y_{k_l} \rangle$$

denotes a subsequence of $\langle y_k \rangle$. (If $f(k)=y$ and $g(l)=k_l$, then $f(g(l))=y_{k_l}$.) For example, if $\langle y_k \rangle = \langle k^2+1 \rangle$ and $\langle k_l \rangle = \langle 2l \rangle$, then $\langle y_{k_l} \rangle$ is $\langle 4l^2+1 \rangle$. This subsequence is obtained from the sequence $\langle y_k \rangle$ by deleting terms as indicated below:

~~2~~, 5, ~~10~~, 17, ~~26~~, 37, ~~50~~, 65, ~~82~~, 101, ...

---

25.19 *Theorem* Let $\langle \mathbf{y}_k \rangle$ be a sequence of points in a metric space $\mathcal{Y}$ with the property that $\mathbf{y}_k \to \boldsymbol{\eta}$ as $k \to \infty$. If $\langle \mathbf{y}_{k_l} \rangle$ is any subsequence of $\langle \mathbf{y}_k \rangle$, then

$$\mathbf{y}_{k_l} \to \boldsymbol{\eta} \quad \text{as} \quad l \to \infty.$$

*Proof* Let $f\colon \mathbb{N} \to \mathcal{Y}$ be defined by $f(k)=\mathbf{y}_k$ and let $g\colon \mathbb{N} \to \mathbb{N}$ be defined by $g(l)=k_l$. We have that $f(k) \to \boldsymbol{\eta}$ as $k \to +\infty$ through $\mathbb{N}$ and that $g(l) \to +\infty$ as $l \to +\infty$ through $\mathbb{N}$ (see exercise 25.22(1)). It follows from exercise 22.33(3ii) that

$$f(g(l)) \to \boldsymbol{\eta} \quad \text{as} \quad l \to +\infty$$

through $\mathbb{N}$ – i.e. $\mathbf{y}_{k_l} \to \boldsymbol{\eta}$ as $l \to \infty$.

---

25.20 *Example* We have that $1/k \to 0$ *as* $k \to \infty$. From theorem 25.19 it follows without further calculation that

$$1/2^k \to 0 \quad \text{as} \quad k \to \infty$$

because $\langle 1/2^k \rangle$ is a subsequence of $\langle 1/k \rangle$.

---

25.21 *Example* Consider the sequence $\langle (-1)^k \rangle$ of real numbers. If $(-1)^k \to l$ as $k \to \infty$, then $(-1)^{2k} \to l$ as $k \to \infty$ and $(-1)^{2k+1} \to l$ as $k \to \infty$. But $(-1)^{2k} = 1$ and $(-1)^{2k+1} = -1$. Hence $l = 1 = -1$. From this contradiction we deduce that $\langle (-1)^k \rangle$ diverges.

---

25.22 *Exercise*

(1) Suppose that $\langle k_l \rangle$ is a strictly increasing sequence of natural numbers. Prove by induction that $k_l \geqq l$ for all $l \in \mathbb{N}$ and deduce that

$$k_l \to +\infty \quad \text{as} \quad l \to \infty.$$

(2) Show that, for any $m \in \mathbb{N}$, the sequence of real numbers

$$\left\langle \left(1 + \frac{m}{k}\right)^k \right\rangle$$

is increasing and bounded above. Deduce that the sequence converges to a real number $\phi(m)$. By considering an appropriate subsequence, prove that

$$\phi(m) = \{\phi(1)\}^m.$$

(3) Let $\langle \mathbf{y}_k \rangle$ be a sequence of points in a metric space $\mathcal{Y}$. If $\boldsymbol{\eta}$ is a cluster point of $E = \{\mathbf{y}_k : k \in \mathbb{N}\}$, prove that $\langle \mathbf{y}_k \rangle$ has a subsequence $\langle \mathbf{y}_{k_l} \rangle$ with the property that

$$\mathbf{y}_{k_l} \to \boldsymbol{\eta} \quad \text{as} \quad l \to \infty.$$

(4) Let $\langle \mathbf{y}_k \rangle$ be a sequence of points in a metric space $\mathcal{Y}$ and let $E = \{\mathbf{y}_k : k \in \mathbb{N}\}$. If $\mathbf{y}_k \to \boldsymbol{\eta}$ as $k \to \infty$, prove that $E \cup \{\boldsymbol{\eta}\}$ is closed. If $\langle \mathbf{y}_k \rangle$ has *no* convergent subsequences, prove that $E$ is closed.

(5) Let $\langle x_k \rangle$ be a sequence of real numbers. If each set $E_n = \{x_k : k > n\}$ has a maximum, prove that $\langle x_k \rangle$ has a decreasing subsequence. If at least one of the sets $E_n$ has no maximum, prove that $\langle x_k \rangle$ has an increasing subsequence.

Deduce that any bounded sequence of real numbers has a convergent subsequence. [*Hint*: Use exercise 25.11(5).]

(6) Let $\langle (x_k, y_k) \rangle$ be a bounded sequence of points in $\mathbb{R}^2$. Prove that $\langle (x_k, y_k) \rangle$ has a convergent subsequence.
[*Hint*: Begin with a convergent subsequence $\langle x_{k_l} \rangle$ of $\langle x_k \rangle$ and consider $\langle y_{k_l} \rangle$.]

---

**25.23 Sequences and compactness**

The Bolzano–Weierstrass theorem (19.6) was central to our discussion of compact sets in chapter 19. We therefore begin by giving a version of the Bolzano–Weierstrass theorem for sequences.

---

25.24 *Theorem* (*Bolzano–Weierstrass theorem*) Any bounded sequence $\langle \mathbf{x}_k \rangle$ of points in $\mathbb{R}^n$ has a convergent subsequence.

*Proof* This may be deduced from theorem 19.6 as follows. If $E = \{\mathbf{x}_k : k \in \mathbb{N}\}$ is infinite, then theorem 19.6 asserts that $E$ has a cluster point $\xi$. This cluster point must be the limit of a subsequence of $\langle \mathbf{x}_k \rangle$ (exercise 25.22(3)). If $E$ is finite, then $\langle \mathbf{x}_k \rangle$ has a subsequence all of whose terms are equal. This subsequence therefore converges.

Alternatively, the theorem may be proved directly as indicated in exercise 25.22(5) and (6).

---

In chapters 19 and 20, we considered a number of different definitions of compactness and the above version of the Bolzano–Weierstrass theorem suggests yet another. We say that a set $K$ in a metric space $\mathcal{X}$ is *sequentially compact* if and only if every sequence of points in $K$ has a subsequence which converges to a point of $K$.

---

25.25 *Theorem* A set $K$ in a metric space $\mathcal{X}$ is compact if and only if each sequence $\langle \mathbf{x}_k \rangle$ of points of $K$ has a subsequence which converges to a point of $K$ – i.e. in a metric space, compactness and sequential compactness are the same.

*Proof* We use the definition of compactness given in §19.5 – i.e. a set $K$ in a metric space $\mathcal{X}$ is compact if and only if each infinite subset $E$ of $S$ has a cluster point $\xi \in S$.

(i) Let $K$ be compact and let $\langle \mathbf{x}_k \rangle$ be a sequence of points of $K$. If $E = \{\mathbf{x}_k : k \in \mathbb{N}\}$ is finite, then $\langle \mathbf{x}_k \rangle$ has a subsequence $\langle \mathbf{x}_{k_l} \rangle$ all of whose terms are equal to one of the terms of $\langle \mathbf{x}_k \rangle$. Hence $\langle \mathbf{x}_{k_l} \rangle$ converges to a point of $K$. If $E$ is infinite, then $E$ has a cluster point $\xi \in K$. By exercise 25.22(3), $\langle \mathbf{x}_k \rangle$ has a subsequence which converges to $\xi$.

(ii) Let $K$ be sequentially compact and let $E$ be an infinite subset of $K$. Since $E$ is infinite, there exists a sequence $\langle \mathbf{x}_k \rangle$ of distinct points of $E$. Because $K$ is sequentially compact, $\langle \mathbf{x}_k \rangle$ has a convergent subsequence whose limit $\xi$ lies in $K$. But, by exercise 25.17(3), $\xi$ is a cluster point of $E$.

---

25.26† *Note* It is important to bear in mind that sequential compactness is *not* the same as compactness in a general topological space. It is true that a countably compact set (§20.7) in a topological space is sequentially compact but even the converse of *this* assertion is false without some subsidiary assumptions.

---

25.27 *Example* Suppose that $\mathcal{X}$ and $\mathcal{Y}$ are metric spaces and that $S \subset \mathcal{X}$. If $f: S \to \mathcal{Y}$ is continuous on the set $S$, then

$$S \text{ compact} \Rightarrow f(S) \text{ compact}.$$

We have already seen two proofs of this very important theorem. (See theorems 19.13 and 21.17.) A third proof may be based on theorem 25.25.

Let $\langle \mathbf{y}_k \rangle$ be any sequence of points in $f(S)$. We seek to show that $\langle \mathbf{y}_k \rangle$ has a subsequence which converges to a point of $f(S)$. Write $\mathbf{y}_k = f(\mathbf{x}_k)$. Since $S$ is compact, $\langle \mathbf{x}_k \rangle$ has a subsequence which converges to a point $\boldsymbol{\xi} \in S$. Suppose $\mathbf{x}_{k_l} \to \boldsymbol{\xi}$ as $l \to \infty$. By corollary 25.9, $f(\mathbf{x}_{k_l}) \to f(\boldsymbol{\xi})$ as $l \to \infty$. It follows that the subsequence $\langle \mathbf{y}_{k_l} \rangle$ converges to the point $f(\boldsymbol{\xi}) \in f(S)$.

---

25.28 *Exercise*

(1) Let $\langle \mathbf{y}_k \rangle$ be a sequence of points in a metric space $\mathcal{X}$ and let $E = \{\mathbf{y}_k : k \in \mathbb{N}\}$. If $\mathbf{y}_k \to \boldsymbol{\eta}$, prove that $E \cup \{\boldsymbol{\eta}\}$ is compact.

†(2) Let $\mathcal{X}$ and $\mathcal{Y}$ be metric spaces and let $f: K \to \mathcal{Y}$ be continuous on the compact set $K$ in $\mathcal{X}$. Use theorem 25.25 to show that $f$ is *uniformly* continuous on $K$ (see §23.20).

[*Hint*: The contradictory of the statement that $f$ is uniformly continuous on $K$ begins with '$\exists \varepsilon > 0 \; \forall \; \delta > 0$'. Use this assertion with $\delta = 1/k$ for each $k \in \mathbb{N}$.]

†(3) Given an example of a bounded sequence in $l^\infty$ (see §20.20) which has *no* convergent subsequence.

---

# 26 OSCILLATION

## 26.1 Divergence

Suppose that $\mathcal{X}$ and $\mathcal{Y}$ are metric spaces and that $f: S \to \mathcal{Y}$ where $S$ is a set in $\mathcal{X}$. Let $\eta \in \mathcal{Y}$ and let $\xi$ be a cluster point of $S$.

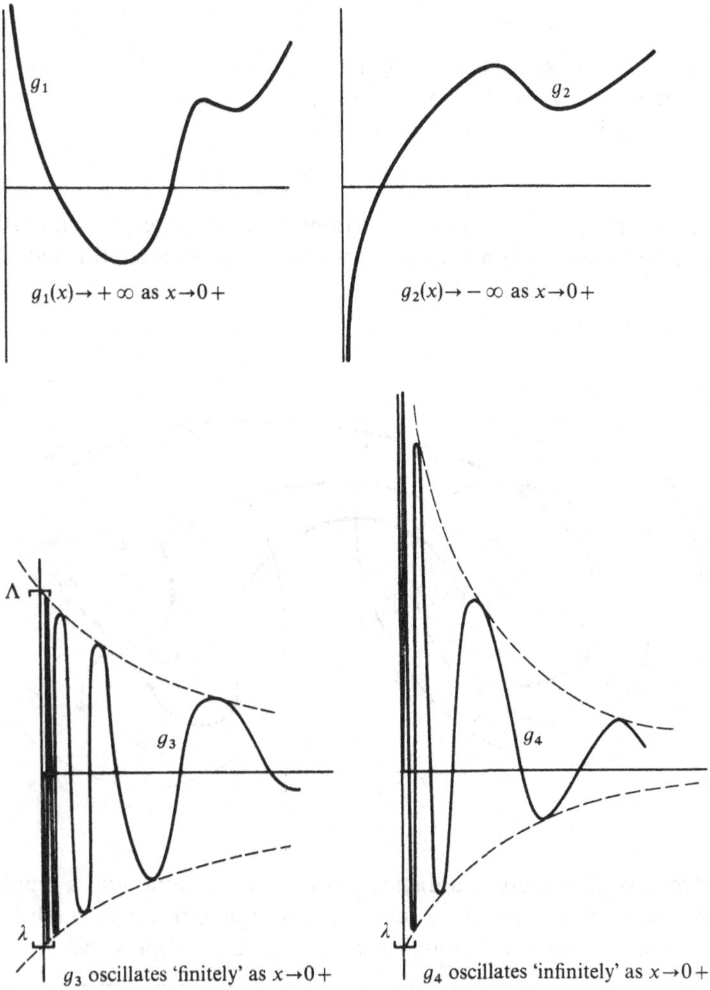

If $f(\mathbf{x}) \to \boldsymbol{\eta}$ as $\mathbf{x} \to \boldsymbol{\xi}$ through the set $S$, we say that the function *converges* as $\mathbf{x}$ approaches $\boldsymbol{\xi}$ through $S$. If the function does not converge as $\mathbf{x}$ approaches $\boldsymbol{\xi}$ through $S$, then it is said to *diverge*. Divergent functions can exhibit a variety of different behaviour as the diagrams on p. 181 illustrate.

We saw in chapter 24 that the behaviour of the functions $g_1$ and $g_2$ can be treated by the same machinery as one uses for convergent functions. Although these functions do not converge as $x$ approaches 0 from the right when regarded as mappings to the space $\mathbb{R}$, they do converge when regarded as mappings to the space $[-\infty, +\infty]$. The functions $g_3$ and $g_4$ are more interesting. These are examples of oscillating functions. To discuss these, we require the notion of a limit point.

## 26.2 Limit points

Suppose that $\mathcal{X}$ and $\mathcal{Y}$ are metric spaces and that $f: S \to \mathcal{Y}$ where $S$ is a set in $\mathcal{X}$. Let $\boldsymbol{\eta} \in \mathcal{Y}$ and let $\boldsymbol{\xi}$ be a cluster point of $S$.

Suppose it is true that, for some $\boldsymbol{\lambda} \in \mathcal{Y}$,

$$f(\mathbf{x}) \to \boldsymbol{\lambda} \quad \text{as} \quad \mathbf{x} \to \boldsymbol{\xi}$$

through the set $T$ where $T$ is a subset of $S$ for which $\boldsymbol{\xi}$ is a cluster point. Then we say that $\boldsymbol{\lambda}$ is a *limit point* of $f$ as $\mathbf{x}$ approaches $\boldsymbol{\xi}$ through $S$.

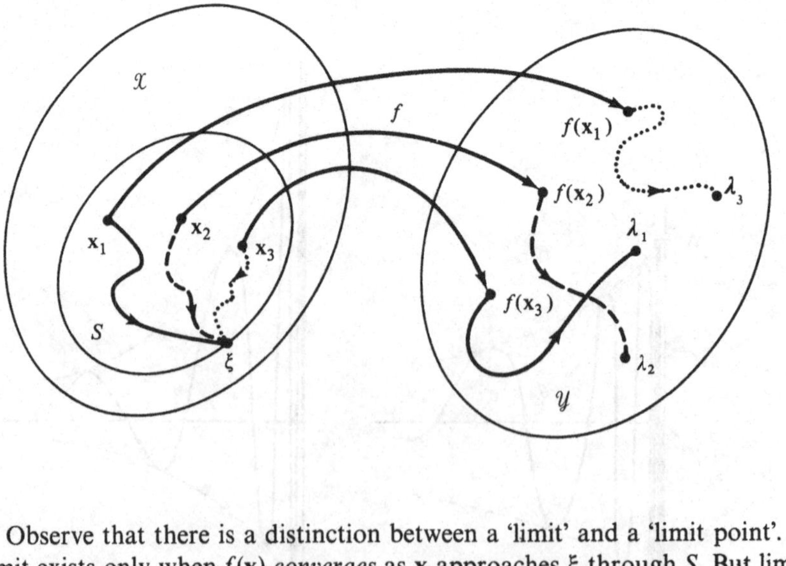


Observe that there is a distinction between a 'limit' and a 'limit point'. A limit exists only when $f(\mathbf{x})$ *converges* as $\mathbf{x}$ approaches $\boldsymbol{\xi}$ through $S$. But limit points may exist even though $f(\mathbf{x})$ *diverges* as $\mathbf{x}$ approaches $\boldsymbol{\xi}$ through $S$.

26.3 *Examples* (i) Consider the function $f: \mathbb{R} \to \mathbb{R}$ defined by

$$f(x)=\begin{cases} 1 & (x \geqq 0) \\ -1 & (x<0). \end{cases}$$

This function diverges as $x$ approaches 0 but $f(x) \to 1$ as $x \to 0+$ and $f(x) \to -1$ as $x \to 0-$. Hence the function has two limit points as $x$ approaches 0. These are 1 and $-1$.

(ii) Consider the function $f: \mathbb{R}^2 \to \mathbb{R}^1$ of example 22.11. We have seen that

$$f(x, y) \to \frac{\alpha^2-1}{\alpha^2+1} \quad \text{as} \quad (x, y) \to (0, 0)$$

along the line $y=\alpha x$. Examining each $\alpha \in \mathbb{R}$ in turn, we find that every point of $[0, 1)$ is a limit point of $f(x, y)$ as $(x, y)$ approaches $(0, 0)$. Also

$$f(x, y) \to 1 \quad \text{as} \quad (x, y) \to (0, 0)$$

along the line $x=0$. Hence 1 is also a limit point.

Since $-1 \leqq f(x, y) \leqq 1$ for all $(x, y) \in \mathbb{R}^2$, the function can have no other limit points. It follows that the set of limit points of $f(x, y)$ as $(x, y) \to (0, 0)$ is equal to $[-1, 1]$.

---

26.4 *Exercise*

(1) Suppose that $\mathcal{X}$ and $\mathcal{Y}$ are metric spaces and that $f: S \to \mathcal{Y}$ where $S$ is a set in $\mathcal{X}$ for which $\boldsymbol{\xi}$ is a cluster point.

Prove that $\boldsymbol{\lambda} \in \mathcal{Y}$ is a limit point of $f$ as $\mathbf{x}$ approaches $\boldsymbol{\xi}$ through the set $S$ if and only if there exists a sequence $\langle \mathbf{x}_k \rangle$ of points of $S$ such that $\mathbf{x}_k \neq \boldsymbol{\xi}$ $(k=1, 2, \ldots)$ and $\mathbf{x}_k \to \boldsymbol{\xi}$ as $k \to \infty$ for which

$$f(\mathbf{x}_k) \to \boldsymbol{\lambda} \quad \text{as} \quad k \to \infty.$$

[*Hint*: Use theorem 25.8].

(2) Find all real limit points of the following sequences of real numbers:

(i) $\langle k \rangle$ (ii) $\langle (-1)^k k \rangle$ (iii) $\langle 1/k \rangle$
(iv) $\langle (-1)^k \rangle$ (v) $\langle 1+(-1)^k \rangle$ (vi) $\langle k+(-1)^k k \rangle$.

These sequences may also be regarded as taking values in $[-\infty, +\infty]$. Find all limit points from $[-\infty, +\infty]$ of the given sequences.

(3) Find all real limit points as $x \to 0+$ of the following functions $f: (0, \infty) \to \mathbb{R}$:

(i) $f(x)=\sin \dfrac{1}{x}$ (ii) $f(x)=x \sin \dfrac{1}{x}$

(iii) $f(x)=\dfrac{1}{x} \sin \dfrac{1}{x}$ (iv) $f(x)=1-\sin \dfrac{1}{x}$

(v) $f(x)=\frac{1}{x}-\frac{1}{x}\sin\frac{1}{x}$ (vi) $f(x)=\frac{1}{x^2}-\frac{1}{x}\sin\frac{1}{x}$.

These functions may also be regarded as taking values in $[-\infty, +\infty]$. Find all limit points from $[-\infty, +\infty]$ of the given functions as $x \to 0+$.

(4) Find all real limit points as $(x, y)$ approaches $(0, 0)$ of the function $f$ of exercise 22.24(3).

(5) Give examples of functions $f$: $(0, \infty) \to \mathbb{R}$ for which the set $L$ of *real* limit points as $x \to 0+$ is as indicated below:

(i) $L=\emptyset$ (ii) $L=\mathbb{R}$ (iii) $L=[0, 1]$
(iv) $L=[0, \infty)$ (v) $L=\{-1, 1\}$ (vi) $L=\mathbb{N}$
(vii) $L=\{0\}$ but $f(x)\not\to 0$ as $x \to 0+$.

Show that in the case of each of your examples *except* (iii) and (v), $+\infty$ or $-\infty$ is also a limit point.

(6) Suppose that $f$: $(0, \infty) \to \mathbb{R}$ is continuous on $(0, \infty)$. Let $L$ be the set of real limit points as $x \to 0+$. Prove that $L$ is connected.

---

## 26.5† Oscillating functions

26.6† *Theorem* Suppose that $\mathcal{X}$ and $\mathcal{Y}$ are metric spaces and let $f$: $S \to \mathcal{Y}$ where $S$ is a set in $\mathcal{X}$ for which $\xi$ is a cluster point. Then the set $L$ of limit points of $f$ as $\mathbf{x}$ approaches $\xi$ through $S$ is given by

$$L=\bigcap_{k=1}^{\infty} \overline{f(S \cap A_k)}$$

where the set $A_k$ is obtained by removing $\xi$ from the open ball with centre $\xi$ and radius $1/k$.

*Proof* (i) Suppose that $\boldsymbol{\lambda} \in L$. Then there exists a sequence $\langle \mathbf{x}_k \rangle$ of points of $S \setminus \{\xi\}$ satisfying $\mathbf{x}_k \to \xi$ as $k \to \infty$ such that $f(\mathbf{x}_k) \to \boldsymbol{\lambda}$ as $k \to \infty$ (exercise 26.4(1)). Given any $J$, there exists a $K$ such that

$$k>K \Rightarrow \mathbf{x}_k \in A_J \cap S \Rightarrow f(\mathbf{x}_k) \in f(A_J \cap S).$$

Hence $\boldsymbol{\lambda} \in \overline{f(A_J \cap S)}$. It follows that

$$L \subset \bigcap_{k=1}^{\infty} \overline{f(A_k \cap S)}.$$

(ii) Suppose that $\boldsymbol{\lambda} \in \overline{f(A_k \cap S)}$ for each $k \in \mathbb{N}$. Then for each $k \in \mathbb{N}$, there exists a sequence $\langle \mathbf{x}_{k,l} \rangle$ of points of $A_k \cap S$ such that $f(\mathbf{x}_{k,l}) \to \boldsymbol{\lambda}$ as $l \to \infty$. For each $k \in \mathbb{N}$, choose $l_k$ so that $d(\boldsymbol{\lambda}, f(\mathbf{x}_{k,l_k}) < 1/k$. Then $\mathbf{x}_{k,l_k} \to \xi$ as $k \to \infty$ and $f(\mathbf{x}_{k,l_k}) \to \boldsymbol{\lambda}$ as $k \to \infty$.

Thus $\lambda \in L$ and so

$$\bigcap_{k=1}^{\infty} \overline{f(A_k \cap S)} \subset L.$$

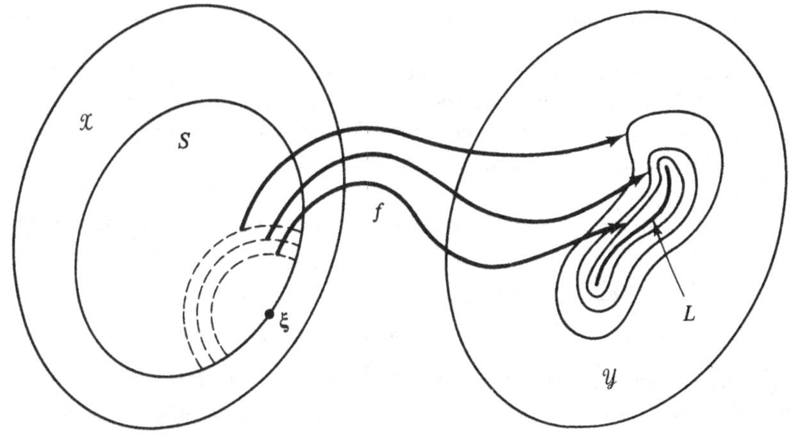


---

26.7† *Theorem* Let $\mathcal{X}$ and $\mathcal{Y}$ be metric spaces and suppose that $f: S \rightarrow \mathcal{Y}$ where $S$ is a set in $\mathcal{X}$ for which $\boldsymbol{\xi}$ is a cluster point. Then the set $L$ of limit points of $f$ as $\mathbf{x}$ approaches $\boldsymbol{\xi}$ through $S$ is *closed.*

*Proof* By the previous theorem, $L$ is the intersection of a collection of closed sets. Thus $L$ is closed by theorem 14.14.

---

26.8† *Theorem* In addition to the assumptions of the previous theorem, suppose that $f(S)$ lies in a compact subset $K$ of $\mathcal{Y}$. Then the set $L$ of limit points of $f$ as $\mathbf{x}$ approaches $\boldsymbol{\xi}$ through $S$ is non-empty.

*Proof* The sets $S \cap A_k$ are non-empty for each $k \in \mathbb{N}$ because $\boldsymbol{\xi}$ is a cluster point of $S$. It follows that $f(S \cap A_k)$ is non-empty for each $k \in \mathbb{N}$. Thus $\langle \overline{f(S \cap A_k)} \rangle$ is a nested sequence of non-empty, closed subsets of a compact set $K$. It follows from the Cantor intersection theorem (19.9) that the sequence has a non-empty intersection – i.e. $L \neq \emptyset$.

---

26.9† *Theorem* With the assumptions of the previous theorem,

$$f(\mathbf{x}) \rightarrow \boldsymbol{\eta} \quad \text{as} \quad \mathbf{x} \rightarrow \boldsymbol{\xi}$$

through $S$ if and only if $L = \{\boldsymbol{\eta}\}$.

*Proof* If $f(\mathbf{x})\to\boldsymbol{\eta}$ as $\mathbf{x}\to\boldsymbol{\xi}$ through $S$, then it is trivial to prove that $L=\{\boldsymbol{\eta}\}$. If it is false that $f(\mathbf{x})\to\boldsymbol{\eta}$ as $\mathbf{x}\to\boldsymbol{\xi}$ through $S$, then there exists a sequence $\langle\mathbf{x}_k\rangle$ of points of $S\setminus\{\boldsymbol{\xi}\}$ such that $\mathbf{x}_k\to\boldsymbol{\xi}$ as $k\to\infty$ but $f(\mathbf{x}_k)\not\to\boldsymbol{\eta}$ as $k\to\infty$. We deduce the existence of an $\varepsilon_0>0$ and a subsequence $\langle\mathbf{x}_{k_l}\rangle$ for which $d(\boldsymbol{\eta}, f(\mathbf{x}_{k_l}))\geqq\varepsilon_0$ $(l\in\mathbb{N})$. But $\langle f(\mathbf{x}_{k_l})\rangle$ has a convergent subsequence by theorem 25.25. Hence $L\neq\{\boldsymbol{\eta}\}$.

---

The assumption that $f(S)$ be a subset of a compact set $K\subset\mathcal{Y}$ which is made in theorems 26.8 and 26.9 is not so restrictive as it may seem at first sight. For example, if $\mathcal{Y}=\mathbb{R}$, one can always begin by replacing $\mathbb{R}$ by the *compact* space $[-\infty, +\infty]$.

---

26.10 *Example* Consider the sequence $\langle(-1)^k k-k\rangle$ of real numbers. We have that

$$(-1)^{2k}2k-2k\to 0 \quad \text{as} \quad k\to\infty$$
$$(-1)^{2k+1}(2k+1)-(2k+1)\to-\infty \quad \text{as} \quad k\to\infty.$$

The sequence therefore has only one *real* limit point namely 0. However, one cannot conclude from theorem 26.9 that the sequence converges because the range of the sequence does not lie in a compact subset of $\mathbb{R}$. Note that, if we regard the sequence as taking values in the *compact* space $[-\infty, +\infty]$, we find that the set $L$ of limit points contains *two* elements namely 0 and $-\infty$. Thus again theorem 26.9 does not apply.

---

Suppose that $\mathcal{X}$ is a metric space and that $f: S\to\mathbb{R}$ where $S$ is a set in $\mathcal{X}$ for which $\boldsymbol{\xi}$ is a cluster point. Let $L$ denote the set of limit points from $[-\infty, +\infty]$ as $\mathbf{x}$ approaches $\boldsymbol{\xi}$ through $S$.

Since $[-\infty, +\infty]$ is compact, we know from theorem 26.8 that $L\neq\emptyset$. If $L$ consists of a single point, then either $f$ converges as $\mathbf{x}$ approaches $\boldsymbol{\xi}$ through $S$ or else $f$ diverges to $+\infty$ or diverges to $-\infty$ as $\mathbf{x}$ approaches $\boldsymbol{\xi}$ through $S$. If $L$ consists of more than one point, we say that $f$ *oscillates* as $\mathbf{x}$ approaches $\boldsymbol{\xi}$ through $S$.

If $f$ oscillates as $\mathbf{x}$ approaches $\boldsymbol{\xi}$ through $S$ and $L$ is a compact subset of $\mathbb{R}$, it is customary to say that $f$ 'oscillates finitely'. (Note that, since $L$ is closed, it follows that $L$ is a compact subset of $\mathbb{R}$ if and only if $L$ is a subset of $\mathbb{R}$ which is bounded in $\mathbb{R}$.) Otherwise $f$ is said to 'oscillate infinitely'.

---

## 26.11† Lim sup and lim inf

Let $\mathcal{X}$ be a metric space and suppose that $f: S\to\mathbb{R}$ where $S$ is a set in $\mathcal{X}$ for which $\boldsymbol{\xi}$ is a cluster point.

The set $L$ of limit points from $[-\infty, +\infty]$ as $\mathbf{x}$ approaches $\boldsymbol{\xi}$ through $S$ is non-empty and closed by theorems 26.7 and 26.8. It follows that $L$ has a maximum element $\Lambda$ and a minimum element $\lambda$ (corollary 14.11). We call $\Lambda$ the *limit superior* (or lim sup) of $f$ as $\mathbf{x}$ approaches $\boldsymbol{\xi}$ through $S$. We call $\lambda$ the *limit inferior* (or lim inf). Sometimes $\Lambda$ is referred to as the *upper limit* and $\lambda$ as the *lower limit*.

The function $g_3$ illustrated in §26.1 is an example of a function for which both $\lambda$ and $\Lambda$ are finite (i.e. $\lambda$ and $\Lambda$ are elements of $\mathbb{R}$). The function $g_4$ of §26.1 is an example of a function for which $\lambda$ is finite but $\Lambda = +\infty$.

---

26.12† *Theorem* Let $\mathcal{X}$ be a metric space and suppose that $f: S \to \mathbb{R}$ where $S$ is a set in $\mathcal{X}$ for which $\boldsymbol{\xi}$ is a cluster point.

Then the function $f$ oscillates as $\mathbf{x}$ approaches $\boldsymbol{\xi}$ through $S$ if and only if $\lambda \neq \Lambda$. Moreover, given any $\eta \in [-\infty, +\infty]$,

$$f(\mathbf{x}) \to \eta \quad \text{as} \quad \mathbf{x} \to \boldsymbol{\xi}$$

through $S$ if and only if $\lambda = \eta = \Lambda$.

*Proof* This is an immediate consequence of theorem 26.9.

---

26.13† *Proposition* With the assumptions of the preceding theorem, $\Lambda$ is the limit superior of $f$ as $\mathbf{x}$ approaches $\boldsymbol{\xi}$ through $S$ if and only if

(i) $\forall l > \Lambda \ \exists \delta > 0 \ \forall \mathbf{x} \in S, \quad 0 < d(\boldsymbol{\xi}, \mathbf{x}) < \delta \Rightarrow f(\mathbf{x}) < l$

(ii) $\forall l < \Lambda \ \forall \delta > 0 \ \exists \mathbf{x} \in S, \quad 0 < d(\boldsymbol{\xi}, \mathbf{x}) < \delta \quad \text{and} \quad f(\mathbf{x}) > l.$

---

Item (i) above asserts that nothing larger than $\Lambda$ is a limit point. Item (ii) guarantees that $\Lambda$ lies in the closure of the set $L$ of limit points. Since $L$ is closed, this means that $\Lambda \in L$.

Similar criteria, of course, hold for the limit inferior $\lambda$.

The notation

$$\Lambda = \limsup_{\mathbf{x} \to \boldsymbol{\xi}} f(\mathbf{x}); \quad \lambda = \liminf_{\mathbf{x} \to \boldsymbol{\xi}} f(\mathbf{x})$$

is often used. The reason for the choice of this notation will be evident from the statement of the following result.

---

26.14† *Proposition* Let $\mathcal{X}$ be a metric space and suppose that $f: S \to \mathbb{R}$ where $S$ is a set in $\mathcal{X}$ for which $\boldsymbol{\xi}$ is a cluster point. Let $A_\delta$ denote the set obtained by removing $\boldsymbol{\xi}$ from the open ball with centre $\boldsymbol{\xi}$ and radius $\delta > 0$. Then

$$\text{(i)} \ \Lambda = \lim_{\delta \to 0+} \left\{ \sup_{\mathbf{x} \in S \cap A_\delta} f(\mathbf{x}) \right\}$$

$$\text{(ii)} \ \lambda = \lim_{\delta \to 0+} \left\{ \inf_{\mathbf{x} \in S \cap A_\delta} f(\mathbf{x}) \right\}.$$

---

The proof of (i) above depends on the observation that the function $F:(0,\infty)\to[-\infty,+\infty]$ defined by

$$F(\delta)=\sup_{\mathbf{x}\in S\cap A_\delta} f(\mathbf{x})$$

increases on $(0,\infty)$. This guarantees the existence of the limit as $\delta\to0+$. The identification of the limit with $\Lambda$ is then achieved with the help of proposition 26.13.

---

26.15† *Exercise*

(1) Determine the lim sup and lim inf for each of the examples given in exercises 26.4(2), (3) and (4).

(2) Suppose that $\langle a_k\rangle$ is any sequence of real numbers. Prove that $\Lambda\in[-\infty,+\infty]$ is the lim sup of the sequence if and only if

(i) For any subsequence $\langle a_{k_l}\rangle$ of $\langle a_k\rangle$,

$$a_{k_l}\to\eta \text{ as } l\to\infty \Rightarrow \eta\leqq\Lambda$$

and (ii) There exists a subsequence $\langle a_{k_l}\rangle$ of $\langle a_k\rangle$ such that

$$a_{k_l}\to\Lambda \quad\text{as}\quad l\to\infty.$$

What is the corresponding result for the lim inf of a sequence?

(3) Suppose that $\langle a_k\rangle$ is any sequence of real numbers. Prove that $\Lambda\in[-\infty,+\infty]$ is the lim sup of the sequence if and only if

(i) $\forall L>\Lambda$, $\{k: a_k\geqq L\}$ is finite

and (ii) $\forall l<\Lambda$, $\{k: a_k\geqq l\}$ is infinite.

What is the corresponding result for the lim inf of a sequence?

(4) Suppose that $\langle a_k\rangle$ and $\langle b_k\rangle$ are any sequences of real numbers. Prove that

$$\limsup_{k\to\infty}(a_k+b_k)\leqq\left(\limsup_{k\to\infty} a_k\right)+\left(\limsup_{k\to\infty} b_k\right)$$

whenever the right-hand side makes sense. Give an example for which the left-hand side is $-\infty$ and the right-hand side is $+\infty$. Show that the two sides are equal whenever $\langle b_k\rangle$ is a *convergent* sequence of real numbers.

What are the corresponding results for the lim infs of the sequences?

(5) Suppose that $\langle a_k\rangle$ and $\langle b_k\rangle$ are any bounded sequences of real numbers. Prove that

$$\liminf_{k\to\infty}(a_k+b_k)\leqq\left(\liminf_{k\to\infty} a_k\right)+\left(\limsup_{k\to\infty} b_k\right)\leqq\limsup_{k\to\infty}(a_k+b_k).$$

Show also that

$$\limsup_{k\to\infty}(-a_k)=-\left(\liminf_{k\to\infty} a_k\right).$$

Deduce that, for any bounded sequence $\langle c_k\rangle$ of real numbers,

$$\liminf_{k\to\infty}(c_k-c_{k+1})\leqq0\leqq\limsup_{k\to\infty}(c_k-c_{k+1}).$$

(6) Suppose that $\langle a_k \rangle$ is any sequence of positive real numbers. If there exists a $K$ such that $a_{k+1} \leqq r a_k$ for any $k > K$, prove that there exists an $H$ such that $a_k < Hr^k$ for all $k \in \mathbb{N}$. Deduce that

$$\limsup_{k \to \infty} a_k^{1/k} \leqq \limsup_{k \to \infty} \frac{a_{k+1}}{a_k}.$$

# 27 COMPLETENESS

## 27.1 Cauchy sequences

A *Cauchy sequence* $\langle \mathbf{x}_k \rangle$ of points in a metric space $\mathcal{X}$ is a sequence with the property that, for any $\varepsilon > 0$, there exists an open ball $E$ of radius $\varepsilon$ and a $K$ such that

$$k > K \Rightarrow \mathbf{x}_k \in E.$$

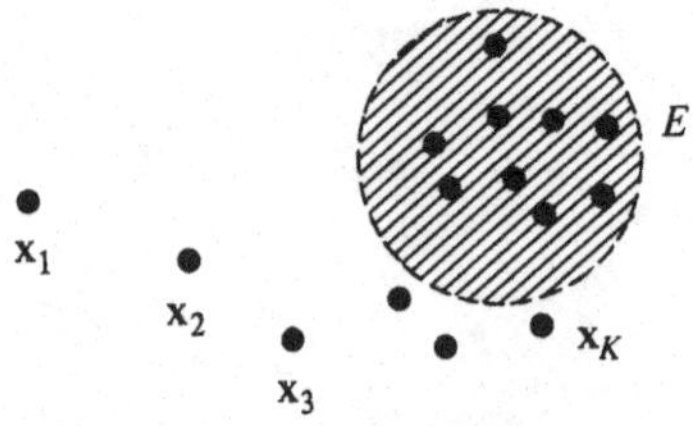


The definition of a convergent sequence asserts that the terms of the sequence $\langle \mathbf{x}_k \rangle$ can be forced as close to the limit $\xi$ as we choose by taking $k$ sufficiently large. The definition of a Cauchy sequence asserts that the terms of the sequence $\langle \mathbf{x}_k \rangle$ can be forced as *close to each other* as we choose by taking $k$ sufficiently large.

This latter point is more evident if the definition is rewritten in the form:

$$\forall \varepsilon > 0 \ \exists K, \quad (k > K \text{ and } l > K) \Rightarrow d(\mathbf{x}_k, \mathbf{x}_l) < \varepsilon.$$

## 27.2 Completeness

A *complete* metric space $\mathcal{X}$ is a metric space in which every Cauchy sequence converges.

Our first task is to link this notion with the ideas introduced in §20.15 by showing that a complete metric space is one in which a suitable analogue of the Bolzano–Weierstrass theorem (19.6) is true.

27.3† *Theorem* A metric space $\mathfrak{X}$ is complete if and only if every totally bounded set in $\mathfrak{X}$ has the Bolzano–Weierstrass property.

*Proof* (i) Suppose that $\mathfrak{X}$ is complete and that $S$ is a totally bounded set in $\mathfrak{X}$. Let $E$ be an infinite subset of $S$ and let $\langle \mathbf{x}_k \rangle$ be a sequence of distinct points of $E$. The definition of a Cauchy sequence is equivalent to the assertion that the range of the sequence is totally bounded. It follows that $\langle \mathbf{x}_k \rangle$ is a Cauchy sequence and hence converges. Its limit is then a cluster point of $E$ (exercise 25.17(3)).

(ii) Suppose that every totally bounded set in $\mathfrak{X}$ has the Bolzano–Weierstrass property. Let $\langle \mathbf{x}_k \rangle$ be a Cauchy sequence. Then its range is totally bounded and hence is either finite or else possesses a cluster point. In either case $\langle \mathbf{x}_k \rangle$ has a convergent subsequence (exercise 25.17(3)) and therefore converges (exercise 27.7(3)).

---

Not only is an analogue of the Bolzano–Weierstrass theorem true in a complete metric space, but we also have the following analogues of the Heine–Borel theorem and the Chinese box theorem.

---

27.4† *Theorem* A set $K$ in a complete metric space $\mathfrak{X}$ is compact if and only if it is closed and totally bounded.

*Proof* See §20.16.

---

27.5† *Theorem* Let $\langle B_k \rangle$ be a nested sequence of non-empty closed balls in a complete metric space $\mathfrak{X}$ whose radii tend to zero. Then

$$\bigcap_{k=1}^{\infty} B_k$$

is non-empty (and in fact consists of a single point).

*Proof* Note that a closed ball need *not* be compact in a general complete metric space $\mathfrak{X}$. (See §20.20.) The Cantor intersection theorem (19.9) therefore does not apply.

Let $\mathbf{x}_k \in B_k$. Given any $\varepsilon > 0$, let $B_K$ be the first ball with radius less than $\varepsilon$. Then

$$k > K \Rightarrow \mathbf{x}_k \in B_K$$

and hence $\langle \mathbf{x}_k \rangle$ is a Cauchy sequence. Since $\mathfrak{X}$ is complete, $\langle \mathbf{x}_k \rangle$ therefore converges. But each of the sets $B_k$ are closed and the limit of $\langle \mathbf{x}_k \rangle$ therefore belongs to each of these sets (corollary 25.15).

---

Theorem 23.17 was concerned with the conditions under which one can reverse the limiting operations in a repeated limit. If $\mathfrak{X}$ is a *complete* metric space, this theorem takes a more satisfactory form.

---

27.6† *Theorem* Let $\mathscr{X}$, $\mathscr{Y}$ and $\mathscr{Z}$ be metric spaces and suppose that $\mathscr{Z}$ is *complete.* Let $\xi \in \mathscr{X}$ and $\eta \in \mathscr{Y}$. Suppose that $A$ is a set in $\mathscr{X}$ for which $\xi$ is a cluster point and that $B$ is a set in $\mathscr{Y}$. Let $f: S \setminus (\xi, \eta) \to \mathscr{Z}$ where $S = A \times B$. Let $l: B \to \mathscr{Z}$ and let $m: A \to \mathscr{Z}$.

Suppose that

(i) $f(x, y) \to l(y)$ as $x \to \xi$ through the set $A$ *uniformly* for $y \in B$, and
(ii) $f(x, y) \to m(x)$ as $y \to \eta$ through the set $B$ *pointwise* for $x \in A$.

Then there exists a $\zeta \in \mathscr{Z}$ such that

(iii) $l(y) \to \zeta$ as $y \to \eta$ through the set $B$, and
(iv) $m(x) \to \zeta$ as $x \to \xi$ through the set $A$.

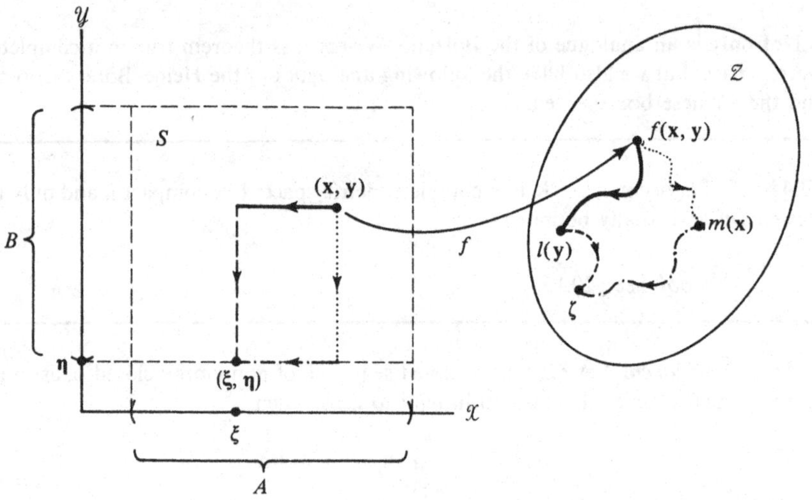


*Proof* Let $\varepsilon > 0$ be given. From (i) we have that there exists a $\delta > 0$ such that for each $x \in A$ and each $y \in B$,

$$0 < d(x, \xi) < \delta \Rightarrow d(l(y), f(x, y)) < \varepsilon/4.$$

From (ii) we have that, for each $x \in B$ there exists a $\Delta_x$ such that

$$0 < d(y, \eta) < \Delta_x \Rightarrow d(m(x), f(x, y)) < \varepsilon/4.$$

It follows that if $x \in A$ satisfies $0 < d(x, \xi) < \delta$ and $y \in B$ satisfies $0 < d(y, \eta) < \Delta_x$, then

$$d(l(y), m(x)) < \varepsilon/2.$$

We may deduce that, if $x_1 \in A$ and $x_2 \in A$ satisfy $0 < d(x_1, \xi) < \delta$ and $0 < d(x_2, \xi) < \delta$, then

$$d(m(x_1), m(x_2)) < \varepsilon.$$

From this it follows that, if $\langle x_k \rangle$ is any sequence of points of $A \setminus \{\xi\}$ such that $x_k \to \xi$ as $k \to \infty$, then $\langle m(x_k) \rangle$ is a Cauchy sequence in $\mathscr{X}$. Item (iv) then follows from theorem 25.8 and the fact that $\mathcal{X}$ is complete. We can then appeal to theorem 23.17 to obtain item (iii).

---

27.7 *Exercise*

(1) Prove that any convergent sequence of points in a metric space $\mathcal{X}$ is a Cauchy sequence. Prove that any Cauchy sequence in a metric space $\mathcal{X}$ is bounded.

(2) Prove that any sequence $\langle \mathbf{x}_k \rangle$ in a metric space $\mathcal{X}$ which satisfies $d(\mathbf{x}_k, \mathbf{x}_{k+1}) \leqq 2^{-k}$ $(k \in \mathbb{N})$ is a Cauchy sequence. Prove that any Cauchy sequence $\langle \mathbf{x}_k \rangle$ in a metric space $\mathcal{X}$ has a subsequence $\langle \mathbf{x}_{k_l} \rangle$ which satisfies.

$$d(\mathbf{x}_{k_l}, \mathbf{x}_{k_{l+1}}) \leqq 2^{-l} \quad (l \in \mathbb{N}).$$

(3) Prove that any Cauchy sequence in a metric space $\mathcal{X}$ which has a convergent subsequence is itself convergent.

†(4) Let $\mathcal{Y}$ be a set in a metric space $\mathcal{X}$. Then $\mathcal{Y}$ may be regarded as a metric subspace of $\mathcal{X}$. Prove the following:
 (i) If $\mathcal{X}$ is complete, then $\mathcal{Y}$ is complete if and only if $\mathcal{Y}$ is closed in $\mathcal{X}$.
 (ii) If $\mathcal{Y}$ is compact in $\mathcal{X}$, then $\mathcal{Y}$ is complete.

†(5) Let $\mathcal{X}$ be a metric space for which it is true that every nested sequence of non-empty closed balls whose radii tend to zero has a non-empty intersection. Prove that $\mathcal{X}$ is complete. (See theorem 27.5.)

†(6) If $\mathcal{X}$ is a metric space, a function $f: \mathcal{X} \to \mathcal{X}$ is called a *contraction* if and only if there exists a real number $\alpha < 1$ such that

$$d(f(\mathbf{x}), f(\mathbf{y})) \leqq \alpha d(\mathbf{x}, \mathbf{y})$$

for each $\mathbf{x} \in \mathcal{X}$ and each $\mathbf{y} \in \mathcal{X}$. Let $f: \mathcal{X} \to \mathcal{X}$ be a contraction. Prove that:
 (i) $f$ is continuous on $\mathcal{X}$.
 (ii) Any sequence $\langle \mathbf{x}_k \rangle$ of points of $\mathcal{X}$ which satisfies $\mathbf{x}_{k+1} = f(\mathbf{x}_k)$ is a Cauchy sequence.
 (iii) If $\mathcal{X}$ is complete, then $f$ has a fixed point – i.e. for some $\xi \in \mathcal{X}$, $f(\xi) = \xi$. (See example 17.13.)

---

## 27.8 Some complete spaces

The results of the previous section show that complete spaces are of some interest. In this section we shall consider a few specific examples of complete spaces.

Since the Bolzano–Weierstrass theorem holds in $\mathbb{R}^n$, it follows from theorem 27.3 that $\mathbb{R}^n$ is complete. The proof of the next theorem establishes this result in a more direct fashion.

---

27.9 *Theorem* The metric space $\mathbb{R}^n$ is complete.

*Proof* Let $\langle \mathbf{x}_k \rangle$ be a Cauchy sequence in $\mathbb{R}^n$. By exercise 27.7(1), $\langle \mathbf{x}_k \rangle$ is bounded and hence, by theorem 25.24, $\langle \mathbf{x}_k \rangle$ has a convergent subsequence. The conclusion therefore follows from exercise 27.7(3).

---

Let $S$ be a non-empty set and let $\mathcal{Y}$ be a metric space. We shall denote the set of all bounded functions $f: S \to \mathcal{Y}$ by $\mathcal{B}(S, \mathcal{Y})$. The set $\mathcal{B}(S, \mathcal{Y})$ becomes a metric space if we define the distance between two 'points' $f$ and $g$ of $\mathcal{B}(S, \mathcal{Y})$ by

$$u(f, g) = \sup_{\mathbf{x} \in S} d(f(\mathbf{x}), g(\mathbf{x})).$$

We call $u$ the *uniform metric* on $\mathcal{B}(S, \mathcal{Y})$.

This is not a new notion. We introduced the same idea in §23.12 in connection with uniform convergence and in §20.20 we studied the special case $l^\infty = \mathcal{B}(\mathbb{N}, \mathbb{R})$.

---

27.10† *Theorem* If $\mathcal{Y}$ is complete, then so is the metric space $\mathcal{B}(S, \mathcal{Y})$.

*Proof* Let $\langle f_k \rangle$ be a Cauchy sequence in $\mathcal{B}(S, \mathcal{Y})$. Then, for each $\mathbf{x} \in S$, $\langle f_k(\mathbf{x}) \rangle$ is a Cauchy sequence in $\mathcal{Y}$. Since $\mathcal{Y}$ is complete, it follows that, for each $\mathbf{x} \in S$, $\langle f_k(\mathbf{x}) \rangle$ converges – i.e. there exists a function $f: S \to \mathcal{Y}$ such that $f_k(\mathbf{x}) \to f(\mathbf{x})$ as $k \to \infty$.

In the language of §23.11 and §23.12, we have shown that $\langle f_k(\mathbf{x}) \rangle$ converges *pointwise* for $\mathbf{x} \in S$. What we need to show is that $\langle f_k \rangle$ converges in the space $\mathcal{B}(S, \mathcal{Y})$ – i.e. $\langle f_k(\mathbf{x}) \rangle$ converges *uniformly* for $\mathbf{x} \in S$.

Given any $\varepsilon > 0$, there exists a $K$ such that for any $k > K$ and any $l > K$,

$$u(f_l, f_k) < \varepsilon/2.$$

It follows that, for each $\mathbf{x} \in S$,

$$d(f_l(\mathbf{x}), f_k(\mathbf{x})) < \varepsilon/2.$$

But, for each $\mathbf{x} \in S$, $f_k(\mathbf{x}) \to f(\mathbf{x})$ as $k \to \infty$. Thus, for any $l > K$

$$d(f_l(\mathbf{x}), f(\mathbf{x})) \leqq \varepsilon/2.$$

(See exercise 16.16(4).) We deduce that, for any $l > K$

$$u(f_l, f) \leqq \varepsilon/2 < \varepsilon.$$

It follows that $f_l \to f$ as $l \to \infty$ in the space $\mathcal{B}(S, \mathcal{Y})$.

---

27.11† *Theorem* Suppose that $\mathcal{Y}$ is a complete metric space. Then the set $C$ of all continuous functions in $\mathcal{B}(S, \mathcal{Y})$ is closed in $\mathcal{B}(S, \mathcal{Y})$.

*Proof* Suppose that $\langle f_k \rangle$ is a sequence of functions in $C$ such that $f_k \to f$ as $k \to \infty$. To prove that $C$ is closed, we need to show that $f \in C$ (corollary 25.15).

Let $\varepsilon > 0$ be given. Then there exists a $K$ such that

$$k > K \Rightarrow u(f, f_k) < \varepsilon/3.$$

Let $\xi \in S$ and let $J > K$. Since $f_J$ is continuous on $S$, there exists a $\delta > 0$ such that for each $\mathbf{x} \in S$

$$d(\xi, \mathbf{x}) < \delta \Rightarrow d(f_J(\xi), f_J(\mathbf{x})) < \varepsilon/3.$$

It follows that, if $d(\xi, \mathbf{x}) < \delta$, then

$$\begin{aligned} d(f(\xi), f(\mathbf{x})) &\leqq d(f(\xi), f_J(\xi)) + d(f_J(\xi), f_J(\mathbf{x})) + d(f_J(\mathbf{x}), f(\mathbf{x})) \\ &< \varepsilon/3 + \varepsilon/3 + \varepsilon/3 = \varepsilon \end{aligned}$$

and so $f$ is continuous on $S$ – i.e. $f \in C$.

---

We shall use the notation $\mathcal{C}(S, \mathcal{Y})$ to denote the space of all continuous functions $f: S \to \mathcal{Y}$. Such functions need not in general be bounded. However, if $S$ is a *compact* set in a metric space $\mathcal{X}$, then a continuous function $f: S \to \mathcal{Y}$ *is* bounded and, moreover,

$$u(f, g) = \max_{\mathbf{x} \in S} d(f(\mathbf{x}), g(\mathbf{x})).$$

To obtain this result we apply theorem 19.14 to the continuous function $F: S \to \mathbb{R}$ defined by $F = d \circ (f, g)$.

---

27.12† *Theorem* Suppose that $\mathcal{Y}$ is a complete metric space and $S$ is compact. Then $\mathcal{C}(S, \mathcal{Y})$ is complete.

*Proof* This follows from theorem 27.11 and exercise 27.7(4i).

---

## 27.13† Incomplete spaces

It is by no means the case that all metric spaces are complete. As a typical example of an incomplete space we shall consider the system $\mathbb{Q}$ of rational numbers (regarded as a metric subspace of $\mathbb{R}$). Some more examples are given in exercise 27.15.

The fact that $\mathbb{Q}$ is not complete follows from the fact that $\mathbb{Q}$ is not a closed subset of $\mathbb{R}$ (see exercise 27.7(4i)). It is instructive, however, to consider some specific examples for which the properties of a complete space fail to hold in $\mathbb{Q}$.

---

27.14 *Examples* Consider the sequences $\langle x_k \rangle$ and $\langle y_k \rangle$ defined by $x_1 = 2$, $y_1 = \frac{1}{2}$ and

$$x_{k+1} = \frac{1}{2}\left(x_k + \frac{2}{x_k}\right); \quad y_{k+1}^{-1} = \frac{1}{2}\left(y_k^{-1} + \frac{1}{2y_k^{-1}}\right).$$

These are both sequences of rational numbers. The sequence $\langle x_k \rangle$ decreases and converges (in the space $\mathbb{R}$) to $\sqrt{2}$. (See exercise 25.11(6).) Similarly, the sequence $\langle y_k \rangle$ increases and converges (in the space $\mathbb{R}$) to $\sqrt{2}$.

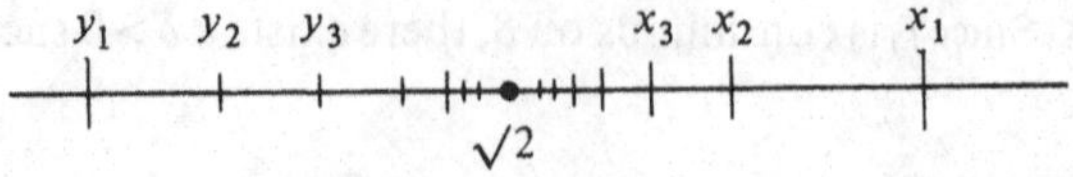


(i) Since $\langle x_k \rangle$ converges in the space $\mathbb{R}$, it is a Cauchy sequence in $\mathbb{R}$ (exercise 27.7(1)). It follows that $\langle x_k \rangle$ is a Cauchy sequence in the space $\mathbb{Q}$. But $\langle x_k \rangle$ does *not* converge if it is regarded as a sequence of points in $\mathbb{Q}$. There is no *rational* number $\xi$ such that $x_k \to \xi$ as $k \to \infty$. In fact, $x_k \to \sqrt{2}$ as $k \to \infty$ and $\sqrt{2}$ is *irrational.*

(ii) Let $I_k = \{r : r \in \mathbb{Q} \text{ and } x_k \leqq r \leqq y_k\}$. Then $I_k = [x_k, y_k] \cap \mathbb{Q}$ and hence is closed in the metric space $\mathbb{Q}$ (corollary 21.12). It follows that $\langle I_k \rangle$ is a nested sequence of closed boxes in $\mathbb{Q}$ (§19.2) but

$$\bigcap_{k=1}^{\infty} I_k = \emptyset$$

– i.e. the Chinese box theorem fails in the space $\mathbb{Q}$.

(iii) Consider the set $S = \{r : r \in \mathbb{Q} \text{ and } 0 \leqq r \leqq \sqrt{2}\}$. This set is closed in the metric space $\mathbb{Q}$. It is also totally bounded. But it is *not* compact in $\mathbb{Q}$. For example, the continuous function $f: \mathbb{Q} \to \mathbb{R}$ defined by $f(x) = x$ does *not* achieve a maximum on $S$.

---

## 27.15† *Exercise*

(1) Find the uniform distance between the functions $f: \mathbb{R} \to \mathbb{R}$ and $g: \mathbb{R} \to \mathbb{R}$ defined by

$$f(x) = \frac{1}{1+x^2}; \quad g(x) = \frac{x}{1+x^2}$$

where these functions are regarded as points in the metric space $\mathcal{B}(\mathbb{R}, \mathbb{R})$.

(2) Let $\mathcal{X}$ be the set of all continuous functions $f: [a, b] \to \mathbb{R}$. Let the metric in $\mathcal{X}$ be defined by

$$d(f, g) = \int_a^b |f(x) - g(x)| dx.$$

Prove that $\mathcal{X}$ is not complete.

(3) Let $S$ be a set in a metric space $\mathcal{X}$ and let $\xi \in \mathcal{X}$. Let $\mathcal{Y}$ be a complete metric space and suppose that $L_\xi(S, \mathcal{Y})$ is the subset of $\mathcal{B}(S, \mathcal{Y})$ consisting of all bounded functions $f: S \to \mathcal{Y}$ such that there exists an $\boldsymbol{\eta} \in \mathcal{Y}$ for which $f(\mathbf{x}) \to \boldsymbol{\eta}$ as $\mathbf{x} \to \boldsymbol{\xi}$ through $S$. Prove that $L_\xi(S, \mathcal{Y})$ is closed in $\mathcal{B}(S, \mathcal{Y})$. [*Hint*: Use theorem 27.6.]

(4) Explain why the open interval (0, 1) (regarded as a metric subspace of $\mathbb{R}$) is not complete. Give an example of a Cauchy sequence in (0, 1) which does not converge in (0, 1).

Let $f: \mathbb{R} \times \mathbb{R} \to (0, 1)$ be defined by $f(x, y) = xy$. Show that $f(x, y) \to y$ as $x \to 1$ through (0, 1) *uniformly* for $y \in (0, 1)$ and that $f(x, y) \to x$ as $y \to 1$ through (0, 1) *pointwise* for $x \in (0, 1)$. Explain the relevance of this result to the assumption in theorem 27.6 that $\mathcal{Z}$ is complete.

(5) Let $P_n: [0, 1] \to \mathbb{R}$ be defined by

$$P_n(x) = 1 + x + \frac{x^2}{2!} + \frac{x^3}{3!} + \ldots + \frac{x^n}{n!}.$$

Prove that $\langle P_n \rangle$ converges in the space $\mathcal{B}([0, 1], \mathbb{R})$.

(6) Let $\mathcal{P}$ denote the metric space of all real polynomials where the metric is defined by

$$u(P, Q) = \max_{0 \leq x \leq 1} |P(x) - Q(x)|.$$

Prove that $\mathcal{P}$ is not complete. [*Hint*: See the previous question.]

---

## 27.16† Completion of metric spaces

When a space lacks a certain desirable property, a natural mathematical response is to seek to fit the space inside a larger space which does have the desirable property. In particular, if $\mathcal{X}$ is an *incomplete* metric space, can we fit $\mathcal{X}$ inside a larger *complete* metric snace? In this section we answer this question in the affirmative by constructing such a complete metric space. We shall, in fact, construct the *smallest* complete metric space inside which $\mathcal{X}$ can be fitted. This is called the *completion* of $\mathcal{X}$ and denoted by $\mathcal{X}^*$.

The construction is very simple. The defect in $\mathcal{X}$ is that it has Cauchy sequences $\langle \mathbf{x}_k \rangle$ which do not converge. We therefore need to *invent* a limit for each such Cauchy sequence. We must be careful, however, not to invent too many objects. In particular, if two Cauchy sequences $\langle \mathbf{x}_k \rangle$ and $\langle \mathbf{y}_k \rangle$ satisfy

$$d(\mathbf{x}_k, \mathbf{y}_k) \to 0 \quad \text{as} \quad k \to \infty \tag{1}$$

then we shall want both sequences to converge to the *same* limit. We therefore begin by introducing an equivalence relation $\sim$ on the set $S$ of Cauchy sequences of points of $\mathcal{X}$ by writing

$$\langle \mathbf{x}_k \rangle \sim \langle \mathbf{y}_k \rangle$$

if and only if (1) holds. The set $\mathcal{X}^*$ is defined to be the set of all equivalence classes of $S$ defined by this equivalence relation. We shall use the notation $[\langle \mathbf{x}_k \rangle]$ to denote the equivalence class containing the Cauchy sequence $\langle \mathbf{x}_k \rangle$. The original space $\mathcal{X}$ is fitted inside the new space $\mathcal{X}^*$ by identifying an element $\mathbf{x}$ of $\mathcal{X}$ with the element $[\langle \mathbf{x} \rangle]$ of $\mathcal{X}^*$.

We need to show that $\mathcal{X}^*$ is a complete metric space. First it is necessary to introduce a metric into $\mathcal{X}^*$ which is consistent with that of $\mathcal{X}$. We therefore define $d: \mathcal{X}^* \to \mathbb{R}$ by

$$d([\langle \mathbf{x}_k \rangle], [\langle \mathbf{y}_k \rangle]) = \lim_{k \to \infty} d(\mathbf{x}_k, \mathbf{y}_k).$$

Note that the limit on the right-hand side exists because $\langle d(\mathbf{x}_k, \mathbf{y}_k) \rangle$ is a Cauchy sequence of real numbers whenever $\langle \mathbf{x}_k \rangle$ and $\langle \mathbf{y}_k \rangle$ are Cauchy sequences in $\mathcal{X}$. (See exercise 27.20(1).) But every Cauchy sequence of real numbers converges because $\mathbb{R}$ is complete (theorem 27.9).

---

27.17† *Theorem* For any metric space $\mathfrak{X}$, the metric space $\mathfrak{X}^*$ is complete.

*Proof* We have to prove that every Cauchy sequence in $\mathfrak{X}^*$ converges to a point of $\mathfrak{X}^*$. We proceed by showing that, if $\langle X_j \rangle$ is a sequence in $\mathfrak{X}^*$ satisfying

$$d(X_j, X_{j+1}) < 2^{-(j+1)}, \tag{2}$$

then there exists an $X \in \mathfrak{X}^*$ such that $X_j \to X$ as $j \to \infty$. Since each Cauchy sequence in $\mathfrak{X}^*$ has a subsequence which satisfies (2) (exercise 27.7(2)), the result will then follow from exercise 27.7(3).

Recall that the elements of $\mathfrak{X}^*$ are equivalence classes of Cauchy sequences in $\mathfrak{X}$. All subsequences of a given Cauchy sequence in $\mathfrak{X}$ clearly belong to the same equivalence class. From exercise 27.7(2) it follows that we may write

$$X_j = [\langle \mathbf{x}_k^{(j)} \rangle],$$

where $\langle \mathbf{x}_k^{(j)} \rangle$ is a sequence in $\mathfrak{X}$ which satisfies

$$d(\mathbf{x}_k^{(j)}, \mathbf{x}_{k+1}^{(j)}) < 2^{-(k+2)}$$

and hence

$$d(\mathbf{x}_k^{(j)}, \mathbf{x}_l^{(j)}) < 2^{-(k+1)} \tag{3}$$

provided that $l \geqq k$.

Now (2) asserts that

$$\lim_{k \to \infty} d(\mathbf{x}_k^{(j)}, \mathbf{x}_k^{(j+1)}) < 2^{-(j+1)}.$$

We may therefore deduce the existence of a strictly increasing sequence $\langle k_j \rangle$ of natural numbers such that, for any $l \geqq k_j$,

$$d(\mathbf{x}_l^{(j)}, \mathbf{x}_l^{(j+1)}) < 2^{-(j+1)}. \tag{4}$$

From (3) we also have that

$$d(\mathbf{x}_{k_j}^{(j+1)}, \mathbf{x}_{k_{j+1}}^{(j+1)}) < 2^{-(k_j+1)} \leqq 2^{-(j+1)}. \tag{5}$$

We are now in a position to define $X \in \mathfrak{X}^*$ by

$$X = [\langle \mathbf{x}_{k_j}^{(j)} \rangle].$$

For this to be an acceptable definition, it is necessary that $\langle \mathbf{x}_{k_j}^{(j)} \rangle$ is a Cauchy sequence in $\mathfrak{X}$. This fact follows from exercise 27.7(2) because

$$\begin{aligned} d(\mathbf{x}_{k_j}^{(j)}, \mathbf{x}_{k_{j+1}}^{(j+1)}) &\leqq d(\mathbf{x}_{k_j}^{(j)}, \mathbf{x}_{k_j}^{(j+1)}) + d(\mathbf{x}_{k_j}^{(j+1)}, \mathbf{x}_{k_{j+1}}^{(j+1)}) \\ &< 2^{-(j+1)} + 2^{-(j+1)} = 2^{-j} \end{aligned}$$

by (4) and (5).

It remains to show that $X_l \to X$ as $l \to \infty$. From (4) we have that, for each $j \geqq l$,

$$\begin{aligned} d(\mathbf{x}_{k_j}^{(l)}, \mathbf{x}_{k_j}^{(j)}) &\leqq d(\mathbf{x}_{k_j}^{(l)}, \mathbf{x}_{k_j}^{(l+1)}) + \ldots + d(\mathbf{x}_{k_j}^{(j-1)}, \mathbf{x}_{k_j}^{(j)}) \\ &< 2^{-(l+1)} + \ldots + 2^{-j} \\ &< 2^{-l}. \end{aligned}$$

But $X_l = [\langle \mathbf{x}_{k_j}{}^{(l)} \rangle]$ and thus it follows that

$$d(X_l, X) \leqq 2^{-l} \to 0 \quad \text{as} \quad l \to \infty.$$

---

## 27.18† Completeness and the continuum axiom

After the previous discussion it is natural to consider first the completion $\mathbb{Q}^*$ of the rational number system. Is $\mathbb{Q}^*$ the same as the real number system $\mathbb{R}$?

Before seeking to answer this question, we should remind ourselves that $\mathbb{Q}$ and $\mathbb{R}$ are not just metric spaces. Both are also ordered fields. The question is therefore only meaningful if we can extend the algebraic structure of $\mathbb{Q}$ to $\mathbb{Q}^*$ (as well as the metric structure). In particular, we need to be able to add, multiply and to order the equivalence classes which constitute the elements of $\mathbb{Q}^*$.

The appropriate definitions are the obvious ones. We define

(i) $[\langle x_k \rangle] + [\langle y_k \rangle] = [\langle x_k + y_k \rangle]$
(ii) $[\langle x_k \rangle] \cdot [\langle y_k \rangle] = [\langle x_k y_k \rangle]$
(iii) $[\langle x_k \rangle] > [\langle y_k \rangle] \Leftrightarrow \exists K (k > K \Rightarrow x_k > y_k)$.

With these definitions $\mathbb{Q}^*$ becomes an *ordered field* and

$$d([\langle x_k \rangle], [\langle y_k \rangle]) = |[\langle x_k \rangle] - [\langle y_k \rangle]|.$$

To prove this is a somewhat tiresome but essentially trivial task and so we shall omit the details.

Recall that $\mathbb{Q}$ is the 'smallest' ordered field. Thus $\mathbb{Q}^*$ is the 'smallest' *complete* ordered field. We know from theorem 27.9 that $\mathbb{R}$ is complete and thus $\mathbb{R}^* \subset \mathbb{R}$. (These remarks, of course, take for granted that structures isomorphic to $\mathbb{Q}$, $\mathbb{Q}^*$ or $\mathbb{R}$ are to be identified with $\mathbb{Q}$, $\mathbb{Q}^*$ or $\mathbb{R}$ respectively – see §9.21.)

The next theorem shows that the system $\mathbb{Q}^*$ and the system $\mathbb{R}$ are the same – i.e. the real number system is the completion of the rational number system. Recall that $\mathbb{R}$ is an ordered field which satisfies the *continuum axiom*. This asserts that every non-empty set which is bound above has a *smallest* upper bound.

---

27.19† *Theorem* The system $\mathbb{Q}^*$ satisfies the continuum axiom.

*Proof* Let $S$ be a non-empty set in $\mathbb{Q}^*$ which is bounded above. Let $a_1 \in S$ and let $b_1$ be an upper bound of $S$.

We construct an increasing sequence $\langle a_k \rangle$ of points of $S$ and a decreasing sequence $\langle b_k \rangle$ of upper bounds of $S$ such that

$$0 \leqq b_{k+1} - a_{k+1} \leqq 2^{-k+1}(b_1 - a_1).$$

The construction is inductive. If $c_k = \frac{1}{2}(a_k + b_k)$ is an upper bound of $S$, we take $b_{k+1} = c_k$ and $a_{k+1} = a_k$. If $c_k$ is not an upper bound we take $b_{k+1} = b_k$ and choose $a_{k+1} \in S$ so that $a_{k+1} \geqq c_k$.

The sequences $\langle a_k \rangle$ and $\langle b_k \rangle$ are both Cauchy sequences (exercise 27.7(2)) and hence converge. They have a common limit $\xi$ which is easily shown to be the smallest upper bound of the set $S$.

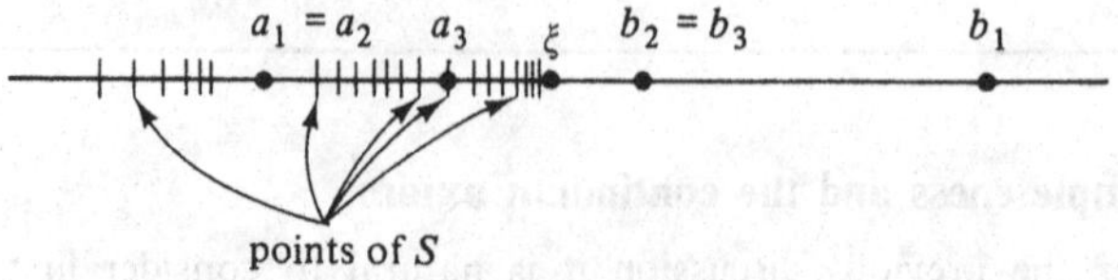


---

**27.20†** *Exercise*

(1) Let $\langle \mathbf{x}_k \rangle$ and $\langle \mathbf{y}_k \rangle$ be Cauchy sequences in a metric space $\mathfrak{X}$. Prove that $\langle d(\mathbf{x}_k, \mathbf{y}_k) \rangle$ is a Cauchy sequence of real numbers and hence converges.

(2) Check that $\mathbb{Q}^*$ is an ordered field with the definitions of addition, multiplication and an ordering given in §27.18.

(3) Every ordered field $\mathfrak{X}$ contains the system $\mathbb{N}$ of natural numbers (or else a system isomorphic to $\mathbb{N}$). We say that $\mathfrak{X}$ is Archimedean if $\mathbb{N}$ is unbounded above in $\mathfrak{X}$. (See §9.16.) The system $\mathbb{R}$ is Archimedean (theorem 9.17) and hence so are $\mathbb{Q}$ and $\mathbb{Q}^*$. Prove that an Archimedean ordered field $\mathfrak{X}$ is complete if and only if it satisfies the continuum axiom (i.e. $\mathfrak{X} = \mathbb{R}$).
(Note that by the assertion that the Archimedean ordered field $\mathfrak{X}$ is complete we mean that $\mathfrak{X}$ is 'complete' with respect to the 'metric' $d: \mathfrak{X} \times \mathfrak{X} \to \mathfrak{X}$ defined by $d(\mathbf{x}, \mathbf{y}) = |\mathbf{x} - \mathbf{y}|$.)

---

# 28 SERIES

## 28.1 Convergence of series

Suppose that $\langle \mathbf{a}_k \rangle$ is a sequence of points in a normed vector space $\mathcal{X}$. (See §13.17.) The $K$th *partial sum* of the sequence is defined by

$$\mathbf{s}_K = \sum_{k=1}^{K} \mathbf{a}_k = \mathbf{a}_1 + \mathbf{a}_2 + \ldots + \mathbf{a}_K.$$

If the limit of the sequence $\langle \mathbf{s}_K \rangle$ exists, then we say that the *series*

$$\sum_{k=1}^{\infty} \mathbf{a}_k = \mathbf{a}_1 + \mathbf{a}_2 + \mathbf{a}_3 + \ldots$$

*converges* and we write

$$\mathbf{s} = \sum_{k=1}^{\infty} \mathbf{a}_k$$

if and only if $\mathbf{s}_K \to \mathbf{s}$ as $K \to \infty$. If the sequence of partial sums does not converge, we say that the series *diverges*.

---

28.2 *Example* If $|x| < 1$, then

$$\sum_{k=0}^{\infty} x^k = (1-x)^{-1}.$$

To prove this result we observe that

$$\sum_{k=0}^{K} x^k = 1 + x + x^2 + \ldots + x^K$$

$$= \frac{1-x^{K+1}}{1-x} \to \frac{1}{1-x} \quad \text{as} \quad K \to \infty.$$

If $|x| \geqq 1$, then the sequence of partial sums diverges and so the infinite series does not exist.

---

28.3 *Theorem* (*Comparison test*) Suppose that $\mathscr{X}$ is a *complete* normed vector space and that

$$\sum_{k=1}^{\infty} b_k \tag{1}$$

is a convergent series of non-negative real numbers. If

$$\|\mathbf{a}_k\| \leqq b_k \quad (k \in \mathbb{N}),$$

then the series

$$\sum_{k=1}^{\infty} \mathbf{a}_k$$

converges.

*Proof* The partial sums of the sequence (1) are a Cauchy sequence (exercise 27.7(1)). Given any $\varepsilon > 0$, it follows that there exists an $N$ such that, for any $K > J > N$,

$$|t_K - t_J| = \sum_{k=J+1}^{K} b_k < \varepsilon$$

where $t_K$ denotes the $K$th partial sum of the series (1). It follows that, for any $K > J > N$,

$$\begin{aligned} \|\mathbf{s}_K - \mathbf{s}_J\| &= \left\| \sum_{k=J+1}^{K} \mathbf{a}_k \right\| \\ &\leqq \sum_{k=J+1}^{K} \|\mathbf{a}_k\| \leqq \sum_{k=J+1}^{K} b_k < \varepsilon \end{aligned}$$

and so $\langle \mathbf{s}_k \rangle$ is a Cauchy sequence. Since $\mathscr{X}$ is complete, we may conclude that $\langle \mathbf{s}_k \rangle$ converges.

---

Many analysis textbooks are stuffed full with tests for establishing the convergence of series. Most of these tests, however, are useful only for somewhat obscure series and we shall therefore restrict our attention to two of the more important.

---

28.4 *Proposition* (*Ratio test*) Let $\langle b_k \rangle$ denote a sequence of positive real numbers. Then the series

$$\sum_{k=1}^{\infty} b_k$$

converges if

(i) $\limsup_{k\to\infty} \frac{b_{k+1}}{b_k} < 1$

and diverges if

(ii) $\liminf_{k\to\infty} \frac{b_{k+1}}{b_k} > 1.$

*Proof* If (i) holds, there exists a $\rho < 1$ and an $H$ such that $b_k < H\rho^k$ $(k \in \mathbb{N})$. The convergence of the series then follows from the comparison test. If (ii) holds, then $b_k \not\to 0$ as $k \to \infty$ and so the series diverges.

---

28.5 *Proposition* (*Root test*) Let $\langle b_k \rangle$ denote a sequence of non-negative real numbers. Then the series

$$\sum_{k=1}^{\infty} b_k$$

converges if

(i) $\limsup_{k\to\infty} b_k^{1/k} < 1$

and diverges if

(ii) $\limsup_{k\to\infty} b_k^{1/k} > 1.$

*Proof* The same remarks apply as in the proof of proposition 28.4.

---

28.6 *Example* Consider the case $b_k = k^{-\alpha}$. Then

$$\lim_{k\to\infty} \frac{b_{k+1}}{b_k} = 1; \quad \lim_{k\to\infty} b_k^{1/k} = 1.$$

It follows that neither the ratio test nor the root test are helpful in determining the convergence or divergence of the series

$$\sum_{k=1}^{\infty} \frac{1}{k^{\alpha}}.$$

As it happens, the series converges for $\alpha > 1$ and diverges for $\alpha \leqq 1$. (See exercise 28.10(3).)

---

## 28.7 Absolute convergence

A series

$$\sum_{k=1}^{\infty} \mathbf{a}_k$$

converges *absolutely* if and only if

$$\sum_{k=1}^{\infty} \|\mathbf{a}_k\|$$

converges. The comparison test shows that in a *complete* space any absolutely convergent series converges.

The ratio test and the root test can therefore be used with $b_k = \|\mathbf{a}_k\|$ to establish the convergence of certain series in a complete space. But note that both tests (and the comparison test) are only able to establish *absolute* convergence. However, series exist which converge but do *not* converge absolutely. (See exercise 28.10(4).)

---

## 28.8 Power series

If $\langle a_k \rangle$ is a sequence of real (or complex) numbers and $\zeta$ is a real (or complex) number, then the series

$$\sum_{k=1}^{\infty} a_k(z-\zeta)^k$$

is called a power series (about the point $\zeta$). For what values of $z$ does this power series converge? The root test is useful here. Suppose that

$$\limsup_{k\to\infty} |a_k|^{1/k} = \rho.$$

Then

$$\limsup_{k\to\infty} |a_k(z-\zeta)^k|^{1/k} = \rho|z-\zeta|.$$

If $0<\rho<+\infty$, it follows that the power series converges when $|z-\zeta|<1/\rho$. If $\rho=0$, the power series always converges and, if $\rho=+\infty$, it converges only when $z=\zeta$.

This argument shows that there exists an $R\in[0, +\infty]$ such that the power series converges when $|z-\zeta|<R$ and diverges when $|z-\zeta|>R$. It follows that the set $S$ of values of $z$ for which the power series converges satisfies

$$\{z : |z-\zeta|<R\} \subset S \subset \{z : |z-\zeta| \leqq R\}.$$

Thus, if $z$ is a real variable, $S$ is an interval with midpoint $\zeta$. We say that $S$ is the *interval of convergence* of the power series. If $z$ is a complex variable, $S$ is a disc with centre $\zeta$. We call $S$ the *disc of convergence* of the power series. For obvious reasons, $R$ is called the *radius of convergence* of the power series.

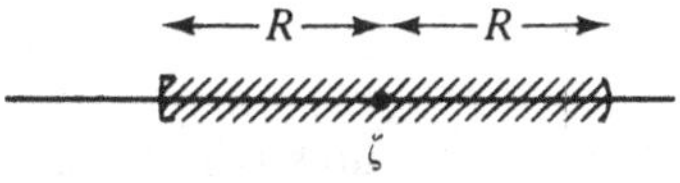


interval of convergence

disc of convergence

A given boundary point $\xi$ of $S$ may or may not be an element of $S$ – i.e. the power series may or may not converge at $\xi$. The above theory yields no information about this question.

Note finally that we have shown that a power series converges *absolutely* at interior points of $S$ (i.e. for $|z-\zeta|<R$). If it converges at a boundary point of $S$, however, there is no reason why the convergence should necessarily be absolute.

---

28.9 *Example* The power series

$$\sum_{k=1}^{\infty} (-1)^{k-1} \frac{z^k}{k}$$

has radius of convergence 1. We have that

$$\lim_{k\to\infty} \left| \frac{(-1)^{k-1}}{k} \right|^{1/k} = 1.$$

The series obtained by writing $z=1$ converges but *not* absolutely. The series obtained by writing $z=-1$ *diverges*.

---

28.10 *Exercise*

(1) Suppose that $\langle \mathbf{a}_k \rangle$ is a sequence of points in a normed vector space $\mathcal{X}$ and that the series

$$\sum_{k=1}^{\infty} \mathbf{a}_k$$

converges. Prove that $\mathbf{a}_k \to 0$ as $k \to \infty$. Give an example of a sequence

$\langle b_k \rangle$ of real numbers for which $b_k \to 0$ as $k \to \infty$ but the series

$$\sum_{k=1}^{\infty} b_k$$

diverges.

(2) Suppose that $\langle a_k \rangle$ is a sequence of *non-negative* real numbers. Prove that the infinite series

$$\sum_{k=1}^{\infty} a_k$$

exists if and only if the sequence of partial sums is bounded above.

(3) Prove that

$$\sum_{k=1}^{\infty} \frac{1}{k} = 1 + \tfrac{1}{2} + \tfrac{1}{3} + \tfrac{1}{4} + \ldots$$

diverges by demonstrating that

$$s_{2^j} \geqq 1 + \frac{1}{2^j} \quad (j \in \mathbb{N})$$

where $s_K$ denotes the $K$th partial sum. Prove that, if $\alpha > 1$, then

$$\sum_{k=1}^{\infty} \frac{1}{k^\alpha} = 1 + \frac{1}{2^\alpha} + \frac{1}{3^\alpha} + \ldots$$

converges by demonstrating that

$$t_{2^j - 1} \leqq 2^{\alpha-1}/(2^{\alpha-1} - 1) \quad (j \in \mathbb{N})$$

where $t_K$ denotes the $K$th partial sum.

(4) If $\langle a_k \rangle$ is any decreasing sequence of non-negative real numbers, prove that

$$a_K - a_{K+1} \leqq \left| \sum_{k=K}^{K+L} (-1)^k a_k \right| \leqq a_K.$$

Show that, if also $a_k \to 0$ as $k \to \infty$, then

$$\sum_{k=1}^{\infty} (-1)^k a_k$$

converges. [*Hint*: Show that the sequence of partial sums is Cauchy.] Deduce that the series

$$s = \sum_{k=1}^{\infty} \frac{(-1)^{k-1}}{k} = 1 - \frac{1}{2} + \frac{1}{3} - \frac{1}{4} + \ldots$$

converges and that $\frac{1}{2} \leqq s \leqq 1$. Prove also that

$$\frac{3}{2}s = 1 + \frac{1}{3} - \frac{1}{2} + \frac{1}{5} + \frac{1}{7} - \frac{1}{4} + \frac{1}{9} + \frac{1}{11} - \frac{1}{6} + \ldots$$

[*Hint*: If $t_K$ denotes the $K$th partial sum of the second series, then

$$t_{3K} = 1 + \frac{1}{3} - \frac{1}{2} + \ldots + \frac{1}{4K-3} + \frac{1}{4K-1} - \frac{1}{2K}.\Bigg]$$

(5) Suppose that $\langle a_k \rangle$ is a sequence of *non-negative* real numbers and that $\mathscr{F}$ is the collection of all *finite* subsets of $\mathbb{N}$. Prove that

$$s = \sum_{k=1}^{\infty} a_k$$

converges if and only if

$$t = \sup_{F \in \mathscr{F}} \left\{ \sum_{k \in F} a_k \right\}$$

is finite and that, in this case, $s = t$.

(6) Suppose that $A$ and $B$ are non-empty sets and that $f: A \times B \to \mathbb{R}$. Prove that

$$\sup_{a \in A} \left( \sup_{b \in B} f(a, b) \right) = \sup_{(a,b) \in A \times B} f(a, b).$$

Let $\langle a_{jk} \rangle$ be a 'double sequence' of *non-negative* real numbers and let $\mathscr{F}$ be the collection of all *finite* subsets of $\mathbb{N}$. Prove that

$$s = \sum_{j=1}^{\infty} \left( \sum_{k=1}^{\infty} a_{jk} \right)$$

exists if and only if

$$t = \sup_{G \in \mathscr{F} \times \mathscr{F}} \sum_{(j,k) \in G} a_{jk}$$

is finite and that, in this case, $s = t$. What can be said of

$$u = \sum_{k=1}^{\infty} \left( \sum_{j=1}^{\infty} a_{jk} \right)?$$

(See exercise 23.19(6).)

**28.11† Uniform convergence of series**

Under what circumstances is it true that

$$\lim_{\mathbf{x}\to\xi} \sum_{k=1}^{\infty} f_k(\mathbf{x}) = \sum_{k=1}^{\infty} \lim_{\mathbf{x}\to\xi} f_k(\mathbf{x})?$$

This is a problem about the circumstances under which two limiting operations 'commute'. We discussed such problems in chapter 23. In particular, we introduced the notion of uniform convergence to provide an appropriate criterion.

Suppose that, for each $k\in\mathbb{N}$, $f_k: S\to\mathcal{Y}$ where $\mathcal{Y}$ is a normed vector space. We say that

$$\sum_{k=1}^{\infty} f_k(\mathbf{x}) \qquad (1)$$

converges *uniformly* for $\mathbf{x}\in S$ if and only if the sequence of partial sums converges uniformly for $\mathbf{x}\in S$. Similarly, (1) converges *pointwise* if and only if the sequence of partial sums converges pointwise for $\mathbf{x}\in S$.

---

28.12† *Theorem* Suppose that $\mathcal{X}$ is a metric space and that $\mathcal{Y}$ is a complete normed vector space. Let $S\subset\mathcal{X}$ and suppose that, for each $k\in\mathbb{N}$, the functions $f_k: S\to\mathcal{Y}$ satisfy

$$f_k(\mathbf{x})\to\eta_k \quad \text{as} \quad \mathbf{x}\to\xi$$

through the set $S$ and that

$$\sum_{k=1}^{\infty} f_k(\mathbf{x})$$

converges *uniformly* for $\mathbf{x}\in S$. Then

$$\sum_{k=1}^{\infty} f_k(\mathbf{x})\to\sum_{k=1}^{\infty}\eta_k \quad \text{as} \quad \mathbf{x}\to\xi$$

through the set $S$.

*Proof* This is an immediate consequence of theorem 27.6.

---

28.13† *Corollary* Suppose that $\mathcal{X}$ is a metric space and that $\mathcal{Y}$ is a complete normed vector space. Let $S\subset\mathcal{X}$ and suppose that, for each $k\in\mathbb{N}$, the function $f_k: S\to\mathcal{Y}$ is continuous on $S$. Then the function $f: S\to\mathcal{Y}$ defined by

$$f(\mathbf{x}) = \sum_{k=1}^{\infty} f_k(\mathbf{x})$$

is continuous on $S$ provided the series converges *uniformly* for $\mathbf{x}\in S$.

---

28.14 *Example* The series

$$f(x)=\sum_{k=0}^{\infty} xe^{-kx}$$

converges *pointwise* for each $x\in[0, \infty)$. For $x>0$, $f(x)$ may be evaluated using the formula for a geometric progression. We obtain that

$$f(x)=\begin{cases} x(1-e^{-x})^{-1} & (x>0) \\ 0 & (x=0). \end{cases}$$

Observe that $f(x)\to 1$ as $x\to 0+$ but $f(0)=0$. Thus $f$ is *not* continuous on $[0, \infty)$ and hence the series does *not* converge *uniformly* for $x\in[0, \infty)$.

---

## 28.15† Series in function spaces

Recall from §27.8 that $\mathscr{B}(S, \mathscr{Y})$ denotes the space of all *bounded* functions $F: S\to\mathscr{Y}$ with metric $u$ defined by

$$u(F, G)=\sup_{\mathbf{x}\in S} d(F(\mathbf{x}), G(\mathbf{x})).$$

If $\mathscr{Y}$ is a normed vector space, then we may regard $\mathscr{B}(S, \mathscr{Y})$ as a normed vector space by defining vector addition and scalar multiplication in $\mathscr{B}(S, \mathscr{Y})$ by

$$(F+G)(\mathbf{x})=F(\mathbf{x})+G(\mathbf{x})$$
$$(\alpha F)(\mathbf{x})=\alpha F(\mathbf{x}).$$

The norm on $\mathscr{B}(S, \mathscr{Y})$ is, of course, defined by

$$\|F\|=\sup_{\mathbf{x}\in S}\|F(\mathbf{x})\|$$

so that $u(F, G)=\|F-G\|$.

We proved in theorem 27.10 that, if $\mathscr{Y}$ is complete, then so is $\mathscr{B}(S, \mathscr{Y})$. In particular, if $\mathscr{Y}$ is a complete normed vector space, then so is $\mathscr{B}(S, \mathscr{Y})$.

Suppose that $\mathscr{Y}$ is a normed vector space and that, for each $k\in\mathbb{N}$, $f_k: S\to\mathscr{Y}$ is *bounded* – i.e. $\langle f_k\rangle$ is a sequence of 'points' in the normed vector space $\mathscr{B}(S, \mathscr{Y})$. Then, as explained in §23.12, to say that the series

$$\sum_{k=1}^{\infty} f_k(\mathbf{x})$$

converges *uniformly* for $\mathbf{x}\in S$ is the same as saying that the series

$$\sum_{k=1}^{\infty} f_k$$

converges in the space $\mathscr{B}(S, \mathscr{Y})$.

This is a useful observation because it allows us to extend many of the preceding results of this chapter to the case of uniformly convergent series. In particular, we obtain the following version of the comparison test.

---

28.16† *Theorem* (*Weierstrass 'M test'*) Suppose that, for each $k \in \mathbb{N}$, $f_k: S \to \mathcal{Y}$, where $\mathcal{Y}$ is a complete normed vector space. Suppose also that $\langle b_k \rangle$ is a sequence of non-negative real numbers for which

$$\sum_{k=1}^{\infty} b_k$$

converges and that, for each $k \in \mathbb{N}$ and each $\mathbf{x} \in S$,

$$\|f_k(\mathbf{x})\| \leqq b_k. \tag{1}$$

Then the series

$$\sum_{k=1}^{\infty} f_k(\mathbf{x})$$

converges *uniformly* for $\mathbf{x} \in S$.

*Proof* Condition (1) implies that

$$\|f_k\| \leqq b_k \quad (k \in \mathbb{N}).$$

Hence the series

$$\sum_{k=1}^{\infty} f_k$$

converges in the space $\mathcal{B}(S, \mathcal{Y})$ by the comparison test.

---

28.17 *Example* Consider the function $f_k: [0, \infty) \to \mathbb{R}$ defined by $f_k(x) = x^2 e^{-kx}$. This achieves a maximum at $x = 2k^{-1}$. Hence, for each $x \in [0, \infty)$ and each $k \in \mathbb{N}$,

$$|f_k(x)| \leqq \left(\frac{4}{e^2}\right)\frac{1}{k^2}.$$

It follows from the Weierstrass 'M test' that the series

$$f(x) = \sum_{k=1}^{\infty} x^2 e^{-kx}$$

converges *uniformly* for $x \in [0, \infty)$.

From corollary 28.13 we may conclude that $f$ is continuous on $[0, \infty)$. In fact, for each $x \in [0, \infty)$, we may evaluate $f(x)$ as in example 28.14 to obtain

$$f(x) = \begin{cases} x^2(1 - e^{-x})^{-1} & (x > 0) \\ 0 & (x = 0). \end{cases}$$

Observe that $f(x) \to f(0)$ as $x \to 0+$.

---

28.18† *Exercise*

(1) Prove that the series

$$f(x)=\sum_{k=0}^{\infty} x\left(\frac{1+x}{1+x+x^2}\right)^k$$

converges pointwise for $x \geqq 0$. Show that $f$ is not continuous on $[0, \infty)$ and deduce that the series does not converge *uniformly* for $x \geqq 0$. Does the series converge uniformly for
(*a*) $0 \leqq x \leqq 1$, (*b*) $1 \leqq x \leqq 2$?

(2) Prove that the series

$$f(x)=\sum_{k=0}^{\infty} \frac{\sin kx}{k^2}$$

converges uniformly for $x \in \mathbb{R}$. Explain why $f$ is continuous on $\mathbb{R}$.

(3) Suppose that $\langle u_k \rangle$ and $\langle v_k \rangle$ are sequences of real numbers and let

$$U_K=\sum_{k=J+1}^{K} u_k \quad (K>J).$$

Prove that

$$\sum_{k=J+1}^{K} u_k v_k = v_K U_K + \sum_{k=J+1}^{K-1} U_k(v_k - v_{k+1})$$

(Abel's 'partial summation' formula).

(4) Suppose that $\langle a_k \rangle$ is a decreasing sequence of positive real numbers which tends to zero and that $\phi_k$: $S \to \mathbb{R}$ satisfies

$$\left|\sum_{k=1}^{K} \phi_k(\mathbf{x})\right| \leqq H$$

for all $K \in \mathbb{N}$ and all $\mathbf{x} \in S$. Prove that

$$\sum_{k=1}^{\infty} a_k \phi_k(\mathbf{x})$$

converges *uniformly* for $\mathbf{x} \in S$. [*Hint*: Use question 3.]

(5) Establish the identity

$$\sin x + \sin 2x + \ldots + \sin nx = \frac{\cos \frac{1}{2}x - \cos(n+\frac{1}{2})x}{2 \sin \frac{1}{2}x}$$

and deduce that the series

$$f(x)=\sum_{k=1}^{\infty} \frac{\sin kx}{k}$$

converges uniformly for $x \in [\delta, 2\pi - \delta]$ provided that $\delta > 0$. [*Hint*: Use question 4.] Deduce that $f$ is continuous on $(0, \pi)$.

(6) Suppose that the real power series

$$f(x)=\sum_{k=0}^{\infty} a_k x^k$$

has interval of convergence $I$ and let $J$ be any *compact* subinterval of $I$. Prove that the series converges *uniformly* for $x \in I$. [*Hint*: To show that the series converges uniformly on $[0, X] \subset I$, take $u_k = a_k X^k$ and $v_k = (xX^{-1})^k$ in question 3.]

Deduce that the function $F: [-1, 1) \to \mathbb{R}$ defined by

$$F(x)=\sum_{k=1}^{\infty} \frac{x^k}{k}$$

is continuous on $[-1, 1)$.

---

## 28.19† Continuous operators

Suppose that $\mathscr{X}$ and $\mathscr{Y}$ are metric spaces and that $T: \mathscr{X} \to \mathscr{Y}$ is *continuous* on $\mathscr{X}$. Then we know from corollary 25.9 that

$$T\left(\lim_{k\to\infty} \mathbf{X}_k\right) = \lim_{k\to\infty} T(\mathbf{X}_k)$$

provided that the left-hand side exists. It follows that

$$T\left(\sum_{k=1}^{\infty} \mathbf{x}_k\right) = \sum_{k=1}^{\infty} T(\mathbf{x}_k)$$

provided that the left-hand side converges.

This is a useful result, particularly in the case when $\mathscr{X}$ is a function space. We then usually call the continuous function $T: \mathscr{X} \to \mathscr{Y}$ a continuous *operator*. We shall illustrate the usefulness of the result by examining the case in which $\mathscr{X}$ is the space of all continuous functions $f: [a, b] \to \mathbb{R}$ (usually denoted by $C[a, b]$) and $T$ is the 'integration operator' on $C[a, b]$.

---

28.20† *Theorem* Let $C[a, b]$ denote the space of all continuous functions $f: [a, b] \to \mathbb{R}$ and define the operator $T: C[a, b] \to \mathbb{R}$ by

$$T(f)=\int_a^b f(t)dt.$$

Then $T$ is continuous on $C[a, b]$.

*Proof* We have that

$$|T(f)-T(g)| = \left|\int_a^b f(t)dt - \int_a^b g(t)dt\right|$$

$$\leq \int_a^b |f(t)-g(t)|dt$$

$$\leq (b-a) \max_{a \leq t \leq b} |f(t)-g(t)|$$

$$= (b-a)\|f-g\|.$$

It follows that $T(f) \to T(g)$ as $f \to g$ and hence that $T$ is a continuous operator on $C[a, b]$.

---

28.21† *Corollary* For each $k \in \mathbb{N}$, let $f_k$: $[a, b] \to \mathbb{R}$ be continuous on $[a, b]$ and suppose that

$$\sum_{k=1}^{\infty} f_k(x) \tag{1}$$

converges uniformly for $x \in [a, b]$. Then

$$\int_a^b \left( \sum_{k=1}^{\infty} f_k(x) \right) dx = \sum_{k=1}^{\infty} \left( \int_a^b f_k(x)dx \right). \tag{2}$$

*Proof* To say that (1) converges uniformly for $x \in [a, b]$ is the same as saying that

$$\sum_{k=1}^{\infty} f_k$$

converges in $C[a, b]$. The result therefore follows immediately from theorem 28.20.

---

28.22 *Example* It is worth noting that, *without* the uniformity condition, corollary 28.21 does *not* hold. Consider, for example, the function $F_k$: $[0, 1] \to \mathbb{R}$ illustrated below.

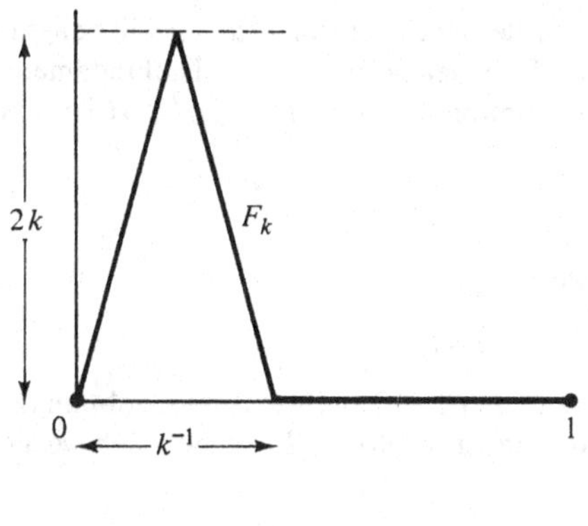


We have that, for *each* $x \in [0, 1]$, $F_k(x) \to 0$ as $k \to \infty$ – i.e. $\langle F_k \rangle$ converges *pointwise* to the zero function. But

$$\int_0^1 F_k(x)dx = 1 \quad (k \in \mathbb{N}).$$

It follows that the series whose sequence of partial sums is $\langle F_k(x) \rangle$ fails to satisfy (2) of corollary 28.21.

---

28.23† *Theorem* Suppose that $\mathcal{X}$ and $\mathcal{Y}$ are metric spaces and that $T\colon \mathcal{X} \to \mathcal{Y}$ is continuous and bijective. Then

$$T^{-1}\left(\sum_{k=1}^{\infty} \mathbf{y}_k\right) = \sum_{k=1}^{\infty} T^{-1}(\mathbf{y}_k)$$

provided that the right-hand side exists.

*Proof* Since $T$ is continuous

$$T\left(\sum_{k=1}^{\infty} T^{-1}(\mathbf{y}_k)\right) = \sum_{k=1}^{\infty} T(T^{-1}(\mathbf{y}_k)) = \sum_{k=1}^{\infty} \mathbf{y}_k.$$

Hence

$$\sum_{k=1}^{\infty} T^{-1}(\mathbf{y}_k) = T^{-1}\left(\sum_{k=1}^{\infty} \mathbf{y}_k\right).$$

---

We illustrate this result by taking $\mathcal{X} = C[a, b]$ and $\mathcal{Y} = C'_\xi[a, b]$ where the latter notation denotes the set of all $F\colon [a, b] \to \mathbb{R}$ which can be written in the form

$$F(x) = \int_\xi^x f(t)dt \quad (a \leq x \leq b)$$

where $f$ is a function in $C[a, b]$. The role of the operator $T\colon \mathcal{X} \to \mathcal{Y}$ will be assumed by the 'integration operator' $I\colon C[a,b] \to C'_\xi[a,b]$ defined by $I(f) = F$. The fundamental theorem of calculus asserts that the 'differentiation operator' $D\colon C'_\xi[a, b] \to C[a, b]$ defined by $D(F) = f$ where

$$f(x) = F'(x) \quad (a < x < b)$$

is inverse to the integration operator – i.e.

$$D = I^{-1}.$$

It is easily shown that $I$ is continuous on $C[a, b]$ and we therefore obtain the following theorem in which $C'[a, b]$ denotes the set of all $f\colon [a, b] \to \mathbb{R}$ which have a continuous derivative on $[a, b]$.

---

28.24† *Theorem* Suppose that, for each $k\in\mathbb{N}$, $g_k\in C'[a, b]$ and that

$$\sum_{k=1}^{\infty} g_k'(x)$$

converges *uniformly* for $x\in[a, b]$ and

$$\sum_{k=1}^{\infty} g_k(\xi)$$

converges for some $\xi\in[a, b]$. Then, for each $x\in(a, b)$

$$D\left(\sum_{k=1}^{\infty} g_k(x)\right)=\sum_{k=1}^{\infty}\left(Dg_k(x)\right). \tag{3}$$

*Proof* Apply theorem 28.23 with $y_k=f_k$, where $f_k$: $[a, b]\to\mathbb{R}$ is defined by

$$f_k(x)=g_k(x)-g_k(\xi).$$

---

28.25 *Example* Consider the function $G_k$: $[0, 1]\to\mathbb{R}$ illustrated below.

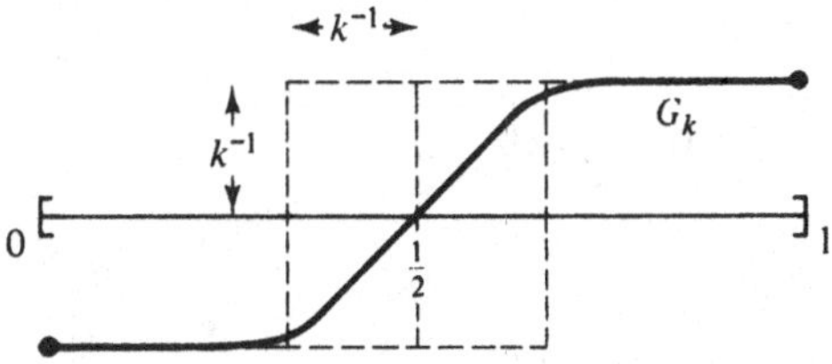


We have that $G_k\in C'[a, b]$ and that $\langle G_k(x)\rangle$ converges *uniformly* to the zero function for $x\in[0, 1]$. But

$$G_k'(\tfrac{1}{2})=1 \quad (k\in\mathbb{N}).$$

It follows that the series whose sequence of partial sums is $\langle G_k(x)\rangle$ fails to satisfy (3) of theorem 28.24.

---

**28.26† Applications to power series**

A trivial consequence of the Weierstrass 'M test' (28.16) is that a real (or complex) power series

$$\sum_{k=0}^{\infty} a_k(z-\zeta)^k$$

converges *uniformly* for $|z-\zeta|\leq r$ provided that the constant $r$ is chosen so that $r<R$ where $R$ is the radius of convergence of the power series. Exercise 28.18(6) is a more subtle result. This demonstrates that a real power series converges uniformly on *any*

compact subinterval of its interval of convergence. We may deduce the following results.

---

28.27† *Proposition* (*Abel's theorem*) The sum of any real power series is continuous on its interval of convergence.

---

28.28† *Proposition* If $c$ and $d$ are any points in the interval of convergence of a real power series

$$f(x)=\sum_{k=0}^{\infty} a_k(x-\xi)^k,$$

then

$$\int_c^d f(x)dx=\sum_{k=0}^{\infty} a_k \int_c^d (x-\xi)^k dx.$$

---

28.29† *Proposition* If $y$ is any *interior* point of the interval of convergence of a real power series

$$f(x)=\sum_{k=0}^{\infty} a_k(x-\xi)^k,$$

then $f$ is differentiable at $y$ and

$$f'(y)=\sum_{k=1}^{\infty} ka_k(y-\xi)^{k-1}.$$

---

The second and third propositions are true also of *complex* power series (provided we replace 'interval of convergence' by 'disc of convergence'). Abel's theorem, however, is *false* for complex power series. The diagram below represents the disc of convergence of a complex power series. If the power series converges at the point $y$, then it is true that $f(z)\to f(y)$ as $z\to y$ along curves like $C_1$ but it is in general false that $f(z)\to f(y)$ as $z\to y$ along 'tangential' curves like $C_2$.

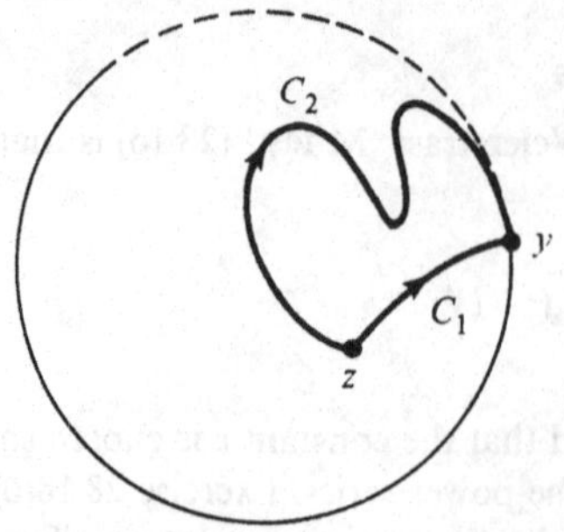


---

28.30 *Example* If $|x|<1$, we have that

$$\sum_{k=0}^{\infty} (-1)^k x^k = \frac{1}{1+x}.$$

By proposition 28.28, if $|y|<1$, then

$$\log(1+y) = \int_0^y \frac{dx}{1+x} = \int_0^y \left( \sum_{k=0}^{\infty} (-1)^k x^k \right) dx$$
$$= \sum_{k=0}^{\infty} (-1)^k \frac{y^{k+1}}{k+1}.$$

The latter power series has interval of convergence $(-1, 1]$ and, by Abel's theorem (28.27), its sum $f(y)$ is therefore continuous on $(-1, 1]$. But $f(y) = \log(1+y)$ for $y \in (-1, 1)$. It follows that $f(1) = \log 2$ – i.e.

$$\log 2 = 1 - \frac{1}{2} + \frac{1}{3} - \frac{1}{4} + \frac{1}{5} - \ldots$$

---

28.31† *Exercise*

(1) Prove that, for $|x|<1$,

$$\frac{1}{1+x^2} = \sum_{k=0}^{\infty} (-1)^k x^{2k}.$$

Hence obtain a power series expansion for arctan $x$ which is valid for $-1 < x \leqq 1$. Deduce that

$$\frac{\pi}{4} = 1 - \frac{1}{3} + \frac{1}{5} - \frac{1}{7} + \frac{1}{9} - \ldots$$

(2) Obtain the general binomial theorem by differentiating

$$(1+x)^{-\alpha} = \sum_{k=0}^{\infty} \frac{\alpha(\alpha-1)\ldots(\alpha-k+1)}{k!} x^k \quad (-1<x<1).$$

(3) The power series

$$f(z) = \sum_{k=0}^{\infty} a_k (z-\zeta)^k; \quad g(z) = \sum_{k=0}^{\infty} b_k (z-\zeta)^k$$

both converge for $|z-\zeta|<r$ and $f(z)=g(z)$ for $|z-\zeta|<r$. Prove that $a_k = b_k$ $(k=0, 1, 2, \ldots)$. [*Hint*: Differentiate.]

---

# 29† INFINITE SUMS

## 29.1† Commutative and associative laws

The operation of addition in a vector space $\mathcal{X}$ satisfies certain laws. (See §13.17.) In this section we wish to concentrate on the commutative law and the associative law.

The *commutative law* asserts that for any $\mathbf{x} \in \mathcal{X}$ and $\mathbf{y} \in \mathcal{X}$,

$$\mathbf{x}+\mathbf{y}=\mathbf{y}+\mathbf{x}.$$

The *associative law* asserts that for any $\mathbf{x} \in \mathcal{X}$, $\mathbf{y} \in \mathcal{X}$ and $\mathbf{z} \in \mathcal{X}$,

$$(\mathbf{x}+\mathbf{y})+\mathbf{z}=\mathbf{x}+(\mathbf{y}+\mathbf{z}).$$

The two laws together imply that, in evaluating a finite sum, it does not matter in which order the requisite additions are performed: the result will always be the same. For example,

$$((1+2)+3)+4=(3+3)+4=6+4=10$$

$$(4+1)+(3+2)=5+5=10.$$

This means that we can write

$$1+2+3+4=10$$

without needing to specify the method of summation which is to be used.

It is natural to ask to what extent these laws extend to the infinite case. We begin with the commutative law. For this purpose we require the notion of a permutation. A *permutation* $\phi$ on a set $A$ is simply a bijective function $\phi: A \rightarrow A$. We shall be concerned with permutations $\phi$ on the set $\mathbb{N}$ of natural numbers. Given such a permutation $\phi: \mathbb{N} \rightarrow \mathbb{N}$ and a sequence $\langle \mathbf{a}_k \rangle$, we call the sequence $\langle \mathbf{a}_{\phi(k)} \rangle$ a *rearrangement* of $\langle \mathbf{a}_k \rangle$ because it has the same terms as $\langle \mathbf{a}_k \rangle$ but placed in a different order.

An appropriate generalisation of the commutative law to series would assert that, for any permutation $\phi$ on $\mathbb{N}$,

$$\sum_{k=1}^{\infty} \mathbf{a}_k = \sum_{k=1}^{\infty} \mathbf{a}_{\phi(k)}$$

provided that one side or the other exists. Unfortunately, this result is *false* in general. Exercise 28.10(4) provides a counterexample. If we write

$$1-\tfrac{1}{2}+\tfrac{1}{3}-\tfrac{1}{4}+\tfrac{1}{5}-\tfrac{1}{6}+\ldots=\sum_{k=1}^{\infty} a_k, \tag{1}$$

then

$$1+\tfrac{1}{3}-\tfrac{1}{2}+\tfrac{1}{5}+\tfrac{1}{7}-\tfrac{1}{4}+\ldots=\sum_{k=1}^{\infty} a_{\phi(k)},$$

where $\phi: \mathbb{N}\to\mathbb{N}$ is the permutation defined by

$$\phi(3k)=2k; \quad \phi(3k-1)=4k-1; \quad \phi(3k-2)=4k-3.$$

But although both infinite series converge, they are *not* equal.

Next we turn to the associative law. This is concerned with the manner in which the terms in a sum are bracketed. We begin with the following instructive example:

$$\begin{aligned} 0=0+0+0+\ldots &=(1-1)+(1-1)+(1-1)+\ldots \\ &=1-1+1-1+1-1+\ldots \\ &=1+(-1+1)+(-1+1)+(-1+1)+\ldots \\ &=1+0+0+0+\ldots \\ &=1. \end{aligned}$$

Clearly something is wrong here. The problem is not hard to locate. The removal of the brackets at the fourth step is invalid. In fact, the series $1-1+1-1+1-1+\ldots$ does not converge. The *removal* of brackets is therefore very definitely *not* in general permissible in an infinite series.

One might hope however that the *insertion* of brackets in a series might be acceptable. This is certainly the case if we choose only to bracket neighbouring sets of terms as indicated below:

$$(a_1+a_2+a_3)+(a_4+a_5)+(a_6+a_7+a_8+a_9)+\ldots$$

In this case, the sequence of partial sums of the new series is $s_3, s_5, s_9, \ldots$. This is a subsequence of $\langle s_K\rangle$ and hence converges to the same limit. But what happens if we choose to bracket sets of non-neighbouring elements? To take an extreme case, we might choose to bracket all the even terms and all the odd terms separately as below:

$$(a_1+a_3+a_5+\ldots)+(a_2+a_4+a_6+\ldots).$$

If we attempt to bracket series (1) in the fashion indicated, we obtain

$$(1+\tfrac{1}{3}+\tfrac{1}{5}+\tfrac{1}{7}+\ldots)+(-\tfrac{1}{2}-\tfrac{1}{4}-\tfrac{1}{6}-\ldots).$$

But neither of these bracketed series converges. (See exercise 28.10(3).)

All these negative results are rather depressing. But no cause for despair exists. As mathematicians, we know that, if matters do not work out very well with one definition, the thing to do is to try another definition. This is the subject of the next section.

## 29.2† Infinite sums

As we saw in the previous section, series do *not* satisfy the appropriate generalisation of the commutative law. It is therefore natural to seek an alternative definition for which the commutative law *does* hold.

Suppose that $\mathfrak{X}$ is a normed vector space and that $a: I \to \mathfrak{X}$ is a function. If $\mathbf{s} \in \mathfrak{X}$. there is, of course, no difficulty in defining

$$\mathbf{s} = \sum_{i \in I} a(i) \tag{1}$$

in the case when the 'index set' $I$ is *finite*. In the case when $I$ is *infinite*, we shall define (1) to mean that

> For any $\varepsilon > 0$, there exists a *finite* set $F \subset I$ such that, for each *finite* set $G$,
>
> $$F \subset G \subset I \Rightarrow \| \mathbf{s} - \sum_{i \in G} a(i) \| < \varepsilon.$$

A vector $\mathbf{s}$ defined in this way will be called an *infinite sum*. (Note incidentally that the convergence notion introduced here differs somewhat from those we have met earlier. It is an example of the convergence of a *net*.)

Since the definition takes no account of any ordering which may exist on the index set $I$, it is clear that infinite sums must satisfy the commutative law. The next theorem puts the proposition in formal terms.

---

**29.3†** *Theorem* Suppose that $\mathfrak{X}$ is a normed vector space and that $a: I \to \mathfrak{X}$. If $\phi$ is a permutation on $I$, then

$$\sum_{i \in I} a(i) = \sum_{i \in I} a(\phi(i))$$

provided one side or the other of the equation exists.

*Proof* We shall suppose that it is the left-hand side which exists. (Otherwise replace $a$ by $b = a \circ \phi$ and $\phi$ by $\phi^{-1}$.)

Let $\varepsilon > 0$ be given. Then there exists a finite set $F \subset I$ such that, for any finite set $G$,

$$F \subset G \subset I \Rightarrow \| \sum_{i \in I} a(i) - \sum_{i \in G} a(i) \| < \varepsilon.$$

Hence, for any finite set $H$,

$$\phi(F) \subset H \subset I \Rightarrow F \subset \phi^{-1}(H) \subset I$$

$$\Rightarrow \| \sum_{i \in I} a(i) - \sum_{i \in \phi^{-1}(H)} a(i) \| < \varepsilon$$

$$\Rightarrow \| \sum_{i \in I} a(i) - \sum_{i \in H} a(\phi(i)) \| < \varepsilon$$

and the result follows.

---

Next it is natural to consider the associative law. The definition of an infinite sum is not tailored to meet the requirements of this law and so we need to work a little harder to obtain a result. As a preliminary we need to relate the idea of an infinite sum to that of a series.

---

## 29.4† Infinite sums and series

We begin with a result for infinite sums which is analogous to the theorem which asserts that, if an infinite series exists, then its terms tend to zero (exercise 28.10(1)).

---

29.5† *Theorem* Suppose that $\mathcal{X}$ is a normed vector space. Let $a: I \to \mathcal{X}$ and suppose that

$$\mathbf{s} = \sum_{i \in I} a(i)$$

exists. Then, given any $\varepsilon > 0$, there exists a finite set $F \subset I$ such that, for each $j \in I$,

$$j \notin F \Rightarrow \|a(j)\| < \varepsilon.$$

*Proof* Let $\varepsilon > 0$ be given. Then $\varepsilon/2 > 0$ and hence there exists a finite set $F \subset I$ such that for any finite set $G$

$$F \subset G \subset I \Rightarrow \| \mathbf{s} - \sum_{i \in G} a(i) \| < \varepsilon/2.$$

Let $j \in I \setminus F$. Then

$$\begin{aligned} \| a(j) \| &= \| \sum_{i \in F \cup \{j\}} a(i) - \mathbf{s} + \mathbf{s} - \sum_{i \in F} a(i) \| \\ &\leqq \| \sum_{i \in F \cup \{j\}} a(i) - \mathbf{s} \| + \| \mathbf{s} - \sum_{i \in F} a(i) \| \\ &< \varepsilon/2 + \varepsilon/2 = \varepsilon. \end{aligned}$$

---

**29.6†** *Corollary* Suppose that $\mathfrak{X}$ is a normed vector space. Let $a: I \to \mathfrak{X}$ and suppose that

$$\mathbf{s} = \sum_{i \in I} a(i)$$

exists. Then the set $H = \{i : a(i) \neq \mathbf{0}\}$ is *countable*.

*Proof* Let $k \in \mathbb{N}$. By the previous theorem the set

$$G_k = \{i : \|a(i)\| \geq 1/k\}$$

is *finite*. But

$$H = \bigcup_{k=1}^{\infty} G_k$$

and so $H$ is the countable union of finite sets. It follows that $H$ is countable by theorem 12.12.

---

The definition of an infinite sum given in §29.2 seems to be one of considerable generality in that the index set may be *any* set whatsoever while the index set in the case of an infinite series is restricted to be $\mathbb{N}$. But the last result shows that this generality is largely illusory. An infinite sum of an *uncountable* collection of non-zero objects never exists. In future, we shall therefore *always* assume that the index set $I$ is *countable*. Indeed, it will often be convenient to assume that $I = \mathbb{N}$. This assumption can be made without loss of generality when $I$ is countable because there then exists a bijection $\phi : \mathbb{N} \to I$. Thus $I$ may be replaced by $\mathbb{N}$ provided that $a : I \to \mathfrak{X}$ is replaced by the sequence $b : \mathbb{N} \to \mathfrak{X}$ where $b = a \circ \phi$.

---

**29.7†** *Theorem* Suppose that $\mathfrak{X}$ is a normed vector space and that $\langle b_k \rangle$ is a sequence of points in $\mathfrak{X}$. Then

$$\sum_{k \in \mathbb{N}} b_k = \sum_{k=1}^{\infty} b_k \tag{1}$$

provided that the left-hand side exists.

*Proof* Let $\varepsilon > 0$ be given. If the left-hand side of (1) exists, then there exists a finite set $F \subset I$ such that for any finite set $G$,

$$F \subset G \subset \mathbb{N} \Rightarrow \left\| \sum_{k \in \mathbb{N}} b_k - \sum_{k \in G} b_k \right\| < \varepsilon.$$

Let $K_0 = \max F$. Then, provided that $K > K_0$, the set $G = \{1, 2, 3, \ldots, K\}$ satisfies $F \subset G \subset \mathbb{N}$. Hence

$$K > K_0 \Rightarrow \left\| \sum_{k \in \mathbb{N}} b_k - \sum_{k=1}^{K} b_k \right\| < \varepsilon$$

and the equation (1) follows.

---

Theorem 29.7 says that, if an infinite sum exists, then so does the corresponding series and the two are equal. Of course, as we know from §29.1 it is in general *false* that the convergence of a series implies the existence of the corresponding infinite sum. There is, however, an important special case for which the convergence of a series *does* imply the existence of the corresponding infinite sum. This is the case of a series with *non-negative* terms.

---

29.8† *Proposition* Suppose that $\langle b_k \rangle$ is a sequence of *non-negative* real numbers. Then

$$\sum_{k\in\mathbb{N}} b_k = \sum_{k=1}^{\infty} b_k$$

provided that the right-hand side converges.

---

The proof follows from exercise 28.10(5). An immediate consequence is that series of *non-negative* real numbers satisfy the commutative law although, as we have seen, this is certainly not the case of series in general.

---

## 29.9† Complete spaces and the associative law

To proceed any further with the study of infinite sums, we need to restrict our attention to the case when $\mathcal{X}$ is a *complete* normed vector space. (Such a space is called a *Banach space.*) This is no surprise since most of the results of the previous chapter required completeness as well.

Recall that a complete metric space is one in which every Cauchy sequence converges (§27.2). Alternatively, it is a metric space in which every totally bounded set has the Bolzano – Weierstrass property. We shall wish to use the fact that a version of the Heine–Borel theorem holds in a complete space. This asserts that every closed and *totally* bounded set is compact (see theorem 27.4). Some examples of complete normed vector spaces were discussed in §27.8. The most notable example of such a space is, of course, the space $\mathbb{R}^n$.

---

29.10† *Theorem* Suppose that $\mathcal{X}$ is a complete normed vector space and that $\mathcal{F}$ is the collection of all finite subsets of $I$. Let $a: I \to \mathcal{X}$ and suppose that

$$\mathbf{s} = \sum_{i\in I} a(i)$$

exists. Then the set

$$S = \left\{ \sum_{i\in H} a(i) : H \in \mathcal{F} \right\}$$

is totally bounded and hence its closure is compact.

*Proof* We have to show that, for each $\varepsilon>0$, there exists a *finite* collection of open balls of radius $\varepsilon$ which covers $S$.

Let $\varepsilon>0$ be given. Then there exists a finite set $F\subset I$ such that, for each finite set $G$,

$$F\subset G\subset I \Rightarrow \|\sum_{i\in G} a(i)-\mathbf{s}\|<\varepsilon.$$

Define the set $C$ by

$$C=\{\mathbf{s}-\sum_{i\in E} a(i): E\subset F\}.$$

Then $C$ is a *finite* set. The collection of all open balls of radius $\varepsilon$ whose centres are elements of $C$ is therefore finite. It covers $S$ because, for each $H\in\mathcal{F}$,

$$\|\sum_{i\in H} a(i)-\mathbf{s}+\sum_{i\in F\setminus H} a(i)\|=\|\sum_{i\in H\cup F} a(i)-\mathbf{s}\|<\varepsilon.$$

---

29.11† *Theorem* Suppose that $\mathfrak{X}$ is a *complete* normed vector space and that $a:I\to\mathfrak{X}$.

Then the infinite sum

$$\sum_{i\in I} a(i) \tag{1}$$

exists if and only if, for any $\varepsilon>0$, there exists a finite set $F\subset I$ such that, for any finite set $H$,

$$H\subset I\setminus F \Rightarrow \|\sum_{i\in H} a(i)\|<\varepsilon. \tag{2}$$

*Proof* It is quite easy to prove that the existence of (1) implies criterion (2). We therefore leave this as an exercise. There then remains the problem of showing that criterion (2) implies the existence of (1).

With the notation of the previous theorem, let

$$S_E=\left\{\sum_{i\in H} a(i): H\in\mathcal{F} \text{ and } E\subset H\right\}.$$

Then $\{\bar{S}_E: E\in\mathcal{F}\}$ is a collection of closed subsets of the *compact* set $\bar{S}$. Since this collection has the finite intersection property (exercise 20.6(3)) it follows that it has a non-empty intersection. Let $\mathbf{s}\in\bar{S}_E$ for each $E\in\mathcal{F}$.

Let $\varepsilon>0$ be given. Then, by (2), there exists a finite set $F_0\subset I$ such that, for any finite set $H$,

$$H\subset I\setminus F_0 \Rightarrow \|\sum_{i\in H} a(i)\|<\varepsilon/2.$$

Also, since $\mathbf{s}\in\bar{S}_{F_0}$, there exists a finite set $F$ satisfying $F_0\subset F\subset I$ such that

$$\|\mathbf{s}-\sum_{i\in F} a(i)\|<\varepsilon/2.$$

Now suppose that $G\supset F$. Then $G\setminus F\subset I\setminus F_0$ and so

$$\|\mathbf{s}-\sum_{i\in G} a(i)\|\leqq\|\mathbf{s}-\sum_{i\in F} a(i)\|+\|\sum_{i\in G\setminus F} a(i)\|$$
$$<\varepsilon/2+\varepsilon/2=\varepsilon.$$

Hence (1) exists.

---

Theorem 29.11 is a variant of the result which says that, in a *complete* space $\mathfrak{X}$, every Cauchy sequence converges. It is not surprising therefore that it has some useful corollaries.

---

29.12† *Corollary* Suppose that $\mathfrak{X}$ is a *complete* normed vector space and that $a: I\to\mathfrak{X}$. If $J\subset I$, then the existence of the infinite sum

$$\sum_{i\in I} a(i)$$

implies the existence of

$$\sum_{i\in J} a(i).$$

---

29.13† *Corollary* Suppose that $\mathfrak{X}$ is a *complete* normed vector space and that $a: I\to\mathfrak{X}$. If $J$ and $K$ are *disjoint* sets with union $I$, then

$$\sum_{i\in I} a(i)=\sum_{i\in J} a(i)+\sum_{i\in K} a(i)$$

provided that one side or other of the equation exists.

---

29.14† *Corollary* Suppose that $\mathfrak{X}$ is a *complete* normed vector space and that $a: I\to\mathfrak{X}$. Then the existence of the infinite sum

$$\sum_{i\in I} a(i)$$

implies that, for any $\varepsilon>0$, there exists a finite set $F\subset I$ such that for *any* set $G$,

$$F\subset G\subset I\Rightarrow\|\sum_{i\in I} a(i)-\sum_{i\in G} a(i)\|=\|\sum_{i\in I\setminus G} a(i)\|<\varepsilon.$$

---

The proofs of the corollaries are easy and we give them as exercises (29.22(1), (2) and (3)). Having obtained these results, we are now in a position to prove a version of the associative law in complete spaces.

---

29.15† *Theorem* Suppose that $\mathcal{X}$ is a *complete* normed vector space and that $a: I \to \mathcal{X}$. Let $\mathcal{W}$ denote any collection of *disjoint* sets with union $I$. Then

$$\sum_{i\in I} a(i) = \sum_{J\in\mathcal{W}} \left\{ \sum_{i\in J} a(i) \right\} \tag{3}$$

provided the left-hand side exists.

*Proof* Let $\varepsilon > 0$ be given. Since the left-hand side of (3) exists, it follows from corollary 29.14 that there exists a finite set $F \subset I$ such that, for any set $G$,

$$F \subset G \subset I \Rightarrow \left\| \sum_{i\in I} a(i) - \sum_{i\in G} a(i) \right\| < \varepsilon.$$

Let $\mathcal{F} = \{ J : J \in \mathcal{W} \text{ and } F \cap J \neq \emptyset \}$ and suppose that $\mathcal{G}$ is a finite collection of sets satisfying $\mathcal{F} \subset \mathcal{G} \subset \mathcal{W}$. Then the set

$$G = \bigcup_{J\in\mathcal{G}} J$$

satisfies $F \subset G \subset I$ and therefore, using corollary 29.13,

$$\left\| \sum_{i\in I} a(i) - \sum_{J\in\mathcal{G}} \left\{ \sum_{i\in J} a(i) \right\} \right\| = \left\| \sum_{i\in I} a(i) - \sum_{i\in G} a(i) \right\| < \varepsilon.$$

Equation (3) follows.

---

Theorem 29.15 asserts that, if an infinite sum exists, then one can bracket its terms in any manner whatsoever without altering its value. The *insertion* of brackets in infinite sums therefore creates no problem. But the problem of *removing* brackets remains – i.e. the existence of the *right*-hand side of (3) does *not* in general guarantee the existence of the left-hand side. The infinite sum

$$\sum_{k\in\mathbb{N}} 0$$

certainly exists. But, if we write $0 = (1-1)$ in this sum and then remove the brackets, we obtain

$$\sum_{k\in\mathbb{N}} (-1)^{k+1},$$

which does not exist.

We noted at the end of §29.4 that, for the special case of series of *non-negative* real numbers, the commutative law does *not* fail. It is also true that the removal of brackets from infinite sums of *non-negative* real numbers creates no problems.

---

29.16† *Proposition* Suppose that $\langle b_k \rangle$ is a sequence of *non-negative* real numbers and that $\mathcal{W}$ is any collection of disjoint sets with union $\mathbb{N}$. Then

$$\sum_{k\in\mathbb{N}} b_k = \sum_{J\in\mathcal{W}} \left\{ \sum_{k\in J} b_k \right\}$$

provided that the *right*-hand side exists.

---

The proof follows trivially from exercise 28.10(6). Propositions 29.8 and 29.16, taken together, mean that the problem of adding up a sequence of non-negative real numbers is very much simpler than the general problem. Roughly speaking, if *any* method of adding up a sequence of non-negative real numbers yields a finite result, then *all* methods will yield the *same* result. Is there a wider class of sequences with this pleasant property? We explore this question in the next section.

---

## 29.17† Absolute sums

We met the notion of an *absolutely* convergent series in §28.7. A similar notion can be defined for infinite sums. If $\mathcal{X}$ is a complete normed vector space and $a: I \to \mathcal{X}$, we say that

$$\sum_{i\in I} a(i)$$

exists *absolutely* if and only if

$$\sum_{i\in I} \|a(i)\| \tag{1}$$

exists. Since the terms of the latter sum are *non-negative* real numbers, the sum exists if and only if any method of adding up its terms yields a finite result. (See propositions 29.8 and 29.16). In particular, in the case when $I=\mathbb{N}$, the infinite sum (1) exists if and only if the series

$$\sum_{i=1}^{\infty} \|a(i)\|$$

converges.

---

29.18† *Proposition* (*Comparison test*) Suppose that $\mathcal{X}$ is a *complete* normed vector space and that $a: I \to \mathcal{X}$ and $b: I \to \mathbb{R}$. Suppose that the infinite sum

$$\sum_{i\in I} b(i)$$

exists and that

$$\|a(i)\| \leqq b(i) \quad (i \in I).$$

Then the infinite sum

$$\sum_{i\in E} a(i)$$

exists.

*Proof* This is the same as the proof of theorem 28.3 except that theorem 29.11 is used instead of the fact that Cauchy sequences converge.

---

29.19† *Corollary* Suppose that $\mathfrak{X}$ is a *complete* normed vector space and that $a: I \to \mathfrak{X}$. Then the existence of

$$\sum_{i\in I} \|a(i)\|$$

implies the existence of

$$\sum_{i\in I} a(i).$$

*Proof* Take $b(i) = \|a(i)\|$ in proposition 29.18.

---

29.20 *Corollary* Suppose that $\mathfrak{X}$ is a *complete* normed vector space and that $a: I \to \mathfrak{X}$. Let $\mathcal{W}$ denote any collection of *disjoint* sets with union $I$. Then the existence of

$$\sum_{J\in\mathcal{W}} \left\{ \sum_{i\in J} \|a(i)\| \right\} \tag{2}$$

implies the existence of the infinite sum

$$\sum_{i\in I} a(i). \tag{3}$$

*Proof* By proposition 29.16, the existence of (2) implies that (3) exists absolutely.

---

An immediate application of this result is to the problem of *removing* brackets from infinite sums.

---

29.21† *Corollary* Suppose that $\mathfrak{X}$ is a *complete* normed vector space and that $a: I \to \mathfrak{X}$. Let $\mathcal{W}$ denote any collection of *disjoint* sets with union $I$. Then

$$\sum_{i\in I} a(i) = \sum_{J\in\mathcal{W}} \left\{ \sum_{i\in J} a(i) \right\}$$

provided the *right*-hand side exists *absolutely*.

*Proof* This follows immediately from corollary 29.20 and theorem 29.15.

---

We shall use this result in §29.23 to study the problem of reversing the order of summation in repeated series.

---

29.22† *Exercise*

(1) Suppose that $\mathfrak{X}$ is *any* normed vector space and that $a: I \to \mathfrak{X}$. If $J$ and $K$ are disjoint sets with union $I$, prove that

$$\sum_{i \in I} a(i) = \sum_{i \in J} a(i) + \sum_{i \in K} a(i)$$

provided the *right*-hand side exists.

(2) Show that the equation of question 1 holds when $\mathfrak{X}$ is a *complete* normed vector space provided the *left*-hand side exists (i.e. prove corollaries 29.12 and 29.13).

(3) Suppose that $\mathfrak{X}$ is a *complete* normed vector space and that $a: I \to \mathfrak{X}$. Show that.

$$\sum_{i \in I} a(i)$$

exists if and only if, for any $\varepsilon > 0$, there exists a finite set $G$ such that for any set $H \subset I \backslash G$

$$\left\| \sum_{i \in H} a(i) \right\| < \varepsilon.$$

(4) Let $\langle a_k \rangle$ be a sequence of real numbers. Write $a_k^+ = \max\{a_k, 0\}$ and $a_k^- = max\{-a_k, 0\}$. Prove that the infinite sum

$$\sum_{k \in \mathbb{N}} a_k$$

exists absolutely if and only if any *two* of the series

$$\text{(i)} \sum_{k=1}^{\infty} a_k \quad \text{(ii)} \sum_{k=1}^{\infty} a_k^+ \quad \text{(iii)} \sum_{k=1}^{\infty} a_k^-$$

converge. If (i) converges but neither (ii) nor (iii) converge, then the series (i) is said to converge *conditionally*. Prove that, in this case, given *any* $\xi \in \mathbb{R}$ there exists a permutation $\phi: \mathbb{N} \to \mathbb{N}$ such that

$$\xi = \sum_{k=1}^{\infty} a_{\phi(k)}.$$

(The latter result is called Riemann's theorem.)

(5) If $a: I \to \mathbb{R}$, prove that the infinite sum

$$\sum_{i \in I} a(i)$$

exists *if and only if* it exists *absolutely*. [*Hint*: Use the previous question.] Obtain the same result when $a: I \to \mathbb{R}^n$.

Let $\langle \mathbf{a}_k \rangle$ be the sequence in $l^\infty$ (see §20.20) whose terms are all zero except for the $k$th which is equal to $1/k$. Prove that

$$\sum_{k \in \mathbb{N}} \mathbf{a}_k$$

exists but *not* absolutely. Deduce that the converse of corollary 29.19 is false in general (although, as question 4 shows, the converse is true for $\mathcal{X} = \mathbb{R}^n$).

(6) Prove that a normed vector space $\mathcal{X}$ is complete if and only if, for every $a: I \to \mathcal{X}$, the existence of

$$\sum_{i \in I} \|a(i)\|$$

implies the existence of

$$\sum_{i \in I} a(i).$$

---

## 29.23† Repeated series

One sometimes has to deal with repeated series of the form

$$\sum_{j=1}^{\infty} \left\{ \sum_{k=1}^{\infty} \mathbf{a}_{jk} \right\}. \tag{1}$$

It follows from corollary 29.21 that if

$$\sum_{j=1}^{\infty} \left\{ \sum_{k=1}^{\infty} \|\mathbf{a}_{jk}\| \right\} \tag{2}$$

exists, then

$$\sum_{j=1}^{\infty} \left\{ \sum_{k=1}^{\infty} \mathbf{a}_{jk} \right\} = \sum_{k=1}^{\infty} \left\{ \sum_{j=1}^{\infty} \mathbf{a}_{jk} \right\}.$$

This is, of course, a special case of a much stronger result. If (2) exists, then *any* method of adding up the terms of (1) will yield the same answer.

---

29.24 *Example* It follows from the binomial theorem (exercise 28.31(2)) that

$$(1-z)^{-2} = 1 + 2z + 3z^2 + 4z^3 + \dots \tag{3}$$

provided that $|z| < 1$. It is instructive to see how this result may be deduced from the theory developed in this chapter.

We begin with the formula for a geometric progression

$$(1-z)^{-1} = 1 + z + z^2 + z^3 + \dots \tag{4}$$

which is valid for $|z| < 1$. Thus

$$(1-z)^{-2}=\left(\sum_{j=0}^{\infty} z^{j}\right)\left(\sum_{k=0}^{\infty} z^{k}\right)=\sum_{j=0}^{\infty}\left(\sum_{k=0}^{\infty} z^{j+k}\right)$$

and so

$$(1-z)^{-2}=\sum_{j=0}^{\infty}\sum_{l=j}^{\infty} z^{l} \tag{5}$$

provided that $|z|<1$. But we know that the power series (4) exists *absolutely* for $|z|<1$. Thus (5) exists absolutely and so we can reverse the order of summation and obtain

$$(1-z)^{-2}=\sum_{l=0}^{\infty}\sum_{j=0}^{l} z^{l}=\sum_{l=0}^{\infty} z^{l}\sum_{j=0}^{l} 1.$$

Thus

$$(1-z)^{-2}=\sum_{l=0}^{\infty}(l+1)z^{l}=1+2z+3z^{2}+\ldots$$

provided that $|z|<1$.

Some explanation of these calculations may be helpful. Given that

$$\sum_{j=0}^{\infty}\sum_{k=0}^{\infty}|a_{jk}|$$

exists, the same result will be obtained regardless of whether the rows or columns in the array below are summed first.

$$\begin{array}{cccccccc}
\vdots & & \vdots & & \vdots & & \vdots & \\
+ & & & & & & & \\
r_3 & = & a_{30} & + & a_{31} & + & a_{32} & + \ldots \\
+ & & & & & & & \\
r_2 & = & a_{20} & + & a_{21} & + & a_{22} & + \ldots \\
+ & & & & & & & \\
r_1 & = & a_{10} & + & a_{11} & + & a_{12} & + \ldots \\
+ & & & & & & & \\
r_0 & = & a_{00} & + & a_{01} & + & a_{02} & + \ldots \\
\| & & & & & & & \\
s & & & & & & &
\end{array}
\qquad
\begin{array}{cccccc}
 & \vdots & \vdots & \vdots & \vdots & \\
 & + & + & + & + & \\
 & a_{30} & a_{31} & a_{32} & a_{33} & \ldots \\
 & + & + & + & + & \\
 & a_{20} & a_{21} & a_{22} & a_{23} & \ldots \\
 & + & + & + & + & \\
 & a_{10} & a_{11} & a_{12} & a_{13} & \ldots \\
 & + & + & + & + & \\
 & a_{00} & a_{01} & a_{02} & a_{03} & \ldots \\
 & \| & \| & \| & \| & \\
s = & c_0 & +c_1 & +c_2 & +c_3 & + \ldots
\end{array}$$

In (5) we are concerned with the special case in which the terms in the array above the main diagonal are all zero as on page 232.

$$
\begin{array}{cccc}
\vdots & \vdots & \vdots & \vdots \\
0 & 0 & 0 & \boxed{a_{33} \quad \cdots} \\
0 & 0 & \boxed{a_{22} + a_{23} \quad \cdots} \\
0 & \boxed{a_{11} + a_{12} + a_{13} \quad \cdots} \\
\boxed{a_{00} + a_{01} + a_{02} + a_{03}} \quad \cdots
\end{array}
\qquad
\begin{array}{cccc}
\vdots & \vdots & \vdots & \vdots \\
0 & 0 & 0 & \boxed{\begin{array}{c}a_{33}\\+\end{array}} \cdots \\
0 & 0 & \boxed{\begin{array}{c}a_{22}\\+\end{array}} & \boxed{\begin{array}{c}a_{23}\\+\end{array}} \cdots \\
0 & \boxed{\begin{array}{c}a_{11}\\+\end{array}} & \boxed{\begin{array}{c}a_{12}\\+\end{array}} & \boxed{\begin{array}{c}a_{13}\\+\end{array}} \cdots \\
\boxed{a_{00}} & \boxed{\begin{array}{c}a_{01}\end{array}} & \boxed{a_{02}} & \boxed{a_{03}} \cdots
\end{array}
$$

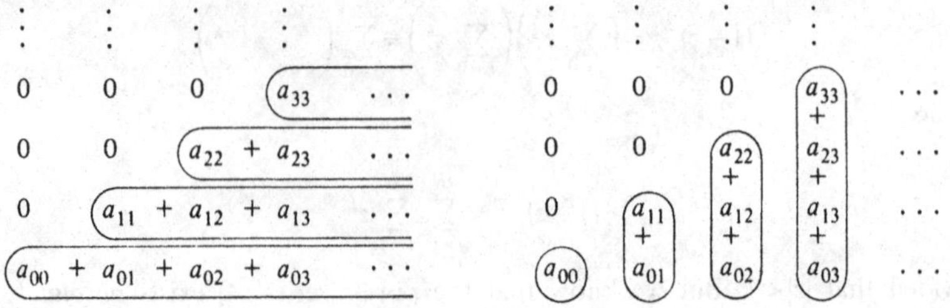

Then

$$
s = \sum_{j=0}^{\infty} r_j = \sum_{j=0}^{\infty} \sum_{l=0}^{\infty} a_{jl} = \sum_{j=0}^{\infty} \sum_{l=j}^{\infty} a_{jl}
$$

$$
s = \sum_{l=0}^{\infty} c_l = \sum_{l=0}^{\infty} \sum_{j=0}^{\infty} a_{jl} = \sum_{l=0}^{\infty} \sum_{j=0}^{l} a_{jl}.
$$

But summing an array by first adding the rows or columns is not the only way in which one can proceed. One can also sum, for example, by diagonals as indicated below.

$$
\begin{array}{ccccc}
\vdots & \vdots & \vdots & \vdots \\
a_{30} & a_{31} & a_{32} & a_{33} & \cdots \\
a_{20} & a_{21} & a_{22} & a_{23} & \cdots \\
a_{10} & a_{11} & a_{12} & a_{13} & \cdots \\
a_{00} & a_{01} & a_{02} & a_{03} & \cdots \\
\end{array}
$$

$$
s \quad = \quad d_0 + d_1 + d_2 + d_3 + \quad \cdots
$$

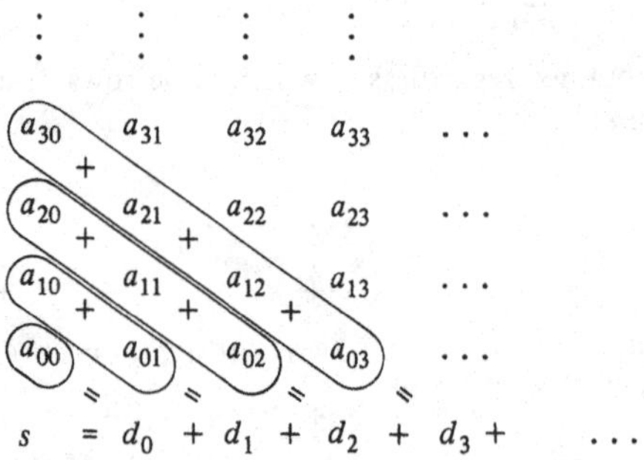

Observe that

$$
d_l = \sum_{j+k=l} a_{jk}.
$$

This method is particularly suitable for multiplying power series. We have that

$$
\left( \sum_{j=0}^{\infty} a_j z^j \right) \left( \sum_{k=0}^{\infty} b_k z^k \right) = \sum_{j=0}^{\infty} \sum_{k=0}^{\infty} a_j b_k z^{j+k}
$$

$$
= \sum_{l=0}^{\infty} \sum_{j+k=l} a_j b_k z^{j+k}
$$

$$
= \sum_{l=0}^{\infty} z^l \sum_{j+k=l} a_j b_k
$$

$$= \sum_{l=0}^{\infty} z^l \sum_{m=0}^{l} a_m b_{l-m}$$

provided that $|z| < R_1$ and $|z| < R_2$ where $R_1$ and $R_2$ are the respective radii of convergence of the two power series.

This method can also be used to obtain (3) from (4). If $|z| < 1$, we have that

$$(1-z)^{-2} = \left(\sum_{j=0}^{\infty} z^j\right)\left(\sum_{k=0}^{\infty} z^k\right)$$

$$= \sum_{l=0}^{\infty} z^l \sum_{m=0}^{l} 1 = \sum_{l=0}^{\infty} (l+1)z^l.$$

---

29.25† *Exercise*

(1) Use the method of example 29.24 to show that, for each natural number $n$,

$$(1-z)^{-n} = \sum_{k=0}^{\infty} \frac{(n+k-1)!}{k!(n-1)!} z^k$$

provided $|z| < 1$.

(2) Prove that

$$\sum_{j=1}^{\infty} \sum_{\substack{k=1 \\ j \neq k}}^{\infty} \frac{1}{j^2 - k^2} \neq \sum_{k=1}^{\infty} \sum_{\substack{j=1 \\ j \neq k}}^{\infty} \frac{1}{j^2 - k^2}.$$

(3) Prove that

$$\sum_{k=0}^{\infty} x(x+1)\ldots(x+k) = e\left\{\frac{1}{x} - \frac{1}{1!}\frac{1}{(1+x)} + \frac{1}{2!}\frac{1}{(2+x)}\cdots\right\}$$

provided that $x \in \mathbb{R} \setminus \mathbb{Z}$.

---

# 30† SEPARATION IN $\mathbb{R}^n$

## 30.1† Introduction

In this chapter we return to the study of the geometry of $\mathbb{R}^n$ which we discussed to some extent in chapter 13. Recall that, in §13.14, we observed that the axioms given by Euclid for his system of plane geometry are incomplete by modern standards. In particular, many of Euclid's proofs tacitly assume that, if $H$ is a line in $\mathbb{R}^2$, then $\mathbb{R}^2 \setminus H$ has precisely two components and that, if $\mathbf{x}$ and $\mathbf{y}$ lie in different components, the line segment joining them cuts $H$.

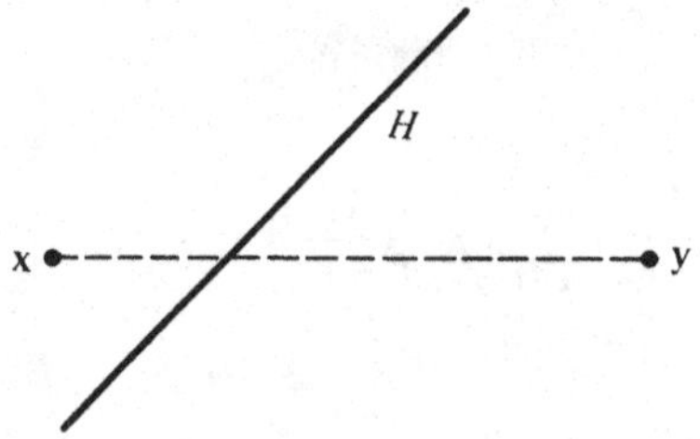


This fact is not hard to prove. Given any hyperplane $H = \{\mathbf{x}: \langle \mathbf{x}, \boldsymbol{\xi} \rangle = c\}$ in $\mathbb{R}^n$, the half-spaces $S_1 = \{\mathbf{x}: \langle \mathbf{x}, \boldsymbol{\xi} \rangle < c\}$ and $S_2 = \{\mathbf{x}: \langle \mathbf{x}, \boldsymbol{\xi} \rangle > c\}$ are convex, open and *disjoint*. It follows that $S_1$ and $S_2$ are the components of $\mathbb{R}^n \setminus H$. (See theorem 15.14 and §17.21.) If $\mathbf{x} \in S_1$ and $\mathbf{y} \in S_2$, then the line segment $L$ joining $\mathbf{x}$ and $\mathbf{y}$ must cut $H$. Otherwise, $L \subset S_1 \cup S_2$ and so $S_1 \cup S_2$ is connected (theorem 17.19).

An analogous argument shows that any hyperplane (§13.14) splits $\mathbb{R}^n$ into two distinct components. It is natural to ask: what other sets have this property? This is one of the questions we shall discuss in this chapter. The chief point that needs to be made is that it is vital not to dismiss the answers to such questions as 'geometrically obvious'. Even when the 'geometrically obvious' results are actually true (which is certainly *not* always the case), they are often exceedingly difficult to prove. The Jordan curve theorem (30.13) is a notable example. It says that any simple 'closed' curve has an 'inside' and an 'outside'. But the proof of the Jordan curve theorem is much too hard to present in this book. Not only is it long and complicated, it also makes use of some quite subtle ideas from algebraic topology. Once one has seen the proof, one becomes very much less ready to dismiss the result as 'obvious'.

It is also worth noting that many results true in $\mathbb{R}^n$ do *not* extend to infinite-dimensional spaces. For example, a hyperplane $H$ in an infinite-dimensional space $\mathfrak{X}$ need not be closed. But it can be shown that, in this case, $\overline{H} = \mathfrak{X}$ and $\mathfrak{X} \setminus H$ is connected! Only when $H$ is closed does $\mathfrak{X} \setminus H$ split nicely into two components.

## 30.2† Separation

The word 'separate' is used with various meanings in different branches of mathematics. In §15.11, we said that two sets $A$ and $B$ are 'separated' if and only if they are not contiguous. This usage is sometimes extended by saying that a topology $\mathcal{T}$ 'separates' points of $\mathcal{X}$ if and only if, for each $\mathbf{x} \in \mathcal{X}$ and $\mathbf{y} \in \mathcal{X}$ with $\mathbf{x} \neq \mathbf{y}$, there exist disjoint open sets $S$ and $T$ for which $\mathbf{x} \in S$ and $\mathbf{y} \in T$. (This is *not* automatically true in a topological space. A topological space for which it is true is called a *Hausdorff space*. See theorem 15.16 and exercise 22.8(3).)

In this chapter we shall use the word 'separate' in a related but not identical sense. We shall say that a set $H$ in a metric (or topological) space $\mathcal{X}$ *separates* $\mathcal{X}$ into two components if and only if $\mathcal{X} \setminus H$ has two components $S$ and $T$. If $A$ and $B$ are two sets in $\mathcal{X}$, we shall say that $H$ *separates* $A$ and $B$ if and only if $H$ separates $\mathcal{X}$ into two components $S$ and $T$ for which $A \subset \bar{S}$ and $B \subset \bar{T}$.

---

30.3 *Example* The line $H = \{(x, 0): x \in \mathbb{R}\}$ separates the sets $A = \{(x, y): (x-1)^2 + y^2 \leqq 1\}$ and $B = \{(x, y): (x+1)^2 + y^2 \leqq 1\}$ in $\mathbb{R}^2$.

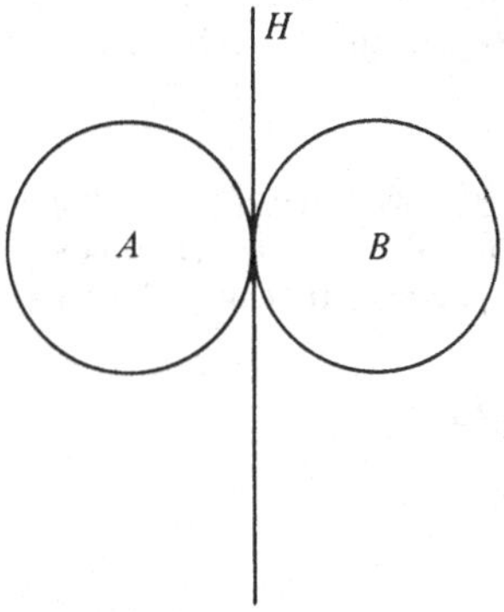


---

## 30.4† Separating hyperplanes

We begin with the observation that, if $Y$ is a non-empty, *closed* set in $\mathbb{R}^n$ and $\xi \in \mathbb{R}^n$, then there exists a point $\mathbf{z} \in Y$ which is *nearest* to $\xi$ – i.e. $\|\xi - \mathbf{z}\| = d(\xi, Y)$. (See corollary 19.16.)

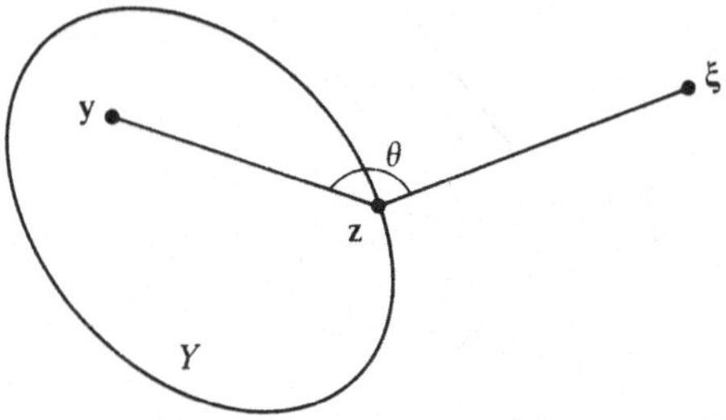

In the special case when $Y$ is *convex* (as in the diagram above), it seems reasonable to expect that $\mathbf{z}$ will be characterised by the fact that the angle $\theta$ is obtuse for all $\mathbf{y} \in Y$. This expectation is confirmed by the following theorem.

---

30.5† *Theorem* (*Projection theorem*) Let $Y$ be a closed, *convex*, non-empty set in $\mathbb{R}^n$. Then for any $\xi \in \mathbb{R}^n$, there exists a *unique* $\mathbf{z} \in Y$ which satisfies $\|\xi - \mathbf{z}\| = d(\xi, S)$. This value of $\mathbf{z}$ is characterised by the fact that

$$\langle \xi - \mathbf{z}, \mathbf{y} - \mathbf{z} \rangle \leqq 0 \tag{1}$$

for all $\mathbf{y} \in Y$.

*Proof* Recall from our discussion of the 'cosine rule' in §13.3 that

$$\|\xi - \mathbf{y}\|^2 = \|\xi - \mathbf{z}\|^2 + \|\mathbf{y} - \mathbf{z}\|^2 - 2\langle \xi - \mathbf{z}, \mathbf{y} - \mathbf{z} \rangle. \tag{2}$$

It follows that, if (1) holds, then $\|\xi - \mathbf{y}\| \geqq \|\xi - \mathbf{z}\|$ for all $\mathbf{y} \in Y$ and hence $\mathbf{z}$ is the nearest point of $Y$ to $\xi$.

Suppose, on the other hand that $\mathbf{z}$ is the nearest point of $Y$ to $\xi$. (The existence of such a $\mathbf{z}$ is guaranteed by corollary 19.16.) If $\mathbf{y} \in Y$, then $\mathbf{u} = \alpha\mathbf{y} + (1-\alpha)\mathbf{z} = \mathbf{z} + \alpha(\mathbf{y} - \mathbf{z}) \in Y$ provided that $0 \leqq \alpha \leqq 1$ because $Y$ is *convex*. Replacing $\mathbf{y}$ in (2) by $\mathbf{u}$, we obtain that

$$\|\xi - \mathbf{u}\|^2 = \|\xi - \mathbf{z}\|^2 + \alpha^2\|\mathbf{y} - \mathbf{z}\|^2 - 2\alpha\langle \xi - \mathbf{z}, \mathbf{y} - \mathbf{z} \rangle.$$

If $\langle \xi - \mathbf{z}, \mathbf{y} - \mathbf{z} \rangle > 0$, then we can ensure that $\|\xi - \mathbf{u}\| < \|\xi - \mathbf{z}\|$ by taking $\alpha$ sufficiently small and positive. Hence $\langle \xi - \mathbf{z}, \mathbf{y} - \mathbf{z} \rangle \leqq 0$ for all $\mathbf{y} \in Y$ – i.e. (1) holds.

It remains to be shown that $\mathbf{z}$ is unique. Observe that, if $\mathbf{y} \in Y$ and $\mathbf{y} \neq \mathbf{z}$, then it follows from (1) and (2) that

$$\begin{aligned} \|\xi - \mathbf{y}\|^2 &\geqq \|\xi - \mathbf{z}\|^2 + \|\mathbf{y} - \mathbf{z}\|^2 \\ &> \|\xi - \mathbf{z}\|^2. \end{aligned}$$

Hence $\mathbf{y}$ cannot be the nearest point of $Y$ to $\xi$.

---

Theorem 30.5 is usually invoked in the case when $Y$ is actually a *vector subspace* of $\mathbb{R}^n$. In this case, the theorem tells us that the nearest point $\mathbf{z} \in Y$ to $\xi$ is characterised by the fact that $\xi - \mathbf{z}$ is *orthogonal* to $Y$.

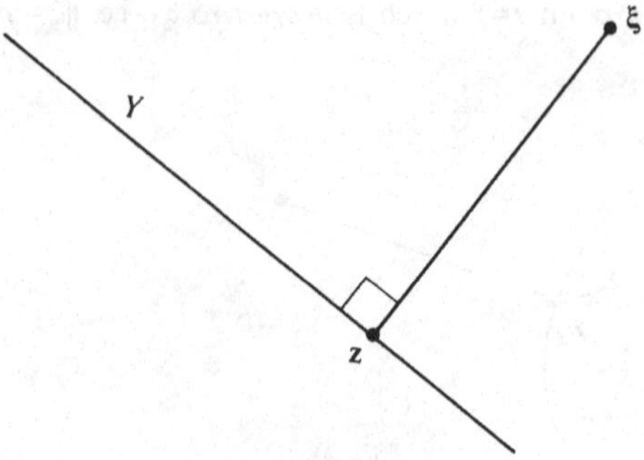

We may then define a function $P: \mathbb{R}^n \to Y$ by writing $P(\boldsymbol{\xi}) = \mathbf{z}$. For obvious reasons, $P$ is said to be the *orthogonal projection* of $\mathbb{R}^n$ onto $Y$.

Suppose that $Y$ is a convex, non-empty set in $\mathbb{R}^n$. Then $\bar{Y}$ is closed and convex. (See exercise 15.10(8).) Suppose that $\boldsymbol{\xi} \notin \bar{Y}$ and let $\mathbf{z}$ be the nearest point in $\bar{Y}$ to $\boldsymbol{\xi}$. We know from theorem 30.5 that, for all $\mathbf{y} \in Y$,

$$\langle \boldsymbol{\xi} - \mathbf{z}, \mathbf{y} - \boldsymbol{\xi} + \boldsymbol{\xi} - \mathbf{z} \rangle \leqq 0.$$

Taking $\mathbf{u} = (\boldsymbol{\xi} - \mathbf{z})/\|\boldsymbol{\xi} - \mathbf{z}\|$, it follows that

$$\langle \mathbf{u}, \mathbf{y} - \boldsymbol{\xi} \rangle \leqq -\|\boldsymbol{\xi} - \mathbf{z}\|$$

for all $\mathbf{y} \in Y$. Thus

$$\langle \mathbf{u}, \mathbf{x} - \boldsymbol{\xi} \rangle = \mathbf{0}$$

is the equation of hyperplane which has the property that $Y$ lies entirely inside one of the two half-spaces determined by the hyperplane. Note that the hyperplane passes through the point $\boldsymbol{\xi}$ and has normal $\mathbf{u}$ where $\|\mathbf{u}\| = 1$. (See §13.14.)

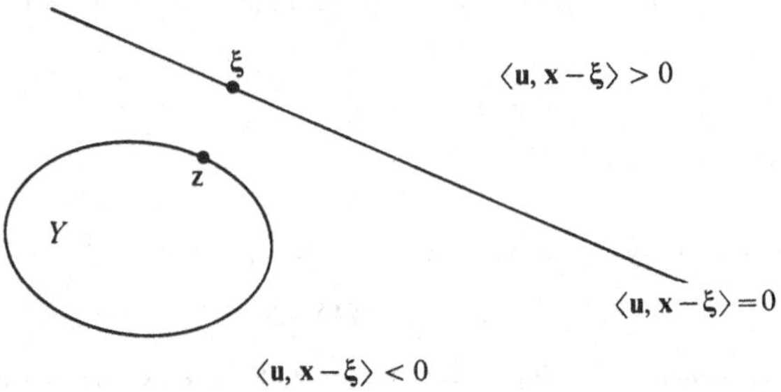


In the next theorem we extend this result to the case when $\boldsymbol{\xi}$ is a boundary point of a convex, non-empty set $Y$. We show that, in this case, there exists a hyperplane through $\boldsymbol{\xi}$ such that $Y$ lies entirely inside one of the two closed half-spaces determined by the hyperplane. Such a hyperplane is called a *supporting hyperplane* of the set $Y$.

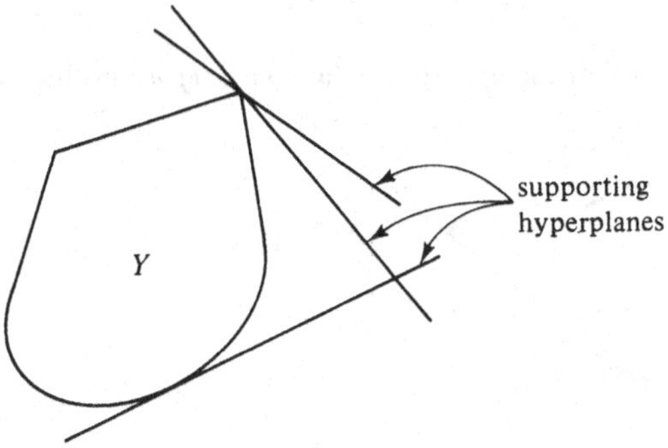


30.6† *Theorem* Let $Y$ be a non-empty, convex set in $\mathbb{R}^n$ and suppose that $\boldsymbol{\xi} \notin \mathring{Y}$.

Then there exists a $\mathbf{u} \neq \mathbf{0}$ such that

$$\langle \mathbf{u}, \mathbf{y} - \boldsymbol{\xi} \rangle \leqq 0$$

for all $\mathbf{y} \in Y$.

*Proof* Since $\boldsymbol{\xi} \notin \mathring{Y}$, each open ball of radius $1/k$ with centre $\boldsymbol{\xi}$ contains a point $\boldsymbol{\xi}_k \in \mathcal{C}(\bar{Y})$ (see exercise 30.10(5)). Let $\mathbf{u}_k$ be chosen so that $\|\mathbf{u}_k\| = 1$ and $\langle \mathbf{u}_k, \mathbf{y} - \boldsymbol{\xi}_k \rangle \leqq 0$ for all $\mathbf{y} \in Y$. Then $\boldsymbol{\xi}_k \to \boldsymbol{\xi}$ as $k \to \infty$ and $\langle \mathbf{u}_k \rangle$ has a convergent subsequence $\langle \mathbf{u}_{k_l} \rangle$ because the unit sphere is closed and bounded and thus is compact by the Heine–Borel theorem. Suppose that $\mathbf{u}_{kl} \to \mathbf{u}$ as $l \to \infty$. Then, from the continuity of the inner product,

$$\langle \mathbf{u}, \mathbf{y} - \boldsymbol{\xi} \rangle \leqq 0$$

as required.

---

30.7† *Theorem* (*Theorem of the separating hyperplane*) Suppose that $A$ and $B$ are non-empty, convex sets in $\mathbb{R}^n$ for which (i) $\mathring{A} \neq \emptyset$ and (ii) $\mathring{A} \cap B = \emptyset$. Then there exists a $\mathbf{u} \neq \mathbf{0}$ such that

$$\langle \mathbf{u}, \mathbf{a} \rangle \leqq \langle \mathbf{u}, \mathbf{b} \rangle$$

for all $\mathbf{a} \in A$ and all $\mathbf{b} \in B$.

*Proof* We apply the previous theorem to the convex set

$$C = \{\mathbf{a} - \mathbf{b} : \mathbf{a} \in A \text{ and } \mathbf{b} \in B\}.$$

Conditions (i) and (ii) ensure that $\mathbf{0} \notin \mathring{C}$. We obtain that there exists a $\mathbf{u} \neq \mathbf{0}$ for which

$$\langle \mathbf{u}, \mathbf{a} - \mathbf{b} - \mathbf{0} \rangle \leqq 0$$

for all $\mathbf{a} - \mathbf{b} \in C$. Hence $\langle \mathbf{u}, \mathbf{a} \rangle \leqq \langle \mathbf{u}, \mathbf{b} \rangle$ for each $\mathbf{a} \in A$ and each $\mathbf{b} \in B$.

---

If the real number $c$ is chosen so that

$$\sup_{\mathbf{a} \in A} \langle \mathbf{u}, \mathbf{a} \rangle \leqq c \leqq \inf_{\mathbf{b} \in B} \langle \mathbf{u}, \mathbf{b} \rangle,$$

then theorem 30.7 tells us that the hyperplane $\langle \mathbf{u}, \mathbf{x} \rangle = c$ *separates* the sets $A$ and $B$.

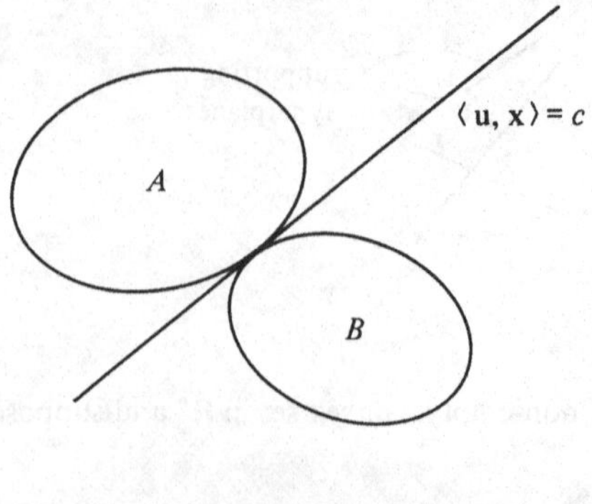

The theorem of the separating hyperplane has many important applications. It is used, for example, in proving Von Neumann's famous maximin theorem of two person zero-sum game theory. As an illustration of these ideas, we give a minor application to the problem of comparing the topologies generated by different norms in $\mathbb{R}^n$.

---

## 30.8† Norms and topologies in $\mathbb{R}^n$

In §13.17 we introduced the idea of a normed vector space. The natural norm to use in the space $\mathbb{R}^n$ is the Euclidean norm

$$\|\mathbf{x}\| = \{x_1{}^2 + x_2{}^2 + \ldots + x_n{}^2\}^{1/2}.$$

In §21.18, however, we came across some other norms which are compatible with the vector space structure of $\mathbb{R}^n$. We shall use the notation

$$\|\mathbf{x}\|_m = \max\{|x_1|, |x_2|, \ldots, |x_n|\}$$

and

$$\|\mathbf{x}\|_l = |x_1| + |x_2| + \ldots + |x_n|$$

to denote these alternative norms. In chapter 23 we found that the first of these alternative norms is sometimes more convenient to use than the Euclidean norm. The reason, of course, that this norm could be substituted for the Euclidean norm in this context is that both norms generate the *same* topology in $\mathbb{R}^n$ and, when considering limits, it is only the topologies in the spaces concerned which matter. (See §21.18 and §23.1.)

There is an infinite collection of norms which are compatible with the vector space structure of $\mathbb{R}^n$ and it is natural to ask whether *all* these different norms generate the *same* topology on $\mathbb{R}^n$. The next proposition answers this question in the affirmative. Note, however, that the corresponding result is very definitely *false* in infinite-dimensional spaces.

---

30.9† *Proposition* All norms on $\mathbb{R}^n$ generate the same topology on $\mathbb{R}^n$.

*Proof* We shall use the usual notation for the Euclidean norm on $\mathbb{R}^n$. We shall suppose that we are also given an alternative norm on $\mathbb{R}^n$ and use the notation $|||\mathbf{x}|||$ to denote this norm. As indicated in §21.18, it is enough if we can show that the set

$$S = \{\mathbf{x} : |||\mathbf{x}||| < 1\}$$

contains an open Euclidean ball $B_r = \{\mathbf{x} : \|\mathbf{x}\| < r\}$ and is contained in another open Euclidean ball $B_R = \{\mathbf{x} : \|\mathbf{x}\| < R\}$.

It follows from the properties of a norm (§13.17) that $S$ is *convex* and that, for each $\mathbf{X} \neq \mathbf{0}$, $\alpha\mathbf{X} \in S$ for small $\alpha > 0$ while $\alpha\mathbf{X} \notin S$ for large $\alpha > 0$.

First suppose that $S$ is unbounded (with respect to the *Euclidean* metric). Each unbounded convex set in $\mathbb{R}^n$ contains a half-line (proof?). Let $\mathbf{X} \neq \mathbf{0}$ be a point on this half-line. Then $\alpha\mathbf{X} \in S$ for *all* $\alpha > 0$ and this is a contradiction. Hence $S$ is bounded and so there exists an $R$ for which $S \subset B_R$.

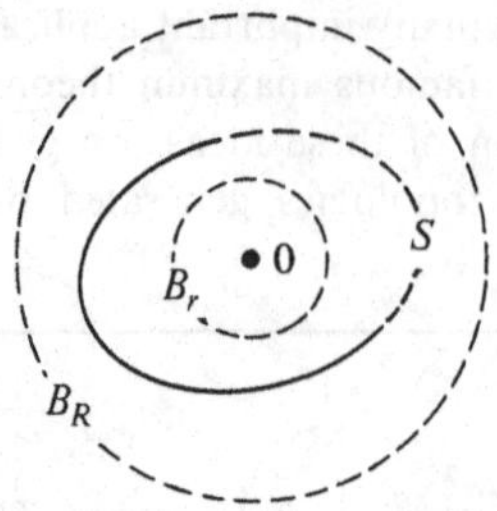


Next suppose that $\mathbf{0}$ is a boundary point of $S$ (with respect to the *Euclidean* metric). Since $S$ is convex, it follows from theorem 30.6 that there exists an $\mathbf{X} \neq \mathbf{0}$ such that $\langle \mathbf{X}, \mathbf{s} \rangle \leqq 0$ for all $\mathbf{s} \in S$. But, if $\alpha \mathbf{X} \in S$ and $\alpha > 0$, then $\langle \mathbf{X}, \alpha \mathbf{X} \rangle = \alpha \|\mathbf{X}\|^2 > 0$. This is a contradiction. We conclude that $\mathbf{0}$ is an interior point of $S$ (with respect to the *Euclidean* metric) and hence there exists an $r > 0$ such that $B_r \subset S$.

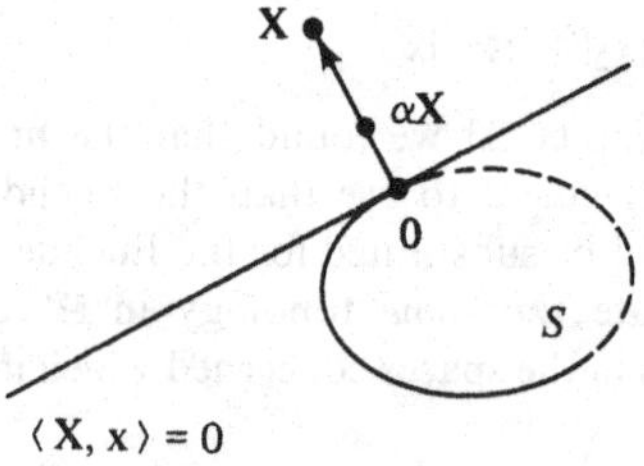


### 30.10† *Exercise*

(1) Prove that any hypersphere (§13.14) separates $\mathbb{R}^n$ into two components.
(2) Prove that any ball, box or line segment $S$ (§13.14) has the property that $\mathbb{R}^n \setminus S$ is connected.
(3) Find hyperplanes which separate the following sets in $\mathbb{R}^2$:
   (i) $A = \{(x, y): x^2 + y^2 \leqq 1\}$; $B = \{(x, y): y \geqq 1\}$
   (ii) $A = \{(x, y): x > 0 \text{ and } xy > 1\}$; $B = \{(x, y): x > 0 \text{ and } xy < -1\}$.
(4) Find two non-empty, convex sets $A$ and $B$ in $\mathbb{R}^2$ with $\mathring{A} \cap B = \emptyset$ which *cannot* be separated by a hyperplane.
(5) Let $S$ be a convex set in $\mathbb{R}^n$.
   (i) If $\xi \in \partial S$, prove that $\xi \in \partial(\mathscr{C}\bar{S})$. Give an example of a non-convex set $S$ for which this result is false.
   (ii) If $S$ is unbounded, prove that $S$ contains a half-line. Give an example of a convex set $S$ in $l^\infty$ for which this result is false.
(6) Prove that in a Hausdorff space (§30.2) every compact set is closed. Prove that $\mathbb{R}^n$ is a Hausdorff space.

## 30.11† Curves and continua

In the remainder of this chapter, we focus our attention on $\mathbb{R}^2$. We have seen that lines and circles separate $\mathbb{R}^2$ into two distinct components. What can be said about other curves?

First, something needs to be said about curves in general. In §17.15, we defined a *curve* in $\mathbb{R}^2$ to be the image $f(I)$ of a compact interval $I$ under a continuous function $f: I \to \mathbb{R}^2$. As we remarked at the time, this definition is not very satisfactory from the intuitive point of view since it fails to classify as a curve various 'one-dimensional' objects such as the example of Brouwer's (§17.18) consisting of a spiral wrapped around a circle.

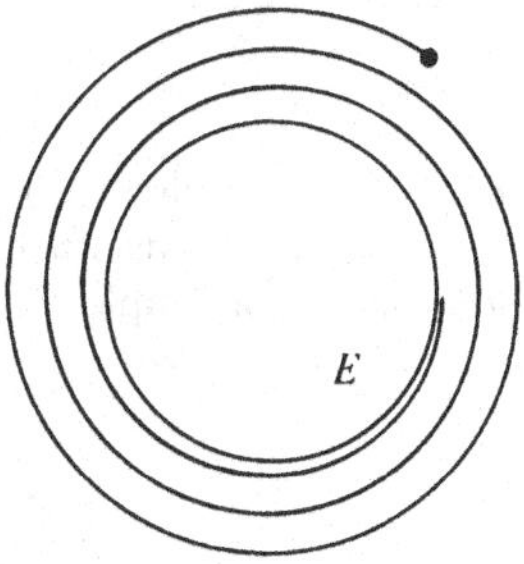


Brouwer's set $E$ is compact and connected. Such a set is called a *continuum* (although this usage is not consistent with the familiar description of $\mathbb{R}$ as 'the continuum'). The fact that $E$ fails to qualify as a curve is something to which one can accommodate oneself. But much worse things can happen than this as Peano discovered in 1890. We describe a version of his construction given by Hilbert.

Let $I = [0, 1]$ and let $S$ be the square $[0, 1] \times [0, 1]$. For each $n \in \mathbb{N}$, Hilbert constructed a continuous function $f_n$: $I \to S$. The curves $f_1(I)$, $f_2(I)$ and $f_3(I)$ are illustrated in the diagram below.

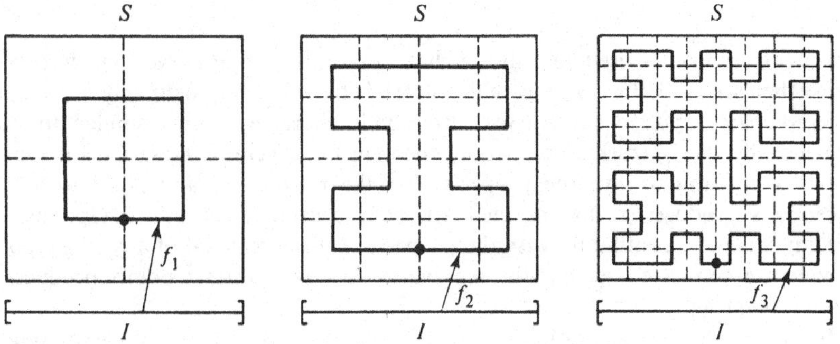


The sequence $\langle f_n \rangle$ of functions converges uniformly to a continuous function $f: I \to S$. As is evident from the construction, the curve $f(I)$ passes through *every* point of $S$ – i.e. $f(I) = S$. We say that $f(I)$ is a 'space-filling curve'.

This construction shows that 'two-dimensional' objects can be curves according to the definition given at the head of this section. This is, of course, highly counter-intuitive. (Even worse, the function $f: I \to S$ has the property that it is continuous on $I$ but differentiable at *no* point of $I$.)

Peano's discovery is by no means an oddity. The Mazurkiewicz–Moore theorem asserts that *every* locally-connected continuum is a curve. (A set $E$ in $\mathbb{R}^2$ is locally-connected if for each $\mathbf{e} \in E$ and each $\varepsilon > 0$ there exists a $\delta > 0$ such that for each $\mathbf{x} \in E$ satisfying $\|\mathbf{e} - \mathbf{x}\| < \delta$ there exists a connected subset of $E$ containing both $\mathbf{e}$ and $\mathbf{x}$ which has diameter at most $\varepsilon$.)

If we are to discuss separation by curves in $\mathbb{R}^2$, it is therefore necessary to begin by restricting the class of curves which we are to study.

---

## 30.12† Simple curves

We shall say that a curve $C$ is a *simple arc* if it is topologically equivalent (§21.1) to the compact interval $[0, 1]$. This means that there exists a homeomorphism $f: [0, 1] \to C$. A curve $C$ will be called a *Jordan curve* (or simple 'closed' curve) if it is topologically equivalent to the unit circle $U$ in $\mathbb{R}^2$.

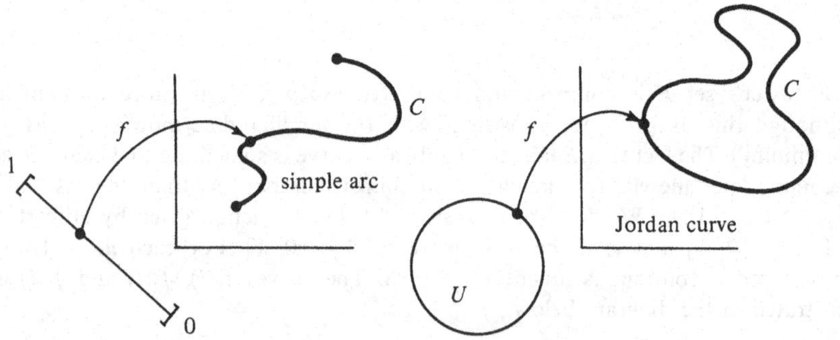


The word 'simple' is used to indicate that we are discussing curves which do not 'cross themselves'. Hilbert's space-filling curve is therefore very definitely *not* simple.

Simple arcs and Jordan curves also admit characterisations similar to that provided by the Mazurkiewicz–Moore theorem for general curves. Thus a simple arc is a continuum with the property that the removal of any point with two exceptions produces a disconnected set. (The exceptions are the endpoints.) A Jordan curve is a continuum with the property that the removal of any single point produces a connected set but the removal of any two distinct points produces a disconnected set.

It is clear what the separation properties of simple arcs and Jordan curves 'ought' to be. If $A$ is a simple arc, then $\mathbb{R}^2 \setminus A$ should be connected while if $J$ is a Jordan curve, then $\mathbb{R}^2 \setminus J$ should have precisely two components. Both results are true but neither is easily proved. The latter result is the more important and we quote it as the *Jordan curve theorem*.

---

30.13† *Proposition* Let $J$ be a Jordan curve in $\mathbb{R}^2$. Then $\mathbb{R}^2\setminus J$ has two components and $J$ is the boundary of both components. One of the components is bounded and the other is unbounded. (The bounded component is called the 'inside' of $J$ and the unbounded component is called the 'outside' of $J$.)

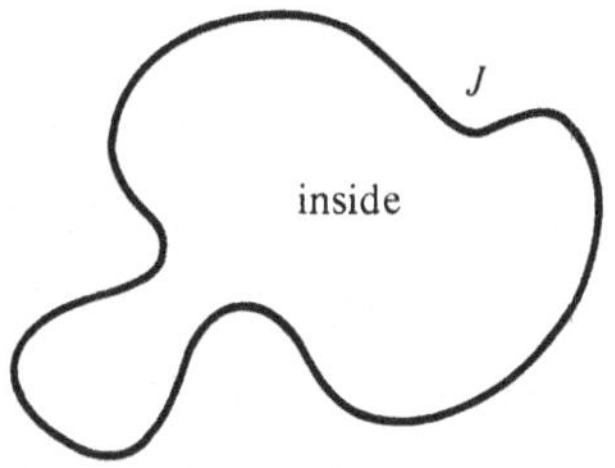


---

Note that the Jordan curve theorem would be trivial if it were known that there exists a homeomorphism $F\colon \mathbb{R}^2\to\mathbb{R}^2$ such that $F(U)=J$. But we are only given that there exists a homeomorphism $f\colon U\to J$. It is a result of Schönflies that each such homeomorphism $f\colon U\to J$ *can* be extended to a homeomorphism $F\colon \mathbb{R}^2\to\mathbb{R}^2$ but the proof of this useful result is harder than the proof of the Jordan curve theorem itself.

---

## 30.14† Simply connected regions

Recall that the Riemann sphere (or Gaussian plane) $\mathbb{R}^{2\#}$ is the one-point compactification of $\mathbb{R}^2$. This notion is particularly important in those cases when one is identifying $\mathbb{R}^2$ with the system $\mathbb{C}$ of complex numbers.

We begin by observing that the Jordan curve theorem remains true with $\mathbb{R}^2$ replaced by $\mathbb{R}^{2\#}$ (except that, if $\infty\in J$, neither component is bounded in $\mathbb{R}^2$).

In complex analysis, one is often concerned with open, connected subsets of $\mathbb{R}^{2\#}$. Such sets are called *regions* (or *domains*). Of particular importance, are simply connected regions. These are regions which have no 'holes'. More precisely, a region $R$ in $\mathbb{R}^{2\#}$ is *simply connected* if and only if $\mathbb{R}^{2\#}\setminus R$ is connected.

simply connected region

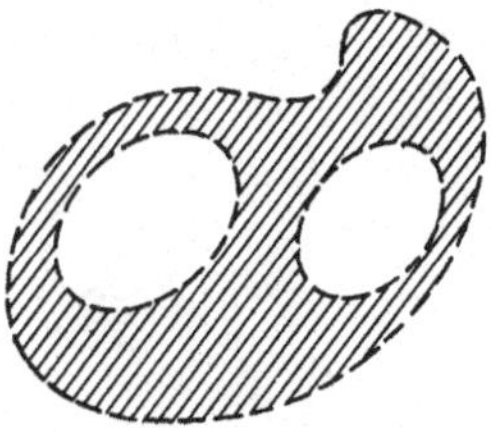

multiply connected region

An important alternative characterisation of a simply connected region $R$ is the fact that if $R$ contains a Jordan curve $J$, then $R$ contains one of the two components of $\mathbb{R}^{2\#}\setminus J$. In particular, if $\infty\notin R$, then $R$ contains the bounded component of $\mathbb{R}^{2\#}\setminus J$.

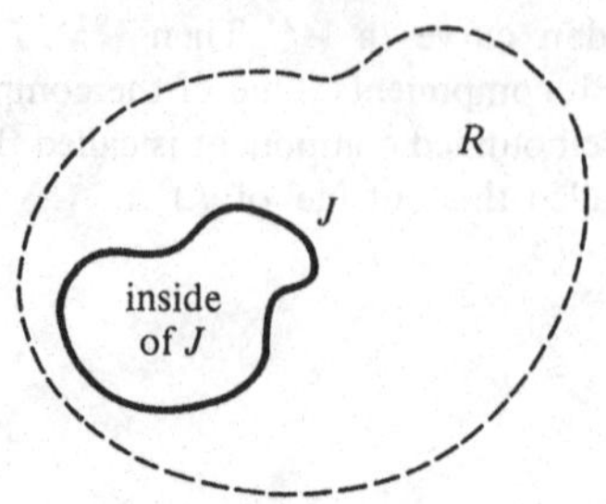


30.15† *Exercise*

(1) Explain why Brouwer's set $E$ (§30.11) separates $\mathbb{R}^2$ into two components. Explain also why $E$ is not a locally connected continuum.

(2) Consider the set $R=\{(x, y): y>x^2\}$ in $\mathbb{R}^2$. Explain why $R$ is simply connected.

(3) Suppose that $\mathcal{X}$ and $\mathcal{Y}$ are topological spaces and that $\mathcal{Y}$ is Hausdorff (§30.2 and exercise 30.10(6)). If $\mathcal{X}$ is compact and $f$: $\mathcal{X} \to \mathcal{Y}$ is continuous and bijective, prove that $f$ is a homeomorphism.

(4) If $f$: $I \to \mathbb{R}^2$ and $g$: $U \to \mathbb{R}^2$ (where $I=[0, 1]$ and $U$ is the unit circle in $\mathbb{R}^2$) are continuous and bijective, prove that $f(I)$ and $g(U)$ are respectively a simple arc and a Jordan curve. Explain why Hilbert's space-filling curve is not simple.

(5) Suppose that $J$ is a Jordan curve in $\mathbb{R}^2$ and that $\xi \in J$. Let $S$ and $T$ denote the inside and the outside of $J$ respectively. Let $\mathbf{a} \in S$ and $\mathbf{b} \in T$. Prove that there exists a simple arc joining $\mathbf{a}$ and $\xi$ which lies entirely inside $S$ and that there exists a simple arc joining $\mathbf{b}$ and $\xi$ which lies entirely inside $T$. [*Hint*: Use the result of Schönflies quoted in §30.12.].

(6) Explain why the result of the previous question is *false* when $J$ is replaced by Brouwer's set $E$.

# NOTATION*



*Numbers in italics refer to pages in Book 1: *Logic, Sets and Numbers.*

# INDEX*



*Numbers in italics refer to pages in Book 1: *Logic, Sets and Numbers.*